ON ORBIT AND BEYOND

SPACE TECHNOLOGY LIBRARY

Published jointly by Microcosm Press and Springer

For further volumes:
www.springer.com/series/6575

Douglas A. Vakoch
Editor

On Orbit and Beyond

Psychological Perspectives on Human Spaceflight

Springer

Editor
Douglas A. Vakoch
California Institute of Integral Studies
San Francisco
CA, USA
and
SETI Institute
Mountain View
CA, USA

ISBN 978-3-642-30582-5 ISBN 978-3-642-30583-2 (eBook)
DOI 10.1007/978-3-642-30583-2
Springer Heidelberg New York Dordrecht London

Library of Congress Control Number: 2012953092

Printed on acid-free paper

Springer is part of Springer Science+Business Media (www.springer.com)

To Tom Pierson, for his steadfast leadership of research and education at the forefront of space exploration and the search for life beyond Earth

Foreword

Space is one of the final frontiers for humankind. This book addresses that final frontier.

We humans have progressed from short orbital flights to landing on the Moon and astronauts living in low Earth orbit. Exploration missions to beyond the Moon such as asteroids and into deep space are in the planning stages.

I am an applied aerospace psychologist who has had the fortune to participate in the psychological portion of US astronaut selection for the past 22 years. The last 13 years I have worked in the Space Medicine Division, Behavioral Health and Performance Group, Johnson Space Center, Houston, Texas. We regularly train, communicate with, and support all US and some International Partner astronauts who have flown on the International Space Station.

One point that comes to my mind was the so-called communication strike of the crew on *Skylab 4*. In November 1973, the final manned mission of this US space station resulted in a conflict between the flight crew and the ground control. The crew's initial task of unloading and stowing the thousands of items needed for their lengthy mission proved to be overwhelming. The schedule for the activation sequence dictated lengthy work periods with a large variety of tasks to be performed, and crew members soon found themselves tired and behind schedule. As the activation of *Skylab* progressed, the astronauts complained of being pushed too hard. Ground crews disagreed; they felt that the astronauts were not working long enough or hard enough. During the course of the mission, this culminated in a radio conference to air frustrations. The space crew took a day off from work. Following this, the workload schedule was modified, and by the end of the mission the crew had completed even more work than had been planned before launch. The experiences of the crew and ground controllers provided important lessons in planning subsequent manned spaceflight work schedules. From that incident, we learned how critical it is to educate and train both ground control personnel and space crews in the social psychological issues of spaceflight.

On Orbit and Beyond is an important book for many reasons. You will read a historical survey of the role of psychologists and psychiatrists in the evolution of space flight, especially the American experience. Surviving and thriving in the extreme space environment is fraught with social challenges as well as simple physical survival. You will discover some of the major social psychological issues of living in low Earth orbit and of future space missions to asteroids and other planets. Advanced technology has led to expensive and complicated space vehicles. Yet we humans have not advanced so quickly. We still have to contend with our narrow margin of life support limits, isolation, confinement, danger, monotony, workload, cultural effects, and personality conflicts.

There are only a dozen or so space behavioral health clinicians in the world. As an applied psychologist daily working with my National Aeronautics and Space Administration, Japanese Aerospace Exploration Agency, Russian Space Agency, and European Space Agency colleagues, I rely on policymakers, operational managers, and scholars in supporting the astronauts' psychological challenges. For many years, Dr. Doug Vakoch has labored to ensure that the public has access to and understanding of the social psychology of living in space. In this book, he has gathered together some of the most renowned applied and research experts whose daily lives have been involved with the space program. These authors hail from Canada, China, Germany, Japan, the Netherlands, Norway, Russia, South Africa, and the United States, making this a truly international effort. You will read about data-based evidence and expert opinions. The authors include space social scientists, psychologists, psychiatrists, and researchers. These are the experts in helping policymakers and managers of the US and other Space Programs make wise decisions.

The first edition of this project was published by NASA in the NASA History Series as *Psychology of Space Exploration: Contemporary Research in Historical Perspective*. I am excited to see the second edition published by Springer with the chapters doubling in number. You will see the progress made since the *Skylab 4* communication incident. Issues addressed in this edition include a thorough history of psychology's role and recent progress in the US Space Program, psychological support, the importance of space analog environments, positive aspects of space flight, the role of astronauts' families, the dynamics of cross-cultural teams in space, autonomy for space crews in spaceflight, and more.

All of us, you the reader, those of us in operational settings, current researchers, space program administrators, and future generations are deeply indebted to Dr. Doug Vakoch for his dedicated commitment to this project. Thank you Doug.

Walter Sipes, Ph.D.
Aerospace Psychologist
NASA Johnson Space Center
National Aeronautics and Space Administration

Preface to the Second Edition

Over half a century ago, humankind first entered space when Yuri Gagarin became the first person to orbit Earth on April 12, 1961. Soon afterward, the United States launched its first manned mission to space, and in 2003 China became the third nation to send a human into space. Over the years, spacefarers from over three dozen countries have flown in space in missions that demonstrate increasing international cooperation. *On Orbit and Beyond* details how the psychological demands of space missions have changed over the decades, requiring increasingly talented and flexible astronauts.

The first missions into space were typically brief, lasting only hours or days, and crews were small. The early years of space exploration were marked by competition between the United States and the Soviet Union, and the astronauts who showed "the right stuff" were not only courageous explorers of this new frontier but heroes back home. As an intensely competitive space race has given way to international cooperation over the decades, the challenges of communicating across cultural boundaries and dealing with interpersonal conflicts have become increasingly important, requiring astronauts to develop additional coping skills and sensibilities.

The psychological stresses of living and working in space are considerable. In the early days of space exploration, short-duration challenges were faced by close-knit crews from the same country. Over the years we have seen increases in both the duration of space missions and the size of crews, as the Soviet Union established permanent space stations, the United States conducted its Space Shuttle program, and a truly transnational effort gave rise to the International Space Station. As crews more frequently included spacefarers from multiple cultures and missions became longer, the challenges of living in space highlighted the need to examine the psychological dimensions of space exploration in even greater depth.

On Orbit and Beyond includes chapters updated since its first incarnation, which was published by NASA as *Psychology of Space Exploration: Contemporary Research in Historical Perspective*. This revised edition has twice as many chapters as the first edition, with new chapters in every section. In short, this new edition provides a more comprehensive analysis of the challenges of surviving and thriving in space, as spacefarers deal with the complexities of working effectively with astronauts from other cultures under interpersonally demanding circumstances.

With an eye to the future, *On Orbit and Beyond* closes with a new section that explores how tomorrow's astronauts will handle the increased autonomy they will face on missions to Mars and beyond. This expansion of attention from Earth's orbit and the Moon to more distant destinations is a response to changing national priorities for space exploration. For example, in 2010, the Obama administration announced a new target for American astronauts as they travel beyond Earth's orbit. Rather than establishing a permanent lunar colony, as had been planned previously, NASA is making plans for a manned mission to an asteroid by 2025. By this timeline, a successful asteroid mission will help prepare for a trip to Mars some years later.

A mission to the red planet would take 2 years, with no chance of quickly returning to Earth if emergencies arise—a prospect that astronauts have never had to face before. In *On Orbit and Beyond*, leading space psychologists and psychiatrists provide guidance for coping with such future missions to asteroids, Mars, and beyond—drawing on lessons learned from past missions to the Moon, long-duration stays on space stations orbiting Earth, and analog environments on Earth.

Preface to the First Edition

Each month, the cover of *Monitor on Psychology*, a magazine sent to over 100,000 members of the American Psychological Association, reflects intriguing new areas of interest to psychologists who work as researchers, clinicians, consultants, and teachers. The importance of human adaptation to space for contemporary psychologists is suggested by the cover of the March 2008 *Monitor*, which featured an astronaut drifting in space, with the tranquil blue Earth in the background and the caption "Deep Space Psych" below.

At one level, the essays in this volume provide an overview and synthesis of some of the key issues in the psychology of space exploration, as well as a sampling of highly innovative empirical research. The characteristic that most clearly sets this collection apart from others, however, is the depth with which the authors have engaged the *history* of the psychology of space exploration.

All psychologists are familiar with the importance of engaging past research and theory while conducting literature reviews in preparation for designing and interpreting new studies. But the contributors to this collection have done much more. They have crafted essays that will be of obvious value to psychologists, psychiatrists, and other behavioral researchers. At the same time, these authors have created a collection with the promise to promote a greater dialogue between psychological researchers and both historians of space exploration and historians of psychology.

Psychologists and historians have quite different criteria for good scholarship and for communicating their findings. These differences make the essays in this volume—meaningful and accessible even to those not formally trained in psychologists' methodologies and mindsets—all the more impressive. With the increasing specialization and isolation of academic disciplines from one another over the past century, these essays serve as a prototype for a broader attempt to bridge the gap between the two cultures of science and the humanities that C. P. Snow identified almost half a century ago—quite fittingly for us, near the beginning of the Space Age. Let us hope that as we prepare once again to send astronauts beyond Earth's orbit, we can do so with the guidance of others equally open to seeing beyond their own specialties.

Acknowledgments

Over the past 15 years, many colleagues from the SETI Institute have shared with me their insights into the space sciences, as well as the ways we can best communicate this work to the broader public. I especially thank Molly Bentley, Linda Bernardi, John Billingham, Jenny Chynoweth, Edna DeVore, Frank Drake, Sophie Essen, Andrew Fraknoi, John Gertz, Gerry Harp, Gail Jacobs, Jane Jordan, Ly Ly, Chris Neller, Tom Pierson, Karen Randall, Jon Richards, John Ross, Pierre Schwob, Seth Shostak, and Jill Tarter.

More recently, I warmly acknowledge the administration, faculty, staff, and students of the California Institute of Integral Studies (CIIS), especially for support from Bryant Welch, Joseph Subbiondo, and Judie Wexler. The work of editing this book was made possible through a generous sabbatical leave from my other academic responsibilities at CIIS in the spring of 2012. In addition, I thank Harry and Joyce Letaw, as well as Jamie Baswell, for their intellectual and financial contributions to promoting space exploration.

For encouraging me to develop this project and then shepherding the book through the editorial process at Springer, I am indebted to Ramon Khanna. Also at Springer, Tamara Schineller has my gratitude for helping to move the book swiftly and efficiently into production. Umamaheswari Gnanamani of SPi Global has my appreciation for faithfully overseeing all aspects of copyediting, layout, and indexing. I thank James Wertz for including *On Orbit and Beyond* in Springer's Space Technology Library.

For providing a foundation for this book by publishing its first edition in the NASA History Series, I thank Bill Barry, Steve Garber, and Steve Dick. For successfully publicizing that edition, I am grateful to Nadine Andreassen of the NASA History Office, as well as Marco Larsen and David Scott of PUBLIC, NYC.

Among the organizations that have fostered discussions about the psychology of space exploration, I especially want to recognize the International Academy of Astronautics (IAA) and the American Psychological Association (APA). Many of the chapters in this book were first presented in nascent form at annual conferences of these organizations, and I thank the leadership of these organizations for creating an institutional structure that supports research in this area.

I especially thank the authors of the chapters that follow for their innovative contributions, providing guidance for the next generation of human space exploration. I also wish to acknowledge the many researchers who, through their review of drafts of these chapters, have greatly strengthened and focused the final work. In particular, I want to recognize Albert Harrison, without whose intellectual leadership this book would never have come into existence. His contributions will be evident in the three chapters he has coauthored; invisible is his extensive work recruiting other contributors, reviewing chapters, and providing long-term as well as last-minute assistance more times than I care to remember.

Finally and most importantly, to my wife Julie Bayless, I am grateful in more ways that I can or will share here. Thank you, forever.

Contents

List of Contributors

Pamela J. Baskin Wyle Laboratories, Houston, TX, USA

Sheryl L. Bishop University of Texas Medical Branch, Galveston, TX, USA

Jennifer E. Boyd San Francisco Veterans Affairs Medical Center, University of California, San Francisco, San Francisco, CA, USA

Joseph V. Brady Institutes for Behavior Resources, Johns Hopkins University School of Medicine, Baltimore, MD, USA

Zabecca S. Brinson Institutes for Behavior Resources, University of Maryland School of Medicine, Baltimore, MD, USA

Hege H. Bye University of Bergen, Bergen, Norway

Lindi Cassel Providence Health Care, Vancouver, BC, Canada

Colleen Cook San Francisco Veterans Affairs Medical Center, University of California, San Francisco, San Francisco, CA, USA

Juris G. Draguns Pennsylvania State University, University Park, PA, USA

Edna R. Fiedler Aerospace Psychology Consultants, Tucson, AZ, USA

Eric D. Gasior Institutes for Behavior Resources, Baltimore, MD, USA

Vadim I. Gushin Institute for Biomedical Problems, Moscow, Russia

Matthew Harris San Francisco Veterans Affairs Medical Center, University of California, San Francisco, San Francisco, CA, USA

Albert A. Harrison University of California, Davis, CA, USA

Robert D. Hienz Institutes for Behavior Resources, Johns Hopkins University School of Medicine, Baltimore, MD, USA

Steven R. Hursh Institutes for Behavior Resources, Johns Hopkins University School of Medicine, Baltimore, MD, USA

Natsuhiko Inoue JAXA Tsukuba Space Center, Tsukuba Science Center, Japan

Bernd Johannes German Aerospace Center (DLR), Köln, Germany

Phyllis J. Johnson University of British Columbia, Vancouver, BC, Canada

Megan A. Kaminski George Mason University, Fairfax, VA, USA

Nick Kanas San Francisco Veterans Affairs Medical Center, University of California, San Francisco, San Francisco, CA, USA

Daniel J. Kealey Centre for Intercultural Learning, Ottowa, ON, Canada

Norbert O. Kraft NASA Ames Research Center, Moffett Field, CA, USA

Jason P. Kring Embry-Riddle Aeronautical University, Daytona Beach, FL, USA

Gloria R. Leon University of Minnesota, Minneapolis, MN, USA

Dietrich Manzey Berlin University of Technology, Berlin, Berlin, Germany

Charles Marmar New York University Medical Center, New York, NY, USA

Ichiyo Matsuzaki University of Tsukuba, Tsukuba, Ibaraki Prefecture, Japan

David Musson McMaster University, Hamilton, ON, Canada

Thomas Neylan San Francisco Veterans Affairs Medical Center, University of California, San Francisco, San Francisco, CA, USA

Regina North NASA Johnson Space Center, Houston, TX, USA

Valerie Olson Rice University, Houston, TX, USA

Lawrence A. Palinkas University of Southern California, Avalon, CA, USA

Julie A. Robinson NASA Johnson Space Center, Houston, TX, USA

Peter G. Roma Institutes for Behavior Resources, Johns Hopkins University School of Medicine, Baltimore, MD, USA

V.P. Salnitskiy Institute for Biomedical Problems, Moscow, Russia

Gro Mjeldheim Sandal University of Bergen, Bergen, Norway

Stephanie Saylor San Francisco Veterans Affairs Medical Center, University of California, San Francisco, San Francisco, CA, USA

Walter Sipes NASA Johnson Space Center, Houston, TX, USA

Kelley J. Slack Wyle Laboratories, Houston, TX, USA

Jack Stuster Anacapa Sciences, Santa Barbara, CA, USA

Peter Suedfeld Department of Psychology, University of British Columbia, Vancouver, BC, Canada

Michael H. Trenchard NASA Johnson Space Center, Houston, TX, USA

Fons J. R. van de Vijver Tilburg University, Tilburg, The Netherlands

North-West University, Potchefstroom, South Africa

Jun Wang Astronaut Center, Beijing, China

Daniel S. Weiss San Francisco Veterans Affairs Medical Center, University of California, San Francisco, San Francisco, CA, USA

Harvey Wichman Claremont McKenna College and Claremont Graduate University, Claremont, CA, USA

Kasia E. Wilk W. J. Stelmaschuk and Associates Ltd, Vegreville, AB, Canada

Kimberly J. Willis NASA Johnson Space Center, Houston, TX, USA

List of Figures

List of Tables

Introduction

Psychology and Space Exploration

Albert A. Harrison and Edna R. Fiedler

Abstract Astronauts, cosmonauts, and others who venture into space live and work in highly unusual and challenging environments where they must withstand multiple stressors. Their abilities to maintain positive psychological outlooks and good interpersonal relations are crucial for personal well-being and mission success. From the inception of the US space program, psychologists, psychiatrists, human factors experts, and other professionals have warned that the psychological stressors of space should be treated as a risk factor and that risks would increase as missions involved larger, more diversified crews undertaking increasingly longer flights. Thus, they called for research leading to the development and application of effective countermeasures. Although psychology played a significant role at the inception of the space program, for many years thereafter certain areas of psychology all but disappeared from NASA. Interest in psychosocial adaptation was rekindled in the mid-1990s when astronauts joined cosmonauts on the Russian space station *Mir*. Focusing on the underutilized areas of personality and social psychology and drawing heavily from psychiatry, the chapters that follow discuss psychology's struggle for acceptance, the history of astronaut selection and psychological support, the use of analogue environments and simulators for research and training, space tourism, the psychological rewards of viewing Earth from space, crew composition, group dynamics, and cross-cultural aspects of multinational missions. This book concludes with a summary, integration, and evaluation of the role of behavioral research and application in space exploration. NASA's recognition of the new field of behavioral health and its links to performance opened the door to many kinds of research that were formerly overlooked.

Albert A. Harrison
University of California, Davis, CA, USA

Edna R. Fiedler
Aerospace Psychology Consultants
Tucson, AZ, USA

> "Once, I was evaluating astronaut applicants" says psychiatrist Nick Kanas. "I asked them to give me some examples of things that might cause stress." One applicant, a test pilot, recalled the time he was flying an experimental aircraft and it spun out of control. As the plane spiraled down, he took out his manual, calmly thumbed through it, and figured out how to pull the plane to safety. "His ability to temporarily control his emotions was striking," laughs Kanas (Science@NASA 2002).

Fully aware of astronauts' remarkable strengths, Kanas also knows that many physical and psychological stressors can pose risks to safety, performance, and quality of life (Kanas 1985, 1991; Kanas et al. 2006). Some of these stressors are associated with flight: riding atop a rocket; rapid acceleration and deceleration; somewhat primitive living conditions; isolation from family and friends; close confinement with other people; and the ever-present specter of a collision, system failure, or other disaster. Other types of stressors come from the astronaut's career. From the earliest days of the space program, astronauts have served as societal exemplars, living under intense public scrutiny; carrying heavy workloads on Earth as in space; and undergoing prolonged absences from home for training, flight, and other purposes. They must withstand the typical hassles of trying to succeed within large bureaucracies, worry over flight assignments, and readjust to their families when they return to Earth (Sipes and Vander Ark 2005). J. Kass, R. Kass, and I. Samaltedinov (1995, p. 658) describe how some of this may seem to an astronaut:

> He knows he has been trained and put into space at great cost and effort, and he has a limited amount of time, especially during a short shuttle mission, to perform the tasks set out for him, efficiently. The precious data of the scientists on the ground, who have dedicated many years for this experiment, can be lost, the equipment can be damaged in such a way that it cannot be repaired in space, or worse still, his blunder can affect the safety of life on the spaceship. Even if such drastic errors are seldom, he is nevertheless under great stress—he has to get the work done quickly, so that the next scheduled event can take place as planned. This kind of stress affects him not only as an individual, but as a member of a team: His peers are watching him, and he knows full well, not only will any mistakes made affect their work as well, but he fails in their eyes in a similar manner as a member of a sports team, whose error can affect the success of the team as a whole (Kass et al. 1995).

This book discusses selected topics in the psychology of space exploration. In this and the following chapters, we and other contributors address the changing role of psychology and psychiatry within US, Russian, European, Chinese, Japanese, and Canadian space programs; review the history of astronaut selection and training; and describe the evolution of techniques for providing spacefarers with psychological support. Contributing authors explain why and how researchers and operational personnel use space-reminiscent settings (such as undersea habitats and polar outposts) for research and training purposes. They trace the not-always-smooth course of the diversification of the astronaut corps from a homogenous collection of white, male test pilots to a more heterogeneous group including women and minorities. They tell about group dynamics and teamwork, as well as occasional friction between crews in flight and people in Mission Control. One of the most dramatic changes over 50 years of crewed flight has been the transition from fiercely competitive national space programs to collaborative efforts with international crews, so cross-cultural issues are addressed in depth. Over the past

50 years, space missions have changed, and so have salient behavioral issues and opportunities for behavioral research and application.

How has psychology fared in the US space program? In his presidential address to the Division of Engineering Psychology on September, 1961, Walter F. Grether affirmed psychology's crucial role in the newly initiated conquest of space, noting that psychologists of that day were responding with creativity and vigor to the enormous behavioral challenges (Grether 1962). Looking back over the history of aviation, Grether remarked that despite a few contributions to military aviation in World War I, for roughly 35 years after the Wright brothers' initial flight at Kitty Hawk, aviation and psychology pretty much went separate ways. Then, beginning with research to benefit civilian aviation in the late 1930s and followed by a powerful military program during World War II, aviation psychology became prominent and influential. "How much different the role of psychology has been in man's early ventures into space!" Grether (1962, pp. 92–93) wrote. Psychological testing, he continued, was prominent in the selection of the initial seven Mercury astronauts, and beyond selection, psychologists were productively engaged in vehicle design, training, task design, and workload management.

Grether pointed out four areas for future research: moving about the interior of spacecraft (once they became large enough for this to occur), conducting extravehicular activities (EVAs) or "spacewalks," performing rendezvous, and living and working under conditions of prolonged isolation and confinement. Highly optimistic about America's future in space, Grether foresaw a strong continuing partnership between psychology and space exploration. One of his few notes of pessimism—that it would not be possible to use the science fiction writer's rocket gun to move from place to place during EVAs—would soon be disproven. Beyond providing psychologists with new opportunities for employment and research support, he felt, space exploration would open new frontiers of knowledge, stimulate thinking about new problems, and lead to new theories, hypotheses, and methods.

Nearly three decades later, celebrants at the 30th anniversary of the 1959 founding of the International Ergonomics Association might conclude that Grether was right. In the field generally known as human factors in the United States and ergonomics in the United Kingdom (UK) and Europe, human factors specialists are interested in the scientific problems of experimental psychology, anatomy, and physiology applied to human work. Classically, human factors addresses people's interaction with physical environments in work settings, but the interests of human factors specialists have broadened over the years (Meister 1989).

In his 1989 presidential address to the association, Alphonse Chapanis could point with pride to rapidly accumulating accomplishments everywhere in the field (Chapanis 1990). Floods of data were appearing in area after area of human activity (work, transportation, leisure-time pursuits), and it was no longer possible to keep abreast of the latest journals and books. The hottest topic of 1989 was computers: how they had revolutionized society, how they spread beyond science and business and were embraced by everyday people, and how they could be humanized through the design of displays and controls. Certainly, much was left to be done—over the lifetime of the association, 71 major railroad disasters had claimed 5,059 persons; 192 major aircraft accidents had killed over 20,000 people; and, in the previous 10 years alone, there had been thousands of nuclear "mishaps," including prominent events at Chernobyl and Three Mile Island. Still, Chapanis's (1990, pp. 276–277) theme was that ergonomics had "come a long way, baby," and that the strongest stimulus for this was America's forays into space:

> Space flights have become so commonplace and so much is known about human performance in space that it is hard to remember the thousands of analyses, studies, and experiments that were done to pave the way for man's leap into these hostile and unknown regions. There were problems of vehicle design involving exotic displays and controls. There were problems of vibration, of *g*-forces, and of weightlessness that had to be explored and solved. For extravehicular activity an entire self-contained environment had to be designed for astronauts and cosmonauts. Torqueless tools had to be designed for use by men who were floating freely and encumbered by space suits with limited mobility. There were problems of nutrition, waste disposal, and work-rest cycles. Nor can we forget the problems of selection, training, and simulator design ... Our leap into space was a significant accomplishment of the past 30 years and the ergonomic findings that helped bring it about have enriched our profession in countless ways (Chapanis 1990).

But other assessments of psychologists' contributions to the US space program were less triumphant. In 1975, Robert L. Helmreich expressed pessimistic views of applying psychology in new areas, stating that prospective customers often respond with profound indifference (Helmreich 1975). In 1983, he elaborated on how data relating to personality and social psychology were underused by the US space program, which (as we shall see in Chap. 1) he considered in contrast to robust use in the Soviet program (Helmreich 1983). In a 1987 conference cosponsored by NASA and the National Science Foundation, psychologist and management consultant Philip R. Harris (1990) observed that

> [a]lthough NASA has been forthright about medical and biological insights gained from previous spaceflights ... the agency has been hesitant on studying or releasing information on the psychosocial experience of its personnel in space. Generally, NASA has limited the access to astronauts by social science researchers, even by its own psychiatrists and psychologists; the agency has failed to capitalize on the data it collected that could improve spaceflight and living for others to follow (Harris 1990, pp. 77–78).

In the early 1990s, outgoing flight surgeon and psychiatrist Patricia Santy concluded that despite an initial flurry of interest, behavioral research all but disappeared from NASA (Santy 1994). For years, she wrote, psychology played a minimal role in astronaut selection, and because the assessment of individual astronaut performance was prohibited, it was not possible to collect normative data for test validation and other purposes. She characterized the application of psychology to space as running 20–30 years behind most areas of medicine and identified formidable organizational barriers to psychology within NASA. Joseph V. Brady, whose research on primate behavior in spaceflight dates back to the 1950s, stated that following John Glenn's flight, there was a dearth of in-flight behavioral experiments (Brady 2005). Brady characterized this as a 30-year hiatus in psychological health research for NASA, a gap that he thought must come to an end given NASA's vision for humans in space. Peter Suedfeld (2005, p. B61) cuts to the heart of the matter: "Through most of NASA's existence, the behavioral sciences have been barely visible on the agency's horizon" (Suedfeld 2005, p. B61).

How can we reconcile such pessimistic views with the optimistic assessments of Grether and Chapanis? Robert Helmreich's point was that, generally, those disciplines that are rooted in biology, engineering, and experimental psychology have found greater acceptance within the space program than disciplines rooted in personality, social, and organizational psychology. Lawrence Palinkas, an anthropologist who has developed an enviable record of hands-on research experience in unusual environments, organized these issues in long-term spaceflight into three "domains": the individual domain (stress and coping), the group dynamics domain (social interaction and intergroup relations), and the organizational domain (management, organizational culture, and behavior) (Palinkas 2001).

From the beginning, physicians, psychologists, and their allies advocated strong behavioral research programs in NASA. Margaret A. Weitekamp points out how interest in high-altitude flight in the 1930s initiated research that evolved into aerospace medicine in the 1940s (Weitekamp 2004). Research to support pilots flying very fast and very high provided a basis for sending astronauts into space. The first conference with "space" in the title was prior to 1950, notes Weitekamp, but some space-oriented research was clandestine or integrated into aviation medicine and psychology in order to avoid the wrath of superiors who thought it wasteful to study "Buck Rogers" issues. In 1961, Bernard Flaherty edited *Psychophysiological Aspects of Space Flight*, which focused on the sensory and biotechnical aspects of spaceflight and simulations, as well as addressing issues of human durability (Flaherty 1961). *Human Factors in Jet and Space Travel* also appeared that year (Sells and Berry 1961). The latter was edited by Saul B. Sells, a NASA consultant who first wrote about astronaut selection and training in 1957, and Charles A. Berry, at one time NASA Director of Life Science and physician to the astronauts. They dealt with performance under conditions of acceleration and deceleration, as well as human adaptation to space. In 1967, Joseph Kubis, along with Edward J. McLaughlin, specifically addressed the psychological aspects of spaceflight (Kubis and McLaughlin 1967). They noted that whereas short-term spaceflight did not have adverse effects on functioning, factors such as emotional stability and group dynamics could prove important in future missions. As would many other writers, they illustrated their points with studies of psychological reactions to isolation and confinement in terrestrial settings.

In the early 1970s, Joseph Kubis addressed issues of group dynamics: group composition, leadership, and teamwork (Kubis 1972). In 1971, the then Air Force psychiatrist Nick Kanas, in collaboration with William E. Fedderson, released an outline of many of the psychological and psychiatric issues that have filtered down and influence discussions today (Kanas and Fedderson 1971).

In 1972, the National Academy of Sciences released the report of a study panel chaired by Donald B. Lindsley of the University of California, Los Angeles (UCLA) (Lindsley 1972). The panel sought "to indicate the blocks of research, roughly in order of priority that will be most fruitful in the years ahead in coming to grips with the problems of long-duration missions . . . In this, there is little doubt in the minds of the study participants that the difficulties are formidable, the unknowns significant, and the prerequisite research extensive ..." (Lindsley 1972, p. 15). Many of the experts were interested in space physiology and medicine, but the panel also included psychologists with expertise in stress, social interaction, and behavior in unusual environments. In addition to recommending basic biomedical and life-support research, the panel urged studies of skilled performance, environmental habitability, group processes, interpersonal interaction, and the relationship of the space crew "microsociety" to the larger flight team. In 1977, partially in response to Lindsley's report, Mary M. Connors, Albert A. Harrison, and Faren R. Aikens began a review of the then current foundations for understanding behavior during anticipated Space Shuttle and space station missions (Connors et al. 1985). Their book, not published until 1985, identified a middle ground between narrowly focused experiments and bold generalizations. They adopted an open systems approach and addressed topics at the individual, small group, and organizational levels.

In the late 1980s, the Committee on Space Biology and Medicine (1987, p. 168) of the National Research Council gave further impetus to psychology, noting that "[a]lthough the evidence is fragmentary it seems likely that behavioral and social problems have already occurred during long-term missions ..... An understanding of the problems and their amelioration is essential if man desires to occupy space for extended periods of time. Even more important from a scientific perspective, it seems likely that significant advances in our basic knowledge of human interaction and processes will emerge from the research needed to ensure effective performance and adjustment in space" (Committee on Space Biology and Medicine 1987).

Revisiting the issue some 10 years later, the next Committee on Space Biology and Medicine (1998, p. 195) reaffirmed the urgency of their predecessors' recommendations: "Despite this [the 1987 panel's] assessment of the importance of behavioral issues, little progress has been made transforming the recommendations for research on human behavior and performance in space into action .... As could be predicted from controlled simulation studies, the history of space exploration has seen many instances of reduced energy levels, mood changes, poor interpersonal relations, faulty decision-making, and lapses in memory and attention. Although these negative psychological reactions have yet to result in disaster, this is no justification for ignoring problems that may have disastrous consequences. Furthermore, there are degrees of failure short of disaster and degrees of success short of perfection; if favorable organizational and environmental conditions can increase the level and probability of success, they are worthy of consideration" (Committee on Space Biology and Medicine 1998).

The 1998 Committee's recommendations included studying the effects of the physical and psychosocial environment of spacecraft on cognitive, psychophysiological, and affective measures of behavior and performance; the development and evaluation of countermeasures for mitigating adverse effects of the physical and social environments on individual and group performance; in-flight studies of the characteristics of sleep during long-duration missions; ground-based studies of change and stability in individual psychophysiological patterns in response to psychosocial and environmental stressors; the effects of individual differences on cognitive, psychophysiological, and affective measures of behavior and performance; improved methods for assessing interpersonal relations and crew compatibility; and improved training [didactic and experiential] in psychological and social adaptation to space. The Committee also urged exploring the effects of crew composition on crew tension, cohesion, and performance; factors affecting ground-crew communication and interactions; and conditions that affect the distribution of authority, decision-making, and task assignments between space crews and ground control.

In 2001, the National Academy of Sciences issued *Safe Passage: Astronaut Care for Exploration Missions* (Ball and Evans 2001), prepared by the Committee on Creating a Vision for Space Medicine During Travel Beyond Earth Orbit of the Institute of Medicine of the National Academy of Sciences. This panel of experts identified some of the medical and behavioral issues that should be resolved immediately in anticipation of a future return to the Moon and a mission to Mars. This far-ranging work covers astronaut health in transit to Earth orbit and beyond, health maintenance, emergency and continuing care, the development of a new infrastructure for space medicine, and medical ethics. Most importantly for present purposes, *Safe Passage* includes a chapter on behavioral health, a topic that we discuss in some detail in Chap. 1.

Different missions raise different questions about human behavior. The most conspicuous questions of the earliest days of spaceflight had to do with life support, the human-machine interface, and the optimization of human performance to ensure mission success. Certainly these topics remain crucial today, but to them we may add many more. Following Apollo and the race to the Moon, NASA entered new eras in 1981, when the Space Shuttle took flight, and again in 1993, when astronauts joined cosmonauts first on Russia's *Mir* space station and then on the International Space Station (ISS) in 2000. Topics such as habitability, loneliness, cultural conflicts, the need to sustain a high level of performance over the long haul, and postflight adjustment gained a degree of immediacy and could no longer be ignored. Consistent with David Meister's views on conceptual changes in human factors, there has been, over the years, a shift from a purely "displays and knobs" orientation to a more holistic approach, with project managers, engineers, and behavioral researchers sharing the goal of a seamless human-machine structure or "system integration" (Meister 1989).

In their discussion of post-Apollo psychological issues, Connors, Harrison, and Akins noted that as missions change, so do behavioral requirements (Connors et al. 1985). Perhaps the most conspicuous trends are in the direction of increased crew size, diversity, and mission duration. The first round of US flights, under Project Mercury, were solo but rapidly gave way to two-person crews with the advent of Project Gemini in 1965, followed by three-person crews during the Apollo program. After Mercury, note Clay Foushee and Robert Helmreich, the test pilot became a less relevant model than the multiengine aircraft commander, who not only requires technical skills but also human relations skills as the leader of a team (Foushee and Helmreich 1998). America's first space station, *Skylab*, provided a "house in space" for three-person crews; apart from occasional emergencies or visitors, three-person crews were also typical for Soviet (1970–89) and then Russian (1990 and onward) space stations and the ISS. Space Shuttles are relatively capacious and usually carry six to eight crewmembers. Other than during brief visits from Shuttle crews, the ISS has been home to crews of two to six people. We suspect that later space stations will house larger crews. Although it is possible to envision huge orbiting platforms and communities on the Moon and Mars, foreseeable missions are unlikely to exceed eight people or so, with the result that crews will remain within the "small group" range.

A second salient trend is toward increasing diversity of crew composition. The initial vision was for a highly diverse pool of astronaut candidates, including mountain climbers, deep sea divers, and arctic explorers, but, as will be explained in the next chapter, it was military test pilots who got the nod. The military remains well represented, but over the years, the astronaut corps has been expanded to include people from many different professions and a greater number of women and minorities. Further complexity was added with the Soviet guest cosmonaut program beginning in the 1970s, the inclusion of international crewmembers on the Shuttle, and international missions on *Mir* and the ISS. Already, tourists have entered the mix, and the first industrial workers in commercial space ventures may not be far behind.

Third, initial spaceflights were measured in hours, then days. (Indeed, within each series of flights, successive Mercury and then Gemini flights were longer and longer, to establish that astronauts could withstand the long trip to the Moon.) The third *Skylab* crew remained on orbit 84 days. *Skylab* was short-lived, but the Soviets set endurance records in this area; the present record of 437.7 days was set by a Russian cosmonaut on *Mir* during a 1984–1985 mission. ISS missions have usually lasted about 3 months, but individuals are staying on the Space Station for up to 6 months, as demonstrated in 2007 and 2008 by Sunni Williams and Peggy Whitson. Extended stays can also result from unexpected circumstances, such as the loss of the Shuttle *Columbia*, which delayed the retrieval of one crew. If and when humans go to Mars, the mission could take over 2 years.

And, of course, technology is advancing in all areas, including space exploration. Over the years, electromechanical gauges that dominated cockpits were replaced first with cathode-ray tubes and now with digital displays. New technology is leading to new human-machine partnerships, with computer-based decision aids, improved communications, and increased availability of automated systems and robotics.

The Space Shuttle Program ended July 2011with the final space flight of *Atlantis*. However, the shuttle concept has been adopted by the USAF in its Boeing X-37, a portable spacecraft. Currently all astronauts and needed cargo are sent to the ISS via the Russian Soyuz. At the time of writing, contracts have been awarded for the development of commercial orbital transportation systems to resupply cargo on the ISS and a NASA deep space exploration spacecraft, the Orion Multipurpose Crew Vehicle. Elsewhere, China has successfully launched a module of its own orbiting space laboratory, Tiangong 1. This module also has successfully docked with *Shenzou 8*, the Chinese spacecraft. The Chinese hope to have a manned docking in the next year.

In the chapters that follow, psychologists, psychiatrists, and others with strong interests in space discuss selected research topics. Section I is organized around surviving and thriving in extreme environments. In Chap. 1, we trace the uneven course of psychology in the space program and describe the history of astronaut selection and psychological support. In Chap. 2, Sheryl Bishop points out that whereas there has been limited opportunity to study astronauts in space, there has been ample opportunity to study people in environments that in some ways resemble that of space. These analogues include polar camps and undersea research vessels that share danger, deprivation, isolation, confinement, and other characteristics with spacecraft, along with simulators intended to imitate or mimic spaceflight conditions. In comparison to studies conducted in "everyday" or laboratory settings, studies set in these more extreme environments offer a balance between accessibility and experimental control on the one hand and a degree of environmental realism on the other. Bishop discusses a wide range of analogues and simulators in the United States and abroad and notes that these are crucial for training purposes.

Spaceflight has positive and rewarding as well as stressful characteristics, and in Chap. 3, Julie Robinson and colleagues discuss one of these psychological benefits: observing Earth. They offer a unique study of taking pictures from space. This is an excellent example of an unobtrusive study, that is, one that does not set up expectations on the part of the research participants or infringe on their privacy. An overwhelming proportion of the photographs taken from the ISS are initiated by crewmembers. What kinds of substitute activities can we devise for some future missions when looking out the window may not be an option?

Space stations are a long way and very different from home. How can occupants preserve at least some semblance of their life on Earth? In Chap. 4, Phyllis J. Johnson tackles this issue, finding answers in astronauts' autobiographical documents, case histories, and pre- and post-mission interviews. NASA, the crew itself, and the crew members' families all contribute to making life on a space station a bit more homelike. NASA contributes through scheduling work and leisure, astronauts through their personalization of the space station and choice of leisure time activities (e.g., by celebrating national and religious holidays), and families through frequent and supportive communications from home.

In Chap. 5, which opens Section II on the interpersonal dimensions of space exploration, Nick Kanas identifies ways that missions can be more successful through careful attention to issues such as communication and leadership. Reviewing research drawn from *Mir* and International Space Station missions as well as a space simulation study, Kanas details the importance of astronauts dealing effectively with the tensions that naturally arise in the stressful environment of space, rather than displacing them on Mission Control. Successful leaders are able to provide support while also helping keep crewmembers on task, each at the appropriate time. Kanas recommends pre-launch training in dealing with interpersonal issues for both astronauts and ground personnel, followed by ongoing instructional review and support during the mission and after astronauts return home. In spite of the challenges of space travel, Kanas notes the positive impacts of being in space, such as an enhanced appreciation for the natural beauty of the Earth.

In Chap. 6, Harvey Wichman points out that soon, spaceflight may no longer be a government monopoly and future spacefarers may include growing proportions of tourists and industrial workers. This development may require departing from the government agency form of organization that has dominated space exploration so far in favor of a private enterprise model of commercial space exploration, and require accommodating people who lack the qualifications of today's astronauts and cosmonauts. In his view, society is at a historical threshold that will require a shift in how engineers, designers, flight managers, and crews perform their tasks. He illustrates some of these points with his industry-sponsored simulation study intended to gauge tourist reactions to spaceflight. Among other findings of the study, the simulation provided insights into ways that harmonious interpersonal relationships during missions can be increased through preflight instruction in conflict resolution.

Group dynamics is a focal point for Jason Kring and Megan A. Kaminski, who, in Chap. 7 explore gender effects on social interaction and the determinants of interpersonal cohesion (crewmember attraction to one another) and task cohesion (commitment to the work at hand). Their review of the basic literature on mixed-gender groups, as well as findings from spaceflight and other extreme environments, points to the conclusion that whereas there are many benefits to mixed-gender crews (typically, men and women bring different skills to the mix), the issue is multifaceted and complex and poses challenges for spaceflight operations. Although researchers are gaining some understanding of the determinants of crew cohesion, the effects on performance depend upon the type of cohesion (interpersonal and task) and the nature of the task. None of this is simplified by findings that cohesion is likely to fluctuate over the course of an extended mission.

In Chap. 8, Gro Mjeldheim Sandal, Hege H. Bye, and Fons J. R. van de Vijver explore the effects of personal values and value change over the course of a mission simulation. Initially they expected their research subjects' personal values in such areas as hedonism (pleasure or sensuous gratification), benevolence (preserving and enhancing the welfare of the group), and tradition (respect, acceptance, and commitment to the ideas and customs of cultural or religious groups) to converge during the course of the 105-day confinement. This increased homogeneity of values would lead to reduced interpersonal tensions. Nothing of that sort happened. Results showed that differences in values increased over time, and that there were greater tensions due to value conflict at the end rather than beginning of the mission.

Section III includes four chapters that discuss the cross-cultural dimensions of space missions. In Chap. 9, a study group appointed by the International Academy of Astronautics and led by Nick Kanas discuss the multifaceted impact of culture on psychological, psychiatric, interpersonal, and organizational issues that have or may arise during long-term (6 month or over) missions. Culture is a broad and pervasive matrix of various factors that affects personal welfare and relationships with other crewmembers and with people back home. Culture affects neuropsychiatric dysfunction, personal coping strategies, psychiatric diagnoses, and treatment. One of the many strengths of this comprehensive overview is its many specific suggestions for research and operations. In Chap. 10, "Flying with

Strangers," Peter Suedfeld, Kasia E. Wilk, and Lindi Cassel draw a distinction between multinational crews, in which "guests" were allowed to participate in US or Soviet/Russian missions, and international crews, which first appeared aboard the International Space Station, which is not owned or operated by any one nation. Through studying the reminiscences of majority and minority participants in multinational and international missions, they test the hypothesis that multinational flights are a source of frustration and annoyance that are not evident in the true partnerships of international flights. In the final chapter in this section, Juris G. Draguns and Albert A. Harrison propose applying a culture assimilator to build cross-cultural awareness and sensitivity. They point out that cross-cultural psychology and spaceflight psychology started at about the same time – and have much to learn from each other.

As missions proceed farther and farther from Earth, real-time communication will become increasingly delayed and on a mission measured in years it will not be possible to anticipate, never mind rehearse, every task to perfection. Crew autonomy is the first focal point for Section IV. In Chap. 12, Kanas and his team report pilot studies from three space-analogous settings: the NEEMO underwater habitat, the Haughton-Mars project, and the pilot phase for the Mars500 simulation. Although findings are tentative due to the limited size of the samples, greater autonomy was associated with increased creativity, improved performance, and higher mood and morale. Increased crew autonomy led to some confusion among mission controllers who became less certain about their work roles. Then in Chap. 13, Peter Roma and his associates describe their experiment where volunteers participated in a game-like simulation of a Mars mission. This required coordination between mission control and an exploration crew roving the red planet's surface. In one variation, exploration proceeded under strict direction and in another, condition subjects could exercise discretion and respond flexibly. Consistent with findings from industrial and organizational psychology, Roma et al. found greater autonomy was associated with improved performance and fewer negative emotional states.

The next two chapters turn to the distant future when space missions might go to the outer reaches of the solar system or beyond. In Chap. 14, Harvey Wichman discusses some of the technology that might increase the plausibility of a round trip to Saturn and its moon Titan by shortening the duration of the mission. He identifies possible "deal-breakers," such as weightlessness and deep space radiation, and what might be done to circumvent them. In Chap. 15, Nick Kanas explores some of the psychological issues that are likely to arise on truly extended, perhaps multigenerational missions. Examples include a severe break-off from Earth and the challenges of re-adapting to Earth after an absence of many decades.

In the book's afterword, Gro Mjeldheim Sandal and Gloria R. Leon present a summary and integration that places the earlier chapters within broader historical, cultural, and organizational contexts. They note that whereas we can point with pride to past accomplishments, missions will continue to change and there will always be a need for more research and new operational procedures. The research that is done—and, perhaps more importantly, that is not done—reflects political as well as scientific and operational concerns. Many of psychology's advances within the American program are recent, and it is not clear if these gains will withstand the test of time. However, sponsors of other space programs, such as the ESA, RSA, CSA, and JAXA understand that psychology and psychiatry are among the many disciplines required to ensure successful spaceflight.

We conclude our introduction with three important caveats. First, although most of the chapters in this book are authored or coauthored by psychologists and psychiatrists, understanding and managing human behavior in space is an interdisciplinary effort. In essence, "spaceflight psychology" includes contributions from architecture and design, engineering, biology, medicine, anthropology, sociology, communications, and organizational studies, as well as many hybrids (such as cognitive science). In a similar vein, the delivery of psychological services to astronauts involves physicians, social workers, and peers, as well as psychologists and psychiatrists.

Second, no one pretends that the chapters in this book are representative of behavioral research as a whole. Our chapters do not provide in-depth treatment of the interface between engineering and psychology, nor do they attend to the interface of biology and behavior, for example, the effects of cumulative fatigue and circadian rhythms on performance and risk. With respect to this, we note a recent chapter by Barbara Woolford and Frances Mount that describes how, over the past 40 years, research on anthropometrics, biomechanics, architecture, and other ergonomics issues slowly shifted from surviving and functioning in microgravity to designing space vehicles and habitats to produce the greatest returns for human knowledge (Woolford and Mount 2006).

Finally, apart from psychological studies of astronauts, we acknowledge many other areas where behavioral science interfaces with NASA. For example, NASA maintains an excellent program in aviation human factors. Even robotic missions, such as those already dispatched to Mars, have a human touch. It is necessary to assemble, organize, and train teams to manage such missions. Considerable preparation is necessary for successful teleoperations, for example, Earth-bound researchers conducting a "glovebox" experiment aboard a satellite thousands of miles away or driving a teleoperated rover on Mars. Satellites devoted to remote sensing must be designed with human sensory, perceptual, and information processing systems in mind. Furthermore, the loss of *Challenger* and *Columbia* reflected organizational and behavioral factors such as miscommunication and faulty judgment as well as technical failures (McConnell 1987; Vaughan 1996). Astronauts in flight and to a lesser extent mission support personnel are the focal point of this book, but there are many areas where behavioral research and application contribute to NASA.

References

Ball, J.R., Evans, C.H. (eds.): Safe Passage: Astronaut Care for Exploration Missions. National Academy Press, Washington, DC (2001)

Brady, J.V.: Behavioral health: the propaedeutic requirement. Aviat. Space Environ. Med. **76**(6), B13–B23 (2005). sect. II

Chapanis, A.: The international ergonomics association: its first 30 years. Ergonomics **33**(3), 275–282 (1990)
Committee on Space Biology and Medicine: A Strategy for Space Medicine and Medical Science for the 1980s and 1990s. National Academy Press, Washington, DC (1987)
Committee on Space Biology and Medicine: A Strategy for Space Biology and Medicine in the New Century. National Academy Press, Washington, DC (1998)
Connors, M.M., Harrison, A.A., Akins, F.R.: Living Aloft: Human Requirements for Extended Spaceflight. NASA SP-483, Washington, DC (1985)
Flaherty, B.E.: Psychophysiological Aspects of Space Flight. Columbia University Press, New York (1961)
Foushee, H.C., Helmreich, R.L.: Group interactions and flight crew performance. In: Wiener, E.L., Nagel, D.C. (eds.) Human Factors in Aviation, pp. 189–228. Academic Press, New York (1998)
Grether, W.F.: Psychology and the space frontier. Am. Psychol. **17**(2), 92–101 (1962)
Harris, P.R.: Personnel deployment systems: managing people in polar and outer space environments. In: Harrison, A.A., Clearwater, Y.A., McKay, C.P. (eds.) From Antarctica to Outer Space: Life in Isolation and Confinement, pp. 65–80. Springer, New York (1990)
Helmreich, R.L.: Applied social psychology: the unfulfilled promise. Pers Soc Psychol Bull **1**, 548–561 (1975)
Helmreich, R.L.: Applying psychology to outer space: unfulfilled promises revisited. Am. Psychol. **38**, 445–450 (1983)
Kanas, N.: Psychosocial factors affecting simulated and actual space missions. Aviat. Space Environ. Med. **56**(8), 806–811 (August 1985)
Kanas, N.: Psychosocial support for cosmonauts. Aviat. Space Environ. Med. **62**(4), 353–355 (August 1991)
Kanas, N., Salnitskiy, V.P., Ritsher, J.B., Gushin, V.I., Weiss, D.S., Saylor, S.A., Kozerenko, O. Po., Marmar, C.R.: Human interactions in space: ISS vs. Shuttle/Mir. Acta Astronaut. **59**, 413–419 (2006)
Kanas, N., Fedderson, W.E., Behavioral, Psychiatric, and Sociological Problems of Long Duration Missions. NASA Technical Memorandum X-58067, Washington, DC (1971)
Kass, J., Kass, R., Samaltedinov, I.: Psychological problems of man in space: problems and solutions. Acta Astronaut. **36**, 657–660 (1995)
Kubis, J.F.: Isolation, confinement, and group dynamics in long duration spaceflight. Acta Astronaut. **17**, 45–72 (1972)
Kubis, J.F., McLaughlin, E.J.: Psychological aspects of spaceflight. Trans. N. Y. Acad. Sci. **30**(2), 320–330 (1967)
Lindsley, D.B. (ed.): Human Factors in Long Duration Spaceflight. National Academy of Sciences, Washington, DC (1972)
McConnell, M.M.: Challenger: A Major Malfunction. Doubleday, New York (1987)
Meister, D.: Conceptual Aspects of Human Factors. Johns Hopkins University Press, Baltimore (1989)
Palinkas, L.A.: Psychosocial issues in long-term spaceflight: an overview. Gravit. Space Biol. Bull. **12**(2), 25–33 (2001)
Santy, P.A.: Choosing the Right Stuff: The Psychological Selection of Astronauts and Cosmonauts. Praeger, Westport (1994)
Science@NASA: How Astronauts Get Along, available at http://science.nasa.gov/science-news/science-at-nasa/2002/21oct_getalong/ (21 Oct 2002). Accessed 13 Mar 2012
Sells, S.B., Berry, C.A. (eds.): Human Factors in Space and Jet Travel: A Medical-Psychological Analysis. Ronald Press, New York (1961)
Sipes, W.E., Vander Ark, S.T.: Operational behavioral health and performance resources for international space station crews and families. Aviat. Space Environ. Med. **76**(6), B36–B41 (2005). Sec. II

Suedfeld, P.: Invulnerability, coping, salutogenesis, integration: four phases of spaceflight psychology. Aviat. Space Environ. Med. **76**(6), B61–B66 (2005). sect. II

Vaughan, D.: The Challenger Launch Decision: Risky Technology, Culture and Deviance at NASA. University of Chicago Press, Chicago (1996)

Weitekamp, M.A.: Right Stuff, Wrong Sex: America's First Women in Space Program. Johns Hopkins University Press, Baltimore (2004)

Woolford, B., Mount, F.: Human spaceflight 3rd ed. In: Salvendi, G. (ed.) Handbook of Human Factors and Ergonomics, 3rd ed. pp. 929–955. Wiley, Hoboken (2006)

Part I
Surviving and Thriving in Extreme Environments

Chapter 1
Behavioral Health

Albert A. Harrison and Edna R. Fiedler

Abstract Experience gained from test pilots, high-altitude balloonists, and animals sent on rocket flights was the starting point for understanding astronaut adaptation and performance in space. Psychology played a significant role in Project Mercury, but before that effort was complete, official interest in such topics as astronaut selection, psychosocial adjustment, group dynamics, and psychological support all but disappeared. Interest was rekindled when astronauts joined cosmonauts on *Mir* and then became full partners on the International Space Station. We review reasons for this period of minimal involvement in the space program and suggest that the "right stuff" image worked against psychology and psychiatry until the mid-1990s, when space station expeditions brought the challenges of long-duration missions into focus. Evidence of renewed interest included the advent of the National Space Biomedical Research Institute, the development of NASA's Bioastronautics Critical Path Roadmap, and the new Human Research Program. In 2001, *Safe Passage: Astronaut Care for Exploration Missions* drew attention to behavioral health, a perspective on psychosocial adjustment that depends not only an absence of neuropsychiatric dysfunction but on the presence of positive interactions with the physical and social environments. We trace the history and current status of astronaut selection and psychological support, two essential ingredients for maintaining behavioral health, from Mercury to the ISS. Behavioral health is important because it reduces risk, helps optimize performance, and contributes to the welfare of astronauts, their coworkers, and their families. We conclude with a brief outline for a comprehensive and continuing program in spaceflight behavioral health.

A.A. Harrison (✉)
University of California, Davis, USA
e-mail: aaharrison@ucdavis.edu

E.R. Fiedler
Aerospace Psychology Consultants, Tucson, USA

D.A. Vakoch (ed.), *On Orbit and Beyond*, Space Technology Library 29,
DOI 10.1007/978-3-642-30583-2_1,

1.1 Introduction

In the 1950s, as America prepared for its first crewed space missions, it was not clear that human performance capabilities could be maintained under the demanding conditions of spaceflight. Where could NASA begin? Much of the research, equipment, and testing procedures used to support test pilots who set successive speed and altitude records transferred easily to the early space program (Wolfe 1979; Weitekamp 2004). Decompression chambers, centrifuges, rocket sleds, and the like made it possible to explore the physiological and performance aspects of conditions that would be encountered in space. Craig Ryan has detailed the contributions of high-altitude ballooning, highlighting the usefulness of gondola designs (which he contends provided a basis for the Mercury spacecraft), flight suits, helmets, and much more (Ryan 1995). Not everything could be "off the shelf"; NASA had to develop elaborate simulators for upcoming space missions. But, on the whole, the same "cast of characters"—engineers, physicians, psychologists and psychiatrists, to mention a few—who brought America to the edge of space brought America into space.

Animal studies offered some reassurance that humans could adapt physiologically and behaviorally to space (Burgess and Dubbs 2007). As early as the late 1940s, biological specimens were launched on balloons and sounding rockets. In 1958, the Russians successfully launched a dog, Laika, who soon died in orbit, but could not be brought back to Earth. Wernher von Braun approached behavioral biologist Joseph V. Brady to see if he would be willing to launch primates, which would leapfrog the Soviets' dogs (Anon 2005). In 1958 and 1959, America's first primate spacefarers, two squirrel monkeys named Able and Baker (known at that time as *Miss* Able and *Miss* Baker) were launched on 15-min flights reaching an altitude of 300 miles on a 1,500-mile trajectory and were successfully recovered following splashdown.

One of the main questions was whether the test animals could do what they had been trained to do during the presumably terrifying rocket rides. Able and Baker were encased in casts to protect them against gravitational changes, but one finger and one toe were exposed so that, after a warning light turned on, the finger could be used to press a lever to avoid a shock to the toe. All the way up and all the way down, they pressed the lever on cue. Later, as a part of the Mercury pretest program, the chimpanzees Ham and Enos received much more elaborate and sophisticated training than did their predecessors (Rholes et al. 1963). They flew in special couches within Mercury capsules; Ham's flight was suborbital, but Enoch completed four orbits. Although acceleration and deceleration forces in excess of 7 g's had an immediate effect on the chimpanzees' performance, once these forces diminished, their performance bounced back to preflight levels. Microgravity did not interfere with visual processes (monitoring the lights), nor did it interfere with eating and drinking. Not only did they perform their assigned tasks in space, but the two chimpanzees also returned to Earth in good health and with their sharply honed skills intact (Brady 2005). Looking back at an episode from this era, Joseph Brady recounted:

> On the recovery ship, after the helicopter had dropped the capsule once or twice before obtaining a good connection on one of these animal pre-test flights—a good reason for practicing before the human flights—the hatch was opened on the flight deck and the chimp came out sputtering and thrashing about. An admiral standing on the deck with several of us said something like "If that chimp could only talk", in response to which I felt required to observe that the best thing that ever happened to us was that the chimp could not talk or the space program might have come to an abrupt end right on the spot (Anonymous 2005, p. 1811).

During the early 1960s, the United States and Soviet Russia were locked in a race to the Moon, and in many ways, the two programs paralleled each other. In the United States, solo missions (Mercury) gave way to two-person missions (Gemini) and then to three-person missions (Apollo) that, in July of 1969, brought astronauts to the Moon. The Apollo Applications Program followed close on the heels of the last astronaut's departure from the Moon. Based on leftover Moon race equipment, the Apollo Applications Program included the Apollo-Soyuz Test Project, where Americans and Soviets joined spacecraft to live together briefly in space, and *Skylab,* America's "house in space" in the mid-1970s (Cooper 1976). By the late 1970s, the U.S. and Soviet programs were following different paths: Americans awaited the orbiter, or Space Shuttle, and Soviets launched a series of space stations. In 1984, President Ronald Reagan approved the development of a U.S. space station, but construction was delayed almost 15 years. President Bill Clinton approved the station as a multinational venture, and it became the International Space Station, or ISS. Prior to its construction, American astronauts joined Russian cosmonauts on *Mir*; later, they worked together as partners on the ISS. As of November, 2012, the ISS reached its 12th anniversary of having multinational crews living and working in space.

Although psychology played a prominent role in the early U.S. space program, some branches had all but disappeared by 1963. To be sure, psychologists did show professional interest in humans in space, and many panels and commissions sought to increase psychology's involvement (see Introduction). Since there were practically no studies of astronauts, researchers relied heavily on studies conducted in Antarctica, submarines and research submersibles, and simulators. Research continues in all three venues; Antarctica took an early lead and remained prominent for many years (Palinkas 2003). A primary reason for this ascendance was that the International Geophysical "Year" (IGY, 1957–1959) stimulated research on human adaptation to isolation and confinement, with authoritative and influential accounts appearing in the early 1970s (Gunderson 1973; Rasmussen 1973).

Other factors that favored Antarctica were the large number of people who ventured there and that, as an international site, it offers opportunities for researchers from many different nations. By picking and choosing research locations, one can find conditions that resemble those of many different kinds of space missions, ranging from relatively luxurious space stations to primitive extraterrestrial camps (Andersen et al. 1990). In 1963, Robert Voas, one of the early space human factors experts, and E. K. Eric Gunderson, a pioneer of psychological research in Antarctica, seriously discussed developing a space mission simulator there, an idea that reemerges from time to time (Gunderson 1990). By the 1980s, it was recognized widely that Antarctica provided a useful meeting ground for people who were interested in adaptation to polar environments and people who were interested in adaptation to space. In 1987, NASA and the National Science Foundation's Division of Polar Programs joined together to sponsor the "Sunnyvale Conference," which brought together researchers from each tradition. Presentations centered on environments (Antarctica and space), theoretical perspectives, isolation and confinement effects, and interventions and outcomes (Harrison et al. 1989, 1991). Antarctic behavioral research became a truly international venture guided in part by the Scientific Committee for Antarctic Research and funded by many sources, including NASA. For example, Des Lugg of NASA Headquarters and Joanna Woods at Johnson Space Center conducted medical and psychological research with the Australian National Antarctic Research Expeditions (Wood et al. 2005; Lugg 2005). The next chapter provides a more detailed discussion of analogue environments.

As noted in the introduction, despite repetitive calls for action, empirical research was slow to accumulate. In the late 1990s, the National Academy of Sciences undertook a comprehensive review of behavioral and medical issues that we need to begin to address right now prepare for future space missions. We consider the Academy's report, *Safe Passage: Astronaut Care for Exploration Missions*, a watershed event (Ball and Evans 2001). Like earlier calls to action, *Safe Passage* drew attention to many biomedical, behavioral, and psychological issues and emphasized their importance for health, performance, and welfare on extended-duration missions. The timing was good because its production and distribution coincided with Americans living on *Mir* and the first missions to the ISS. Although future-oriented, *Safe Passage* was developed in the context of unfolding events on then-contemporary extended-duration missions. Most importantly, this work also introduced the concept of behavioral health, an idea that may be particularly useful because of its breadth and relative lack of pejorative connotations. According to one recent definition, "Compared with earlier formulations (such as mental health), behavioral health is less limited in that it recognizes that effective, positive behavior depends on an interaction with the physical and social environments, as well as an absence of neuropsychiatric dysfunction. Behavioral health is evident not only at the level of the individual, but also at the levels of the group and organization" (Harrison 2005, p. B3).

NASA's recognition of the field of behavioral health and linking of it to performance opened the door for many of the kinds of research that earlier were thought to be too "soft" to be useful to the space program (Bioastronautics Critical Path Roadmap 2012). Today, NASA has shown increased recognition of shared perspectives, privacy, leisure-time activity, family separation and reunification, cultural awareness, the satisfying properties of windows and view ports, and many other topics that were formerly overlooked if not seen as irrelevant or frivolous. From NASA's perspective, the significance of these factors is less in the fact that they can help people "feel good" (although many psychologists would argue that this is a major benefit) than in their potential positive impact on risk and performance. This research, in turn, has implications for organizing and staging space missions. Thus, a combination of maturing social science and interest sparked by space station and exploration missions has opened the door, at least partially, for new kinds of psychological research within the U.S. space program.

1.2 The Right Stuff

For decades, expanding the role of psychology in the U.S. space program was an uphill battle with psychologists' pleas generally falling on deaf ears. Among the more obvious interpretations, it might be tempting to think of NASA managers and engineers as "thing" people rather than "people" people, so the behavioral side of spaceflight is of little interest to them. Perhaps mission managers were simply unaware of the significance of behavioral factors. Or maybe, as "hard" scientists, they saw the behavioral and social sciences as fuzzy if not bumbling efforts that lead to qualitative recommendations that are difficult to implement and unlikely to work. The sociologist Charles Perrow has discussed how resistance to human factors within complex organizations has strong structural and cultural underpinnings and is not overcome easily (Perrow 1983).

Psychologists make contributions to human welfare in such diverse areas as environmental design, problem-solving, decision-making, leadership, and group performance, but many people strongly associate psychology with mental illness and long-term psychotherapy. If such attitudes explained NASA's ambivalence about behavioral factors, education would be the antidote; but for many years, educational efforts had little visible impact on research or mission operations.

The stereotype of clinical psychologists and psychiatrists working with troubled clients may have threatening implications for NASA administrators who need to maintain good public relations and build government support. The historian Roger Launius (Launius 2005, p. 4) points out that from the moment they were introduced to the public in 1959, America was enthralled by the "virtuous, no nonsense, able and professional astronauts" who "put a very human face on the grandest technological endeavor in history" and "represented the very best that we had to offer" (Launius 2005). From the beginning, the press was never motivated to dig up dirt on the astronauts; rather, reporters sought confirmation that they embodied America's deepest virtues: "They wanted to demonstrate to their readers that the Mercury seven strode the Earth as latter-day saviors whose purity coupled with noble deeds would purge this land of the evils of communism by besting the Soviet Union on the world stage" (Launius 2005, p. 4). Today, Launius continues, people look back longingly to a simpler era when good was good and evil was evil, and, at least in memory, heroes did not disappoint. Psychological research or, worse yet, the faintest possibility that a mission would be compromised by psychological factors could be a public relations nightmare.

For project managers and engineers, faith in the right stuff helps cut costs because the person can be engineered out of the equation. Such conviction simplifies and speeds the design process as there is no need to waste time consulting behavior experts. Sliding by psychological issues preserves managerial autonomy and decision-making power. If behavioral professionals were to serve as consequential advisors, mission directors would have to share control, or at least seriously consider the opinion of behavioral experts. Why should managers complicate their task by bringing more players—psychologists, psychiatrists, anthropologists, human factors experts—to the table?

For astronauts, the stereotype of the right stuff helps maintain flight status (Santy 1994). It deters snooping and prying that might suggest a real or imagined blemish that could lead to mission disqualification, a most dreaded result. After all, part of the heroic myth is that under the greatest of adversities, people with the right stuff can still get the job done! Why risk all by getting involved in a research program that could lead to new reasons for disqualification? George Low, manager of the Apollo program, advised subordinates that identity issues, past or present, were off-limits and that personal hang-ups should be put aside in favor of the mission (McQuaid 2007). Michael Collins and his colleagues liked the John Wayne–type image created for the early astronauts and did not want it tarnished (Weitekamp 2004). Flying in space was a macho, masculine endeavor, and there were those who made an effort to reserve the term "astronaut" for men, referring to women who sought to fly in space as "astronautix," "astro-nettes," "feminauts," and "space girls" (Weitekamp 2004).

Marc Shepanek points out that today's astronauts are very much aware of the possible effects of stress, boredom, and many other stressors on safety, performance, and quality of life in space (Shepanek 2005). He notes that while many of them favor research on these topics, not all stand ready to volunteer as test subjects. The concern is that despite strong assurances of confidentiality, one of the results of their participation could be disqualification. This means that operational psychologists cannot also conduct research: the role of the therapist or consulting organizational psychologist must remain sacrosanct with no hints of dual allegiance to research (Flynn 2005). Many kinds of workers, including those in the military and law enforcement, worry about breaches of confidentiality that have adverse repercussions on their careers. Worries about breaches of confidentiality are periodically reinforced by officials who release information despite assurances to the contrary.

Efforts to protect the astronauts' image are evident in the cordon that NASA public relations and legal teams establish to prevent outsiders from obtaining potentially damaging information, the micromanagement of astronauts' public appearances, and the great care with which most astronauts comport themselves in public. Even today, there are topics that are considered "too hot" to be included in otherwise comprehensive and informed discussions.

"The right stuff" is an abstraction or ideal type that living, breathing human astronauts approximate but do not fully attain. By the beginning of the twenty-first century, cracks began to appear in this image. Researchers had long noted behavioral problems in spaceflight-like environments and worried about what might happen during future space missions. Hints of problems came from the Russian space program, which seemed more attuned to the significance of psychological issues. For Americans, conditions that had been heralded since the 1960s became realities in the 1990s when U.S. astronauts joined Russian cosmonauts on *Mir*, living and working in space for prolonged periods of time with peers from a very different culture. A few astronauts described some of the behavioral challenges that they encountered in space: maintaining high performance in the face of extreme danger, loneliness, and minor conflicts with other crewmembers (Burrough 1998; Holland 2000). On the debit side of the balance sheet, members of isolated and confined groups frequently report sleep disturbances, somatic complaints (aches, pains, and a constellation of flu-like symptoms sometimes known as the "space crud"), heart palpitations, anxiety, mood swings including mild depression, inconsistent motivation, and performance decrements. Although the usual approach among crewmembers and with Mission Control is business-like, friendly and positive, there have been recorded incidents of frustration. Crewmembers may sometimes withdraw from one another, get into conflicts with each other or in disputes with Mission Control. Eugene Cernan reports an example of an early conflict between the Apollo 7 crew and Mission Control. The conflicts were so severe that the astronauts never flew again (Cernan and Davis 1999). Both Bryan Burrough and Al Holland have described some of the difficulties that U.S. astronauts experienced on *Mir* (Burrough 1998; Holland 2000). Burrough writes that Soyuz 21 (1976), Soyuz T-14 (1985), and Soyuz TM-2 (1987) were shortened because of mood, performance, and interpersonal issues. Brian Harvey wrote that psychological factors contributed to the early evacuation of a *Salyut 7* crew (Harvey 1996). U.S. researchers and flight surgeons have acknowledged instances of fear, anxiety, depression, sleep disorders, cognitive changes, somatization, impulsive behaviors, social withdrawal, cultural misunderstandings, interpersonal frictions, and anger directed toward Mission Control. After their return, some astronauts reported depression, substance abuse issues, marital discord, and jealousy (Flynn 2005; Shepanek 2005; Suedfeld 2005). Astronauts are highly competent, task-oriented people, who, like other highly functional adults, have the normal ups and downs in their moods and social relationships. And, as in the case of other highly functional adults, these ups and downs can sometimes reduce their effectiveness and impair personal relationships.

It is not only the normal ups and downs of the individual astronaut that affect the teams and their work, but also the pressures and occasionally dysfunctional dynamics of the organization and Mission Control. The Mercury astronauts lobbied aggressively to fly as pilots rather than to ride as mere passengers ("Spam in a can") whose spacecraft were controlled from the ground (Wolfe 1979). H. S. F. Cooper wrote a well-publicized account of conflict between the *Skylab 4* crew and Mission Control (Cooper 1976). At the heart of the matter was the over-programming of the astronauts' time. As psychologist Karl Weick described the situation:

> To get the most information from this final trip in the Apollo program, ground control in Houston had removed virtually all the slack from the astronauts' schedule of activities and had treated the men as if they were robots. To get everything in, ground control shortened meal times, reduced setup times for experiments, and made no allowance for the fact that previous crews aboard Skylab had stowed equipment in an unsystematic manner. The astronauts' favorite pastimes—watching the sun and earth—were forbidden (Weick 1977, p. 31).

Thus, on 27 December 1973, the Skylab 4 astronauts conducted a daylong "sit-down strike." Cooper described the crew pejoratively as hostile, irritable, and downright grumpy, while other writers have described the "strike" as a legitimate reaction to overwork (Connors et al. 1985). William K. Douglas, a NASA flight surgeon, lamented both Cooper's emotionally toned reporting and people's willingness to focus on others' real or imagined failures while overlooking greatness (Douglas 1990). Whatever the "spin" on this particular event, the lessons are clear: the same rapid pace that can be sustained for brief sprints cannot be sustained for marathons. Give astronauts the flexibility to schedule their own activities, and allow time to look out the windows. NASA appears to have taken the lesson to heart. In 2002, Space.Com's Todd Halvorson conducted an interview with enthusiastic ISS astronaut Susan Helms. "It's not that the crew isn't busy maintaining the station, testing the remote manipulator and conducting science, it's that there remains enough time to look out the window, do somersaults in weightlessness, watch movies, and sit around chatting" (Halvorson 2010).

Spaceflight also offers opportunities for psychological growth and development (Harrison and Summit 1991). Training for and working in space allows people to develop their abilities, gain a strong sense of accomplishment, and feel worthwhile. There is unparalleled challenge, the opportunity to redefine one's place in the cosmos. There is the exhilarating feeling, as Harrison Schmitt (1996, p. 37) wrote, of actually "being there" (Schmidt 1996). Walter Cunningham (1977, p. 27) wrote, "It has caused me to seek a challenge wherever I can find one, to charge ahead and never look back . . . that feeling of omnipotence is worth all that it takes to get there." Many of the two dozen or so astronauts and cosmonauts interviewed by Frank White reported "overview effects," truly transformative experiences including senses of wonder and awe, unity with nature, transcendence, and universal brotherhood (White 1987). More testimonials concerning the psychological benefits of life in space come from Apollo 14 astronaut Edgar Mitchell (Mitchell and Williams 1996) and Shuttle-*Mir* astronaut-cosmonaut Jerry Linenger (2000). Astronauts and cosmonauts like the sense of adventure, camaraderie, and grandeur in space (Suedfeld 2005).

We find hints of long-term physical and mental health benefits to life in challenging environments. For example, a long-term follow-up study of Navy personnel who had wintered in Antarctica revealed that following their return, they had undergone fewer hospitalizations than their peers who had identical qualifications but whose orders to go to the South Pole were rescinded as the result of an arbitrary administrative decision (Palinkas 1990). Studies of the mental health of cosmonauts conducted two or three years after their return to Earth found that they were less anxious, depressed, or aggressive and showed reduced hypochondriasis (Myasnikov and Zamaletdinov 1996). The most plausible explanation is that during their stay in tough environments, people develop coping skills, that is, ways of dealing with challenge and stress that continue to serve them well long after they have returned from their expedition.

It was about the time astronauts began traveling on *Mir* and the ISS that greater evidence of psychology began to show in the U.S. space program. NASA's Bioastronautics Critical Path Roadmap (BCPR) is one piece of evidence. Bioastronautics was NASA's shorthand for life in space, and the BCPR was a framework for identifying the knowledge that NASA needs for future space missions (Bioastronautics Critical Path Roadmap 2012). It identified and assigned priorities to the biomedical and behavioral questions that must be addressed (and the kinds of countermeasures that must be designed) for Space Station, lunar, and Mars missions. The BCPR represented a major investment of time and energy, of soliciting and responding to expert advice, and of building consensus. It recognized that NASA's organizational chart was not isomorphic with the way that research is traditionally organized and tried assiduously to address crucial gaps. The BCPR was a useful mechanism for organizing biomedical and behavioral research and fostered research that yielded operationally relevant results. Most importantly, it represented a higher level of "buy-in" to behavioral research on the part of the space agency. Later, the BCPR evolved into the Human Research Program (HRP) (Human Research and Safety 2012). This has both study areas and an education and outreach arm. The three broad areas of study for the HRP are physiology, environment, and technology. Physiology encompasses bone health, muscle function, cardiovascular response, senorimotor, immunology, and behavioral health and performance. Environment includes human factors and habitability, lunar dust, microbiology, and radiation. Technology has three components: exercise, food and nutrition, and exploration medical capability.

Also coincident with turn-of-the-millennium space station missions was the initiation of the National Space Biomedical Research Institute (NSBRI), a consortium of universities and businesses dedicated to solving the problems of astronauts who are undertaking long-duration missions. The NSBRI is best viewed as tightly networked centers of excellence. Members of affiliated organizations form interdisciplinary teams that cut across organizational boundaries and draw strength from one another.

In 2011, NASA chose the Center for the Advancement of Science in Space (CASIS), a non-profit organization to manage the ISS US National Laboratory. CASIS will maximize the ISS US lab through research pathways, marketplace opportunities, investigator support, academic advancement, and advocacy. CASIS also has educational and advocacy outreach elements. CASIS is a portal for investigators interested in research within the extreme environment of space and microgravity (NASA 2012).

In 2003, NASA commissioned a workshop on spaceflight behavioral health. The primary purpose of this workshop was to bring together researchers and practitioners in an effort to identify research gaps and produce an archival record for use by managers, established behavioral health researchers, and newcomers to the field (Harrison 2005). Also, astronauts have begun to respond to questionnaires on such topics as noise levels and communication (Kelley and Kanas 1992, 1993a, b). Astronauts have participated in flight studies involving sleep and circadian rhythms and have taken self-administered tests of cognitive ability, maintained diaries, and provided other information from orbit (Mallis and DeRoshia 2005; Kane et al. 2005). Compared to those of earlier years, many of today's astronauts are more willing to participate in ground-based and in-flight studies, given proper assurances of confidentiality.

We suggest that the NASA-*Mir* missions opened a window of opportunity for fruitful reevaluation of the role of behavior, including psychosocial adaptation, in

U.S. space missions. When extended-duration missions moved from the abstract and theoretical to the real and some astronauts broached topics like risk, loneliness, and culture conflicts, psychological factors were brought into sharp focus. In policy studies, a window of opportunity opens when a major, unexpected catastrophe (known as a focusing event) becomes known to policy-makers and the public at the same time (Birkland 1997). Certainly, minor problems on *Mir* were far removed from catastrophic, but behavioral issues gained salience and became known to NASA officials and the public at the same time. The astronauts' experiences on *Mir* opened a window that generated interest in spaceflight behavioral health.

In 1983, Robert Helmreich pointed out that in contrast to Americans, the Russians seemed to have always maintained a certain degree of interest in psychosocial adaptation (Helmreich 1983; Santy 1994). He reprinted several quotes from cosmonauts showing interest in psychosocial adjustment, group dynamics, and related topics, and he pointed to the publication of a collection of papers on space psychology by Petrov et al. (1979). Nick Kanas and his associates have written extensively on the role of psychology in the Soviet and then Russian space programs and have highlighted the potential value of this research for NASA (Kanas 1985, 1991; Kanas et al. 2006). By the mid-1980s, Oleg Gazenko, head of Soviet Space Medicine, concluded that the limitations of living in space are not medical, but psychological (Oberg and Oberg 1986). Quotes from cosmonaut diaries and Soviet/Russian reports remain popular for illustrating the importance of stress, mental health, crew dynamics, and the like, in part because for a long time, neither NASA support personnel nor astronauts themselves freely commented on such issues.

In the early 1970s, after only three crewed missions, America's "House in Space," *Skylab,* was abandoned. The United States invested in the Shuttle, which supports fairly large crews, but for only short times in space. America expected a space station, but it was not approved until 1984, and the station itself underwent several iterations (Space Station, Space Station Alpha, and Space Station Freedom) before being abandoned for the ISS. The Soviets, on the other hand, moved directly into the era of Salyut and *Mir* space station missions. For them, extended-duration missions—and focusing events in the area of behavioral health—became a reality decades ago. As Mary M. Connors, Albert A. Harrison and Faren R. Aikens wrote in 1986, "The Russians have experienced longer spaceflights than their American counterparts and have given considerable attention to ways of maintaining individuals' psychological health and high morale in space . . .. In the Soviet Union, the Group for Psychological Support is an acknowledged and welcomed component of the ground team. Concern over such issues as intragroup compatibility and the effects of boredom on productivity seem to be actively studied by cosmonauts and psychologists alike. There appears to be little if any loss of status associated with confirmation of psychological or social problems associated with confinement in space" (Connors et al. 1986, pp. 910–911).

Thus, Russians had to confront in the 1970s issues that became pressing for Americans two decades later. As a result, when looking for models for a psychological support program, NASA turned to Russian efforts to support cosmonauts on *Mir* (Sipes and Vander Ark 2005). It is interesting that America's international partners in space—European as well as Japanese—share the Russians' interest in spaceflight psychology (Santy 1994).

1.3 Astronaut Selection

NASA, chartered as a civilian space agency, initially intended to select Mercury astronauts from a relatively broad range of explorers: military and commercial aviators; mountain climbers; polar explorers; bathysphere operators; and other fit, intelligent, highly motivated individuals who had demonstrated capabilities for venturing into dangerous new areas. Strong pressure from the White House limited the pool to military test pilots (Santy 1994). This was a group of accomplished fliers, many of whom had braved death during war. They brought with them the sharp wits, relentless motivation, and strong emotional control that characterize pilots who are willing to push themselves and their aircraft to (and sometimes beyond) the limits. Furthermore, because they were under military command, they were used to taking orders and were already cleared for top-secret technology. Mercury candidates had to be under 40 years of age, have graduated from college with a bachelor's degree in science and engineering, have logged at least 1,500 h flying jet planes, and have graduated from test pilot school. Of course, they were expected to be free of disease or illness and to demonstrate resistance to the physical stressors of spaceflight, such as temperature extremes and rapid acceleration and deceleration. To fit in the cramped confines of the Mercury capsule, their height could not exceed 5 ft 9 in.. The first astronauts had five duties: survive, perform effectively, add reliability to the automated system, complement instrument and satellite observation with scientific human observation, and improve the flight system through human engineering capabilities (Link 1965).

The initial Mercury project used two psychological approaches to selection. One was the industrial-organizational model of select-in characteristics emphasizing astronaut proficiencies needed to successfully complete mission tasks. The second was the psychiatric-clinical psychology model of select-out characteristics. As Robert Voas and Raymond Zedekar pointed out, psychological qualifications fell into two categories: abilities and personality (Voas and Zedekar 1963). Under aptitude and ability, they include high intelligence, general scientific knowledge and research skills, a good understanding of engineering, knowledge of operational procedures for aircraft and missiles, and psychomotor skills such as those used to operate aircraft. As regards personality, astronauts were to demonstrate a strong motivation to participate in the program, high tolerance for stress, good decision-making skills, emotional maturity, and the ability to work with others.

At that time, of 508 military test pilots, 110 met the general requirements and 69 were considered highly qualified. These were invited to the Pentagon for a briefing and interviews. Then, 32 were sent to the Lovelace clinic for an extraordinary physical exam and, after certification at Lovelace, to Wright Air Development Center in Dayton, Ohio, for tests of performance under stress. Here, the candidates were subjected to vibration, acceleration and deceleration, sitting with their feet in tubs of ice water, and numerous psychological and psychiatric evaluations. They completed 13 tests on personality and motivation, and another dozen or so on intelligence and aptitudes. NASA historians offer the following observation:

> Two of the more interesting personality and motivation studies seemed like parlor games at first, until it became evident that how profound an exercise in Socratic introspection was implied by conscientious answers to the test questions "Who am I" and "Whom would you assign to the mission if you could not go yourself?" . . .. Candidates who proceeded this far in the selection process all agreed with the one who complained "Nothing is sacred any more" (Mercury Program Overview).

After five Mercury flights, NASA officials decided that, given the absence of serious performance deficits to date, there was no need to continue exhaustive testing procedures. Although ongoing research would have provided an excellent basis for refining selection methods, by the end of 1962, NASA had prohibited research teams from collecting data on astronaut job performance, thus making it impossible to validate selection methods. At that point, according to Patricia Santy's (1994, p. 29) authoritative work, *Choosing the Right Stuff: The Psychological Assessment of Astronauts and Cosmonauts*, normal reluctance to participate in psychological research was transformed into "outright hostility" (Santy 1994). Psychiatric and psychological data from the Mercury program were confiscated, and researchers were told that apart from incomplete information that had already appeared in an obscure interim report, nothing could be published about astronaut psychology. The reasons for this are not entirely clear—for example, confidentiality was a growing concern, and data that could provide a basis for invidious comparisons could work against crew morale—but Santy (1994, p. 29) favors the view that "NASA became fearful that information on the psychological status and performance of their astronauts would be detrimental to the agency" (Santy 1994).

She also documents the minimal role that psychiatrists and psychologists played in the selection process from Gemini until well into the early Shuttle missions (Santy 1994). In the beginning of the astronaut program, original psychological selection attempted to pick the best-qualified candidates from a very capable group of experienced pilots, but by the 1980s, the selection process simply made sure that candidates were qualified based on the evaluator's opinion. Thus in 1983, Jones and Annes could claim that no psychological testing was involved. Rather, the approach had evolved into an entirely psychiatric process completed by two psychiatrists who separately interviewed each candidate. Whereas the original examination sought the best-qualified candidates, later procedures simply ensured that each candidate met the minimum qualifications (Jones and Annes 1983).

Candidates were no longer rated against one another, but they were screened for various psychopathologic conditions that could be detrimental or unsafe in a space environment. This screening, although conducted by expert aviation psychiatrists, did not have specific and objective criteria by which to rate each candidate. The emphasis was on selecting-out those candidates whose psychological structure would be detrimental in a space environment. Neuroses, personality disorders, fear of flying, disabling phobias, substance abuse, the use of psychotropic medications, or any other psychiatric conditions that would be hazardous to flight safety or mission accomplishment were among the grounds for rejection.

Thus, a selection program that began in 1959 as a model rooted in psychiatry and clinical psychology, and in industrial and organizational psychology, had been reduced to subjective evaluation. Santy provides more detail on how psychiatric evaluations were conducted by two psychiatric consultants who did not collaborate, use a standardized psychiatric interview, or keep detailed documentation, and who used their own subjective set of psychological criteria in the course of the evaluation (Santy 1994). She reviewed the percentage of female and male candidates disqualified psychiatrically. She found that one of the two psychiatrists hired to help in the screening process between 1977 and 1985 psychiatrically disqualified 40.7% of the female candidates and 7.5% of the male candidates. However, since no specific documentation existed, there was no way to know the reasoning behind his decisions (Santy 1994). This is not to say that the psychiatric consultants did a poor job of selecting-out; because no validation studies were completed, there is no evidence by which to evaluate their work.

Under the leadership of psychiatrist Patricia Santy and psychologist Al Holland in the 1980s, and then, in the 1990s, psychiatrist Christopher Flynn, there was a gradual return to evidence- and normative-based astronaut selection. In 1988, a biobehavioral research laboratory was formed within the Space Biomedical Research Institute (SBRI), which at that time was a branch of NASA's Medical Sciences Division, along with Medical Operations. Michael Bungo headed SBRI; Patricia Santy was the director of the laboratory; and psychologist Al Holland became her deputy. The Biobehavioral Laboratory was to develop a new working group of professionals to make recommendations on both the operational and research needs in the areas of the behavioral sciences. At that time, operations were expanding beyond helping to choose astronauts to providing psychological support for the astronaut corps.

The development of standardized, semi-structured interviews and diagnostic criteria, aided by the work done by the Working Group on Psychiatric and Psychological Selection of Astronauts, resulted in a rewrite of NASA psychiatric standards based on the then-current American Psychiatric Association's *Diagnostic and Statistical Manual III* and recommendations for a select-in process (Santy 1994). The reasoning behind the select-in process harkened back to the original logic of 1959, hypothesizing that certain psychological traits were associated with effective astronaut performance. Commencing in 1989, validation work on the select-in criteria was begun. In describing the selection process, Laura Galarza and Al Holland note that selection starts at the time of entry into the astronaut corps, then should continue through the training process and selection for specific missions (Galarza and Holland 1999a, b).

In the 1990s, Galarza and Holland began developing a scientifically defensible select-in process that would screen for personal abilities to help people live and work within a small teams under conditions of isolation and confinement (Galarza and Holland 1999a, b). By using highly qualified subject-matter experts, job analysis, and documented validation techniques, they sought to meet the high standards for selection established by the Society for Industrial and Organizational

Psychologists (SIOP) (Society of Industrial and Organizational Psychology 2003). Although these researchers developed a profile of needed knowledge, skills, and abilities, NASA's prohibition against obtaining in-training or on-the-job performance ratings effectively killed any longitudinal or predictive validation of the proposed astronaut select-in procedures. Today, all astronaut candidate applicants spend several hours completing psychological tests and then undergo extensive psychological and psychiatric interviews. To prevent coaching, the specific tests and interview content are not publicly available. The current selection process resembles the selection procedures for other high-risk jobs and incorporates highly validated tests that are quantitatively scored, along with in-depth, semistructured interviews.

Well before Apollo astronauts set foot on the Moon, there were political pressures to increase the diversity of the astronaut corps by including women and representatives of different racial and ethnic groups. Accommodating people with different cultural backgrounds became a practical matter in the Apollo-Soyuz rendezvous, in the course of the Russian "guest cosmonaut" program, in Shuttle missions with international crews, and, of course, aboard the ISS. Successfully managing cultural, occupational, and other differences in space is likely to become even more crucial as highly trained professionals are joined by industrial workers and tourists.

Margaret Weitekamp recounts how, at the inception of Project Mercury, an Air Force flight surgeon, Don Flickenger, helped initiate a program known as WISE—Women in Space Earliest (Weitekamp 2004). Women offered certain potential advantages over men; one of the most notable of these was their smaller size (and reduced life-support requirements), which would make them easier to lift into orbit and keep alive at a time when engineers had to fret every extra pound of weight. After word of the program's existence leaked, it was abandoned by the Air Force and taken over by Dr. Randall Lovelace, of the same Lovelace Clinic that conducted the physicals for project Mercury. Aviatrix Jackie Cochran and her wealthy philanthropist husband, Floyd Odlum, provided funding so that Lovelace could put the women through the same rigorous evaluation. Of the 25 women who took the physical, 13 passed. The next step in the process, which involved centrifuges and jet flights, depended on the availability of military facilities and equipment. Although it appeared that the procedures could be done at the Naval Air Station in Pensacola, Florida, actually doing so depended on NASA's officially "requiring" and then reimbursing the testing. Since the program was unofficial (despite widespread perceptions that it was connected with NASA), the space agency did not intervene on the women's behalf. Some of the women continued to press for further testing and flight training, and, eventually, there was a congressional hearing, but public clamor and aggressive lobbying got no results. Kennedy's decision to place a man on the Moon before the decade was finished was interpreted by NASA to mean that it could not divert resources to sending women to orbit.

But there were other barriers to women's participation in space exploration, including the inability of some of the people in NASA's white-male-dominated culture to conceive of women in the "masculine" role of astronaut. Weitekamp (2004, p. 3) writes:

> At a very basic level, it never occurred to American decision makers to seriously consider a woman astronaut. In the late 1950s and early 1960s, NASA officials and other American space policy makers remained unconscious of the way their calculations implicitly incorporated postwar beliefs about men's and women's roles. Within the civilian space agency, the macho ethos of test piloting and military aviation remained intact. The tacit acceptance that military jet test pilots sometimes drank too much (and often drove too fast) complemented the expectation that women wore gloves and high heels—and did not fly spaceships (Weitekamp 2004).

At that time, lack of diversity at NASA was by no means limited to the astronaut corps. In 1974, Congress held a hearing on NASA's Equal Employment Opportunity Program. The chairman's introductory remarks included the statement "It is clear that the NASA equal employment opportunity effort over the years has been inadequate…." (House Committee on the Judiciary, Subcommittee on Civil Rights and Constitutional Rights 1974, p. 1). In the congressional report, NASA admitted that as of the end of fiscal year (FY) 1971, only 16.6% were women and 4.6% minorities (House Committee on the Judiciary, Subcommittee on Civil Rights and Constitutional Rights 1974). Only 3% of the supervisors and 2.4% of the engineers were women.

Kim McQuaid points out that many forces worked against increasing the proportion of women and blacks at NASA (McQuaid 2007). Nationally, efforts to increase diversity through new employment strategies began at about the same time as NASA flourished in the late 1960s and early 1970s. Special hurdles at NASA included an organizational culture that was built on the white-male stereotypes of the time and demanded prior training and experience in science and engineering at a time when very few women or minorities were earning (or were allowed to earn) degrees in science and engineering. In 1973, then–NASA Administrator James Fletcher hired Ruth Bates Harris as a high-level deputy director to oversee NASA's equal opportunity employment processes—but, when it turned out that she would be a fearless leader rather than a compliant bureaucrat, he fired her and then, under pressure, attempted to rehire her at a lower level. This initiated bad press, conflicts with Congress, and a series of internal struggles that brought about diversification. In the 1990s, Administrator Dan Goldin could complain that NASA was still too male, pale, and stale, although, two decades earlier, NASA had responded to new domestic political issues by changing from a civil rights sham to the beginnings of a demonstrably effective, if imperfect, affirmative action program.

Aside from the 1965 selection cycle, when the Academy of Science handled selection and allowed women to apply (none were accepted), it was not until the Shuttle era that women entered the astronaut corps. On 16 January 1978, the first female and black candidates were selected; only a few years later, in 1983, the public wildly acclaimed mission specialist Sally Ride's orbital flight aboard *Challenger*. Some of the women who had participated in the informal women's astronaut selection program of the early 1960s felt vindicated in 1995, when they watched pilot Eileen Collins lift off, carrying their dreams with her (Weitekamp 2004). Today, female astronauts routinely participate in Shuttle and Space Station missions in many different roles. Despite the long road that American women and minorities traveled to prove their worth, the U.S. experience has shown that talented women and minorities, given no special treatment because of gender or ethnicity, are as adept as their white, male colleagues in the world of space.

1.4 Psychological Support

Initially, psychological support for astronauts came from helpful flight surgeons, flak-catchers who tried to minimize interference on the part of the media and the public, and cheering family and friends. By means of shortwave radio, astronauts on the ground encouraged astronauts in orbit. It is clear from Wolfe's *The Right Stuff* that the astronauts' wives provided strong support for one another, as well as for their husbands (Wolfe 1979). The larger community of astronauts and their families still provides psychological support for astronauts before, during, and after their flights.

Professional psychological support for the astronauts and their families evolved over time and gained momentum in the early space station era (Fiedler and Carpenter 2005; Flynn 2005; Kanas and Manzey 2003). Today, psychological support is provided in three stages: preflight, in-flight, and postflight (Sipes and Fiedler 2007). The Operational Psychology team, currently under the leadership of Al Holland and Walter Sipes of NASA and of Steven Vander Ark of Wyle Laboratories, offers preflight training and briefings in such diverse areas as self-care, conflict management and cultural awareness, and field training. Family readiness is addressed in a briefing focused on the astronauts' spouses to explain processes such as the training flow (normally two years in length), crew care packages and private family conferences. Crew care packages are containers of personal items from family and friends that are sent to astronauts residing on the ISS. Favorite foods, surprise gifts from the family, and holiday decorations are a few of the items that have been sent to the ISS in these shipments.

During the flight stage, in addition to the crew care packages and private weekly videoconferences with families, psychological support services include extensive communication with people on the ground (including Mission Control personnel, relatives, and friends), psychological support hardware and software, special events such as surprise calls from celebrities, and semimonthly videoconferences with behavioral health experts and operational psychologists. Astronauts in flight have e-mail accessibility and can use an Internet protocol phone on board the ISS to call back to Earth. As in the past, ham radio allows contact between the ISS and schools throughout the world. Astronauts have also used Facebook and Twitter to stay in contact with family, friends and the public.

A month before their return to Earth, ISS astronauts are briefed on the stresses and joys of returning home following the deployment. Following the flight, astronauts undergo a series of debriefings intended to benefit the astronaut and fine-tune the psychological support program. The astronaut's spouse is given the opportunity to meet with psychological support personnel to provide the latter with feedback on the psychological support provided during the mission. Of course, astronauts and their families can use counseling services at any time. While this briefly covers the current state of the art of psychological support for astronauts on the ISS, psychological support for lunar, asteroid, and Mars missions may have greater constraints and force a return to the mindset of earlier explorers and their families.

1.5 Conclusion

Spaceflight is both demanding and rewarding, and for many years, psychologists focused on the demanding environment and stressful effects. Throughout the history of spaceflight, professionals have expressed concern that the physical, psychological, and interpersonal stressors of spaceflight could endanger a crew, undercut performance, and lower the quality of life. Episodes in spaceflight-analogous environments and a few incidents in space suggest that although no astronauts have been recalled to Earth on the basis of psychological and social challenges, adaptation must be taken into account. Astronaut participation in extended-duration missions, the prospects of a return to the Moon, continuing public enthusiasm for a mission to Mars, the reformulation of research questions following the publication of *Safe Passage*, and the coevolution of NASA's Bioastronautics Critical Path Roadmap and the National Space Biomedical Research Institute initiated a new era for psychology. Additional opportunities are now available with the evolution of the Bioastronautics Critical Path Roadmap into today's Human Research Program and the establishment of the Center for the Advancement of Science in Space for research on the US ISS laboratory.

According to our analysis, since the dawn of the modern space station era, there has been an increase in both research and operational interest in spaceflight behavioral health. Slowly, and perhaps painfully, psychology has gained greater recognition within the U.S. space program, and there is a growing convergence of interests to target research at operational problems (Harrison 2005).

Current NASA administration has mandated that human research be operationally relevant. This is partly driven by funding shortages and partly by needs to meet NASA performance standards and requirements when astronauts once again venture beyond low-Earth orbit. The new Human Research Program provides the bases for defining, documenting, and allocating human research program requirements as they have evolved from the older Bioastronautics Critical Path Roadmap and the new NASA standards and requirements that emphasize future missions. The purpose is to deliver health and performance research, technologies and countermeasures for safe and productive space exploration on timescale of many decades.

A comprehensive program in spaceflight behavioral health wil have to be broad based; be interdisciplinary; and address issues at the individual, small-group, and organizational levels. It will require multiple, convergent methods including archival research, field observations, and both field and laboratory experiments. Research falling under this umbrella must meet high scientific standards, achieve flight certification, and be palatable to astronauts. Only with continued interest and support from NASA—and from psychologists—will spaceflight behavioral health flourish. Long-term success will require accessible, peer-reviewed publications and efforts to target young investigators to replace those who retire.

Contemporary psychology is in a better position to be of help. Many of the theories and tools that are proving useful today were not available at the dawn of the Space Age. New (relative to 1960) resources include cognitive models, which emphasize our information processing power, and humanistic or "positive psychology" models that stress people's positive, striving nature (Suedfeld 2005). These new models have allowed psychologists a fresh take on many important issues. Human factors psychologists benefit from modern computer modeling technologies and increasing evidence of the importance of taking the person into account when developing a human or human-robotic system.

Research technology has changed dramatically over the past 50 years, and the new technology has also been useful for increasing psychology's contributions to NASA. These changes are evident wherever we look, from data collection to analysis. Today, miniaturization and computer technology enable psychological assessments and evidence-based countermeasures that would have been impossible in the 1960s.

Minimally intrusive techniques are particularly useful, and one of these is based on nonintrusive computer monitoring of facial expression (Dinges et al. 2005). Another approach is monitoring cognitive functioning through computer analysis of speech (Lieberman et al. 2005). Encouraging astronauts to monitor their own behavior reduces the threat that performance lapses could lead to flight disqualification. This self-monitoring has been accomplished by means of computers and personal digital assistants (PDAs) that are programmed to measure several dimensions of cognitive functioning (attention, information processing, and recall). Astronauts may use the results of these tests to gauge their own preparedness to engage in a particular activity (Shephard and Kosslyn 2005).

While we see evidence of an expanding role, psychology's future in spaceflight is by no means assured. NASA's resistance to psychology is by no means fully overcome. NASA Administrators must still concern themselves with public relations. Project managers and engineers must still get on with their tasks within the real constraints of cost and practicality. Astronauts remain sensitive to possible threats to flight assignments and careers. The focusing events of *Mir* and the ISS were less than two decades ago, and it is too early to tell if the new interest and infrastructure can withstand the vagaries of funding variations or national and organizational politics.

In the late 1950s the human space program began selecting, training and supporting astronauts. Long before that, governments and wealthy individuals funded explorers to systematically survey and discover new areas of the world. Historians, explorers, and academics have all contributed to an understanding of what it is like to embark on a mission to the unknown and what qualities are needed by the individual, the leader, and the crew (Stuster 2004). Less has been written about the infrastructure that supported explorers and or how their families survived. Today we have formal programs for psychological services. The responsibilities that psychologists feel today were either ignored or, or more likely performed by other professionals (such as ship's doctor or priest) during centuries of earth based exploration. In the future, psychological services may be performed by professionals and technicians that we cannot begin to imagine today.

References

Andersen, D.T., McKay, C.P., Wharton Jr., R.A., Rummel, J.D.: An Antarctic research as a model for planetary exploration. J. Br. Interplanet. Soc **43**, 499–504 (1990)

Anonymous: Journal interview 64: conversation with Joseph V. Brady. Addiction **100**, 1805–1812 (2005)

Ball, J.R., Evans, C.H. (eds.): Safe Passage: astronaut Care for Exploration Missions. National Academy Press, Washington, DC (2001)

Bioastronautics Critical Path Roadmap. http://bioastroroadmap.nasa.gov/index.jsp. Accessed 13 Mar 2012

Birkland, T.A.: Focusing events, mobilization, and agenda setting. J. Public. Policy. **18**(1), 53–74 (1997)

Brady, J.V.: Behavioral health: the propaedeutic requirement. Aviat. Space Environ. Med. **76** (6, sect. II), B13–B24 (2005)

Burgess, C., Dubbs, C.: Animals in Space: from Research Rockets to the Space Shuttle. Springer Praxis, Chichester (2007)

Burrough, B.: Dragonfly: NASA and the Crisis On Board Mir. Harper Collins, New York (1998)

Cernan, E., Davis, D.: The Last Man on the Moon: Astronaut Eugene Cernan and America's Race to Space. St. Martin's Press, New York (1999)

Connors, M.M., Harrison, A.A., Akins, F.R.: Living Aloft: Human Requirements for Extended Spaceflight. NASA SP-483, Washington, DC (1985)

Connors, M.M., Harrison, A.A., Akins, F.R.: Psychology in the resurgent space program. Am. Psychol. **41**(8), 906–913 (1986)

Cooper Jr., H.S.F.: A House in Space. Bantam Books, New York (1976)

Cunningham, W.: The All-American Boys. Macmillan, New York (1977)

Dinges, D.F., Rider, R.L., Dorrian, J., McGlinchey, E.L., Rogers, N.L., Cizman, Z., Goldenstein, S.K., Vogler, C., Venkartamarian, S., Metaxas, D.N.: Optical computer recognition of facial expressions associated with stress induced by performance demands. Aviat. Space Environ. Med. **76**(6, sect. II), B172–B182 (2005)

Douglas, W.K.: Psychological and sociological aspects of manned spaceflight. In: Harrison, A.A., Clearwater, Y.A., McKay, C.P. (eds.) From Antarctica to Outer Space: Life in Isolation and Confinement, pp. 81–88. Springer, New York (1990)

Fiedler, E., Carpenter, F.E.: Evolution of the behavioral health sciences branch of the space medicine and health care systems at the Johnson Space Center. Aviat. Space Environ. Med. **76**(6, sect. II), B31–B35 (2005)

Flynn, C.F.: An operational approach to long-duration mission behavioral health and performance factors. Aviat. Space Environ. Med **76**(6, sect. II), B42–B52 (2005)

Galarza, L., Holland, A.W.: Selecting astronauts for long-duration missions. SAE international document 1999-01-2097, presented at the international conference on environmental systems, Denver, July 1999a

Galarza, L., Holland, A.W.: Critical astronaut proficiencies required for long duration spaceflight. SAE Document 1999-01-2096, presented at the international conference on systems, Denver, July 1999b

Gunderson, E.K.E.: Human Adaptability to Antarctic Conditions. American Geophysical Union, Washington, DC (1973)

Gunderson, E.K.E.: Preface. In: Harrison, A.A., Clearwater, Y.A., McKay, C.P. (eds.) From Antarctica to Outer Space: Life in Isolation and Confinement, p. 1. Springer, New York (1990)
Halvorson, T.: ISS Astronaut Susan Helms: space is More than a Nice Place to Visit. http://www.space.com/missionlaunches/mission/iss_freetime_010615.html. Accessed 23 June 2010
Harrison, A.A.: Behavioral health: integrating research and application in support of exploration missions. Aviat. Space Environ. Med. **76**(6, sect. II), B3–B12 (2005)
Harrison, A.A., Summit, J.E.: How third force psychology might view humans in space. Space. Power **10**, 85–203 (1991)
Harrison, A.A., Clearwater, Y.A., McKay, C.P, The human experience in Antarctica: applications to life in space. Behav. Sci. **34**(4), 253–271 (1989)
Harrison, A.A., Clearwater, Y.A., McKay, C.P. (eds.): From Antarctica to Outer Space: Life in Isolation and Confinement. Springer, New York (1990)
Harvey, B.: The New Russian Space Program: From Competition to Cooperation. Wiley Praxis, Chichester (1996)
Helmreich, R.L.: Applying psychology to outer space: unfulfilled promises revisited. Am. Psychol. 445–450 (1983)
Holland, A.W.: Psychology of Spaceflight. J. Hum. Perform. Extreme Environ. **5**(1), 4 (2000)
House Committee on the Judiciary, Subcommittee on Civil Rights and Constitutional Rights, *NASA's Equal Opportunity Program*, hearings before the Subcommittee on the Judiciary, 93rd Congress, 2nd session, 13–14 March 1974.
Human Research and Safety. http://www.nasa.gov/exploration/humanresearch/index.html. Accessed 13 Mar 2012
Human Research Roadmap. http://humanresearchroadmap.nasa.gov/. Accessed 13 Mar 2012
Jones, D.R., Annes, C.A.: The evolution and present status of mental health standards for selection of USAF candidates for space missions. Aviat. Space Environ. Med. **54**, 730–734 (1983)
Kanas, N.: Psychosocial factors affecting simulated and actual space missions. Aviat. Space Environ. Med. **56**(8), 806–811 (1985)
Kanas, N.: Psychosocial support for cosmonauts. Aviat. Space Environ. Med. **62**(4), 353–355 (1991)
Kanas, N., Manzey, D.: Space Psychology and Psychiatry. Kluwer, Dordrecht (2003)
Kanas, N., Salnitsky, V.P., Ritsher, J.B., Gushin, V.I., Weiss, D.S., Saylor, S.A., Kozerenko, O.P., Marmar, C.R.: Human interactions in space: ISS vs. shuttle/MIR. Acta Astronaut **59**, 413–419 (2006)
Kane, L., Short, P., Sipes, W.E., Flynn, C.F.: Development and validation of the spaceflight cognitive assessment tool for windows (WinSCAT). Aviat. Space Environ. Med. **76**(6, sect. II), B183–B191 (2005)
Kelley, A.D., Kanas, N.: Crewmember communications in space: a survey of astronauts and cosmonauts. Aviat. Space Environ. Med. **63**, 721–726 (1992)
Kelley, A.D., Kanas, N.: Leisure time activities in space: a survey of astronauts and cosmonauts. Acta. Astronaut. **32**, 451–457 (1993a)
Kelley, A.D., Kanas, N.: Communication between space crews and ground personnel: a survey of astronauts and cosmonauts. Aviat. Space. Environ. Med. **64**, 795–800 (1993b)
Launius, R.D.: Heroes in a vacuum: the Apollo astronaut as cultural icon. In: American Institute of Aeronautics and Astronautics [AIAA] Aerospace Sciences Meeting and Exhibit, Reno, 13 Jan 2005
Lieberman, P., Morey, A., Hochstadt, J., Larson, M., Mather, S.: Mount Everest: a space analogue for speech monitoring of cognitive deficits and stress. Aviat. Space Environ. Med. **76**(6, sect. II), B198–B207 (2005)
Linenger, J.M.: Off the Planet. McGraw-Hill, New York (2000)
Link, M.M.: Space Medicine in Project Mercury. NASA, Washington, DC (1965)
Lugg, D.J.: Behavioral health in Antarctica: implications for long-duration space missions. Aviat. Space Environ. Med. **76**(6, sect. II), B74–B78 (2005)

Mallis, M.M., DeRoshia, C.W.: Circadian rhythms, sleep, and performance in space. Aviat. Space Environ. Med. **76**(6, sect. II), B94–B107 (2005)

McQuaid, K.: Race, gender and space exploration: a chapter in the social history of the space age. J. Am. Stud. **41**(2), 405–434 (2007)

Mercury Program Overview. http://www.nasa.gov/mission_pages/mercury/missions/astronaut.html. Accessed 13 Mar 2012

Mitchell, E., Williams, D.: The Way of the Explorer. Putnam, New York (1996)

Myasnikov, V.I., Zamaletdinov, I.S.: Psychological states and group interaction of crew members in flight. In: Huntoon, C.L., Antipov, V., Grigoriev, A.I. (eds.) Humans in Spaceflight, vol. 3, bk. 2, pp. 419–431. AIAA, Reston (1996)

NASA Names CASIS to Manage Space Station National Laboratory. http://www.nasa.gov/home/hqnews/2011/sep/HQ_11-294_ISS_NPO.html. Accessed 13 Mar 2012

Oberg, J.E., Oberg, A.R.: Pioneering the Space Frontier. McGraw-Hill, New York (1986)

Palinkas, L.A.: The psychology of isolated and confined environments: understanding human behavior in Antarctica. Am. Psychol. **58**(3), 353–363 (2003)

Palinkas, L.A., Group adaptation and individual adjustment in antarctica: a summary of recent research, In: Harrison, A.A., et al. (ed.) From Antarctica to Outer Space, pp. 239–252 Springer, New York (1990)

Perrow, C.E.: The organizational context of human factors engineering. Adm. Sci. Q. **28**(4), 521–541 (1983)

Petrov, B.N., Lomov, B.F., Samsonov, N.D. (eds.): Psychological Problems of Spaceflight. Nauka Press, Moscow (1979)

Rasmussen, J.E. (ed.): Man in Isolation and Confinement. Aldine, Chicago (1973)

Rholes Jr., F.H., Grunzke, M.E., Reynolds, H.H.: Chimpanzee performance during the ballistic and orbital project mercury flights. J. Comp. Physiol. Psychol. **86**(1), 2–10 (1963)

Ryan, C.: The Pre-Astronauts: Manned Ballooning on the Threshold of Space. Naval Institute Press, Annapolis (1995)

Santy, P.A.: Choosing the Right Stuff: The Psychological Selection of Astronauts and Cosmonauts. Praeger/Greenwood Publishing Group, Westport (1994)

Schmidt, H.: The millennium project. In: Stoker, C., Emmart, C. (eds.) Strategies for Mars: a Guide for Human Exploration. American Astronautical Society/Univelt, San Diego (1996)

Shepanek, M.: Human behavioral research in space: quandaries for research subjects and researchers. Aviat. Space Environ. Med. **76**(6, sect. II), B25–B30 (2005)

Shephard, J.M., Kosslyn, S.M.: The MiniCog rapid assessment battery: a 'blood pressure cuff' for the mind. Aviat. Space Environ. Med. **76**(6, sect. II), B192–B197 (2005)

Sipes, W.E., Vander Ark, S.T.: Operational behavioral health and performance resources for international space station crews and families. Aviat. Space Environ. Med. **76**(6, sect. II), B36–B41 (2005)

Sipes, W.E., Fiedler, E.: Current psychological support for US astronauts on the international space station. Paper presented at tools for psychological support during exploration missions to mars and moon, European Space Research and Technology Centre [ESTEC], Noordwijk, 26 Mar 2007

Society of Industrial and Organizational Psychology: Principles for the Validation and Use of Personnel Selection Procedures. SIOP, Washington, DC (2003)

Stuster, J.W.: Lessons from previous expeditions for the human exploration of mars. J. Br. Interplanet. Soc. **57**, 3–4 (2004). 126–134

Suedfeld, P.: Invulnerability, coping, salutogenesis, integration: the four phases of space psychology. Aviat. Space Environ. Med. **76**(6, sect. II), B61–B66 (2005)

Voas, R., Zedekar, R.: Astronaut selection and training, chap. 10. In: Mercury Project Summary Including the Results of the Fourth Manned Orbital Flight, May 15 and 16, 1963. Office of Scientific and Technical Information, NASA, Washington, DC (1963)

Weick, K.E.: Organizational Design: organizations as Self-Designing Systems. Organ. Dyn. **6**, 30–46 (1977)

Weitekamp, M.A.: Right Stuff, Wrong Sex: America's First Women in Space Program. Johns Hopkins University Press, Baltimore (2004)
White, F.: The Overview Effect. Houghton Mifflin, Boston (1987)
Wolfe, T.: The Right Stuff. Farrar, Strauss and Giroux, New York (1979)
Wood, J., Schmidt, L., Lugg, D., Ayton, J., Phillips, T., Shepanek, M.: Life, survival and behavioral health in small closed communities: 10 years of studying small Antarctic groups. Aviat. Space Environ. Med. **76**(6, sect. II), B89–B94 (2005)

Chapter 2
From Earth Analogues to Space: Learning How to Boldly Go

Sheryl L. Bishop

Abstract The need to find relevant terrestrial substitutes, that is, analogues, for teams operating in extraterrestrial and microgravity environments is driven by extraordinary demands for mission success. Unlike past frontiers where failure on the part of various groups to succeed represented far more limited implications for continued progress within these environments, accidents like *Challenger* in 1986 and *Columbia* in 2003 underscored the magnified cost of failure for space missions. Where past human frontiers were characterized by centralized decisions to engage in exploration and development largely under the dictates of authoritarian governments or individual sponsors, the exploration of space has been significantly influenced by the general public's perception of "acceptable risk" and fiscal worthiness. To date, space missions have failed due to technological deficiencies. However, history is replete with examples of exploration and colonization that failed due to human frailties, including those that reflect failures of the group. Both historical literature and research on teams operating within extreme environments, including space, have clearly indicated that psychological and sociocultural factors are components critical for individual and group success. Given the limited access to the space frontier and the investment in collective effort and resources, our ability to study individual and group functioning in the actual space environment has been, and will continue to be, severely limited. Thus, studying groups in terrestrial extreme environments as analogues has been sought to provide *predictive* insight into the many factors that impact group performance, health, and well-being in challenging environments.

This chapter provides an overview of the evolution of research utilizing terrestrial analogues and addresses the challenges for selecting, training, and supporting teams for long-duration space missions. An examination of how analogue environments can contribute to our knowledge of factors affecting functioning

S.L. Bishop (✉)
University of Texas Medical Branch, Galveston, TX, USA
e-mail: sbishop@utmb.edu

D.A. Vakoch (ed.), *On Orbit and Beyond*, Space Technology Library 29,
DOI 10.1007/978-3-642-30583-2_2, © Springer-Verlag Berlin Heidelberg 2013

and well-being at both the physiological and the psychological levels will help define the focus for future research.

2.1 Introduction

Humans have long speculated about, studied, and striven to explore the heavens. Many of our earliest myths, such as the flight of Daedalus and Icarus too close to the Sun on wings made of wax, expressed our desire to explore beyond the boundaries of Earth as well as our willingness to push current technology to its limits. Considerations by the earliest philosophers and scientists, including Archimedes, Galileo Galilei, Nicolaus Copernicus, Leonardo da Vinci, Sir Isaac Newton, Jules Verne, H. G. Wells, or Percival Lowell, eventually generated a whole new genre of fictional literature built upon scientific extrapolations, dubbed "science fiction," and gave voice to their speculations about the nature of extraterrestrial environments. Modern scientists and pioneers led by the Wright brothers, Robert Goddard, Konstantin Tsiolkovsky, Hermann Oberth, Wernher von Braun, Sergey Korolev, Yuri Gagarin, and Neil Armstrong pushed the boundaries of knowledge about flight and extended human inquiry beyond our terrestrial boundaries into our local and extended galactic neighborhood. For serious considerations of how humans will fare in space, we have had to extrapolate from human experience on Earth in environments that challenge us in, ideally, similar ways. However, the search for space analogue environments in which to systematically study individual and group adaptation has had to grapple with some significant limitations, i.e., the impossibility of a substitute for a microgravity or reduced-gravity environment or environments that holistically mimic radiation profiles and their inherent danger for those beyond Earth's magnetic field. Since there is no direct equivalent for space, all analogue environments are simulations of greater or lesser fidelity along varying dimensions of interest. Some analogue environments provide extremely good characterizations of expected challenges in testing equipment or hardware, e.g., environmental chambers such as the Space Shuttle mockups of the various decks or the cargo bay in NASA's Weightless Environmental Training Facility (WET-F), but lacks any relevance to assessing how human operators will fare psychologically or as a team. Others, like chamber studies, address important components of human adaptation, e.g., confinement, but fail utterly to incorporate true environmental threats. Others allow for the impact of true dangerous, unpredictable environments but lack any way to systematically compare across specific environments. The spectrum of fidelity to space among terrestrial analogues ranges from laboratory studies where the impact of environmental threat and physical hardship, as well as true isolation and confinement, are limited and, even, sometimes absent, to real teams in real, extreme environments characterized by very little control over extraneous variables.

This, then, is the challenge. Unlike the testing of hardware where various components can be reliably evaluated separately, the study of humans and teams, in particular, is a dynamic endeavor requiring in situ study of the collective.

To develop reliable protocols based on empirical evidence to select, monitor, and support teams effectively in space necessarily involves the demand to study teams in analogue environments that replicate a wide range of physiological, psychological, and psychosocial factors interacting both with the environment and within the team. The high degree of reliance on technology for life support, task performance, and communication must be integrated with new measurement methodologies to overcome heretofore intrusive measurement modalities. The growing frequency of multinational and multicultural teams and the demand for longer-duration missions both further compound the complexity of the challenge. While the primary goal has been the assurance of human health and well-being, the expectation has been that such priorities will naturally lead to improved chances for performance and mission success. Yet achieving this goal depends largely on how well our analogues prepare us for living and working in space.

Analogues for human individual and group performance in space has involved two basic approaches: (1) constructing an environment within a laboratory setting with maximum control over extraneous variables and utilizing volunteer research subjects or (2) studying naturally occurring real-world groups in real environments characterized by a number of confounds (Haythorn and Altman 1966; Zubek 1969). Each comes with its own limitations and strengths. In any assessment of the value of the analogue, the pros and cons of each environment need to be kept in mind. This is especially true when assessing the generalizability of insight of psychosocial factors from substitute environments for space.

Before we began deliberately constructing controlled laboratory environments, there were the records of early expeditionary explorations into various places on Earth (Greely 1986; Stefansson 1925; Pearce 1930). The tradition of publishing personal diaries and mission recounts has been similarly observed by the earliest explorers of space (Lebedev 1988; Lovell and Kluger 1994). Secondary analyses of historical expeditions have become increasingly popular in recent years (Stuster 1996). The very character of natural environments typically guarantees that there will be at least some, if not substantial, periods of inaccessibility, lack of communication or contact, little accessibility of real-time support, and great demands on individuals and groups to engage in autonomous decision-making, problem-solving, conflict resolution, self-monitoring, and self-regulation. These demands inherently build in the potential for conflict with external mission support personnel and researchers who find adherence to mission schedules and timelines far easier to maintain than do those actually on the mission. Shared perspective between these groups becomes increasingly difficult to promote as mission duration, distance, and environmental demands play larger roles in daily decisions of the teams than do seemingly arbitrary mission schedules.

Measurement of these factors is compromised as teams become preoccupied with dealing with the environment, become antagonistic to external evaluation, become noncompliant with schedules that become unimportant to participants, and engage in a general reprioritization of activities that emphasizes near-term, more salient goals (e.g., personal comfort, leisure) over and above long-term mission goals (e.g., study data). Such difficulties have raised questions about the worth of studying groups in real-world environments. In actuality, these conditions are *exactly* what is needed to simulate space missions that have grown in duration, distance from Earth, complexity, and challenge. However, space missions will also be, at least for the foreseeable future, characterized by an extraordinary degree of control, from selecting who goes to establishing the daily details of mission tasks and schedules—elements that are far more variable in real-world groups, such as those in Antarctica or part of polar or mountaineering expeditions. In real-world groups that have higher degrees of structure and control, such as military teams, the command and control structure is distinctly different from the current scientist-astronaut organizational structure of space missions. Fundamental differences in group structures, such as leadership and authority, represent significant elements in whether findings from terrestrial analogues translate to future space crews.

The need for control over the inherent chaos of real-world environments in order to definitively identify critical factors that affect individual and group performance was the driver behind the development of constructed environments of various complexities. Useful data from such artificial environments depend on whether participants are truly immersed in the fiction of a simulation and are responding in the same way they would if the environment were real. This is the paradox researchers in analogue environments face: In laboratory studies, the very attributes of the environment that have the greatest impact on performance are removed (e.g., real danger, uncontrolled events, situational ambiguity, uncertainty, or the interaction with the extreme environment itself). If these features are compromised, as many have argued, then is there value in conducting such laboratory studies? (Palinkas 2003a; Suedfeld 1998). On the pro side, laboratory chamber studies have provided opportunities to evaluate methods of monitoring psychological and interpersonal parameters for subsequent application during real flights and have identified issues that might cause psychological and interpersonal problems in space. They have also provided empirical evidence for a number of behavioral issues anecdotally reported from space, e.g., the tendency of crews to direct aggression toward personnel at Mission Control (Kanas et al. 2000; Sandal, Vaernes, Ursin 1995; Gushin et al. 1996, 1997). They are well suited to first-line inquiry when there is a need to investigate the characteristics of a particular phenomenon suspected of being present. However, complexity is a key defining trait of stressed operational environments. Total reliance on laboratory studies and the presumption of broad generalizability, particularly for research on high-stress, high-risk environments, is highly likely to lead to dissociation between actual operational findings and laboratory and experimental studies (Baddeley 1972; McCarthy 1988; Mears and Cleary 1980; Wilson, Skelly, Purvis 1989). Conversely, data on real-world groups situated in extreme environments has provided insight into a host of factors that impact group performance, health, and well-being emergent from the interaction between the individual, the team, and the environment. The differences found between studies conducted in experimentally controlled chambers and those conducted in messy, noisy, in situ real environments appears to be due to the critical presence of real environmental threat and physical hardship, as well as true isolation and confinement, which have proven to be key factors in individual and group coping. Additionally, when comparing extreme environments with non-extreme natural environments in which people normally operate, the level, intensity, rate of change, and diversity of physical and social stimuli, as well as behavior settings and possible behaviors within an extreme environment, are far more restricted (Suedfeld 1998).

Thus, real teams in extreme environments have validated or corrected findings from chamber studies where critical environmental factors are typically absent or blunted. Real extreme environments allow us to examine various aspects of the psychophysiological relationship that are essential to fully understanding the adaptation of humans to the stresses of these environments and, ultimately, to space. Space, of course, will be the final testing ground for our accumulated knowledge. But are we stuck with choosing between chamber studies and naturally occurring opportunistic teams in real extreme environments? A more recent, hybrid approach of situating research facilities within extreme environments offers a good compromise between the artificial conditions of the laboratory and the open-ended, full access of an expeditionary mission. When teams or individuals operate in extreme environments, their responses are more purely a product of either situational drivers or internal personal characteristics. To the extent that an extreme environment is well characterized and known, it gains in fidelity and allows more accurate inferences about key phenomena to be drawn. For these very reasons, Palinkas has strongly argued that the cumulative experience with year-round presence in Antarctica makes it an ideal laboratory for investigating the impact of seasonal variation on behavior, gaining understanding about how biological mechanisms and psychological processes interact, and allowing us to look at a variety of health and adaptation effects (Palinkas 2003a).

2.2 Psychology and Space

> One important fact, which has emerged during decades of research, is that in the study of capsule environments there are few main effect variables. Almost every outcome is due to an interaction among a host of physical and social environmental variables and personality factors. Thus, although we conceptually deconstruct the situation into particular sources of variance, we must remember that how people experience an environment is more important than the objective characteristics of the environment (Suedfeld and Steel 2000, p. 230).

Investigations into psychological and psychosocial adaptation to extreme environments as substitutes for space are recent phenomena. Expeditions and forays into these environments have historically been for the purposes of exploration, and the primary metric of successful adaptation was survival. One could argue that chronicles such as the *Iliad* and the *Odyssey* were early examples of the more familiar diaries such as those that recounted the historic race to reach the South Pole between modern polar expeditions lead by Roald Amundsen, who reached the South Pole in 1911, and Robert F. Scott, who reached the South Pole in 1912. Humans have been periodically living and working in Antarctica, one of the most challenging environments on Earth, for over a 100 years. The first winter-over in Antarctica occurred during 1898–1899 on board an icebound ship, the *Belgica*, on which Amundsen served as a second mate. A continuous presence on our further-most southern continent has only been in place since the International Geophysical Year of 1956–1957. Systematic research on isolated, confined environments can arguably be dated as beginning as recently as the late 1950s by the military, and much of the early work focused on purely physiological parameters. In their seminal collection of papers dealing with isolated environments from Antarctica to outer space, A. A. Harrison, Y. A. Clearwater, and C. McKay pointed out that early work on psychological factors in extreme environments is often recounted as beginning with C. S. Mullin's research on states of consciousness; E. K. E. Gunderson and colleagues' comprehensive work on adaptation to Antarctica; and classic laboratory studies on group dynamics conducted by I. Altman, W. W. Haythorn, and associates (Harrison et al. 1991a; Mullin 1960; Gunderson 1973, 1974; Altman 1973; Haythorn 1973).

Regardless of which analogue is used to understand what helps or hinders individuals and groups in functioning well under extreme environmental challenges, it is necessary to characterize what we need to know for space. Although specific conditions of the setting vary, most extreme environments share common characteristics: (1) a high reliance on technology for life support and task performance; (2) notable degrees of physical and social isolation and confinement; (3) inherent high risks and associated costs of failure; (4) high physical/physiological, psychological, psychosocial, and cognitive demands; (5) multiple critical interfaces (human-human, human-technology, and human-environment); and (6) critical requirements for team coordination, cooperation, and communication (Bishop 2006a). This last is not insignificant. The accumulated knowledge to date is still fairly rudimentary, given the short historical emergence of the "Space Age." Drawing on research from a number of fields (e.g., social psychology, human factors, military science, management, anthropology, and sociology), researchers easily identified a number of factors that need further investigation. As early as the 1980s, psychological and sociocultural issues had been acknowledged by the National Commission on Space (1986), the National Science Board (1987), and the Space Science Board (1987) to be critical components to mission success, as robust evidence from Antarctica clearly showed psychological issues to impact human behavior and performance significantly in most challenging environments, especially those characterized by isolation and confinement (National Science Board 1987). The most recent roadmap detailing critical strategic and tactical research needed is NASA's Human Research Program Integrated Research Plan whose risk reduction strategy relies on the use of low cost, high fidelity ground analogues to characterize and assess three identified key risks (i.e., sleep/circadian rhythms, workload and fatigue; team adaptation, cohesion and performance; and behavioral symptoms and psychiatric disorders); validate approaches and proof-of-concepts already tested and proven in laboratory studies; and to test and validate countermeasure tools, technologies and protocols prior to spaceflight (NASA 2011). Studies in a variety of analogue environments, e.g., Antarctica, underwater capsules, submarines, caving and polar expeditions, and chamber studies, have confirmed that mission parameters have a significant influence upon the type of "best-fit" crew needed and have isolated a number of psychosocial issues that may negatively affect crewmembers during multinational space missions (Palinkas Gunderson, Burr 1989; Palinkas 1989, 1990, 1991, 2003; Anderson 1991; Ursin 1991; Palinkas, Gunderson, Holland 2000; Bishop and Primeau 2001; Palinkas 2003b; Gilluly 1970; Sexner 1968; Sandal et al. 1999; Bishop, Santy, Faulk 1998; Sandal et al. 1996; Bishop, Grobler, Schjøll, 2001; Kanas 1985). These issues include (1) tension resulting from external stress, (2) factors related to crew heterogeneity (e.g., differences in personality, gender, and career motivation); (3) variability in the cohesion of the crew; (4) improper use of leadership roles (e.g., task/instrumental versus emotional/supportive); (5) cultural differences; and (6) language differences. Of particular uniqueness to challenging environments is the fact that successful performance requires *competent* team interaction, including coordination, communication, and cooperation. The functioning of the operational team often determines the success or failure of the mission. Experience in spaceflight, aviation, polar, and other domains indicates that the stressors present in extreme environments, such as fatigue, physical danger, interpersonal conflict, automation complexity, risk, and confusion, often challenge team processes. The contribution of interpersonal and intrapersonal factors is substantial. For instance, a robust body of evidence from both civilian and military aviation identifies the majority of aircraft accidents as due to human and crew-related performance factors (The Boeing Company 1994; Raymond and Moser 1995; Ricketson, Brown, Graham 1980; Weiner, Kanki, Helmreich 1993; Wiegmann and Shappel 1997; Yacovone 1990). Analyses of critical incidents in medical operating rooms indicate that 70–80% of medical mishaps are due to team and interpersonal interactions among the operating room team (Sexton et al. 1996; Williamson et al. 1993). From pilot to surgeon, firefighter, polar expeditioner or astronaut, we need to know if the characteristics that define adaptable and functional individuals and teams have commonalities across various environments. A comprehensive review of over 50 years of research on team effectiveness conducted by Kozlowski and Ilgen in 2006 identified key team processes involving cognitive, motivational and behavioral processes that contributed substantially to team effectiveness as well as identifying three primary types of interventions (team training and development, team leadership and dynamic regulation of team member behavior) with the potential to improve team processes and team effectiveness (Kozlowski and Ilgen 2006). It is therefore critical that teamwork in these environments be examined and understood. A fundamental need to enable these investigations is developing reliable, minimally intrusive and valid methodologies for assessing individual and group responses to these stressors and identifying dysfunctional and functional coping responses (NASA 2011). The use of extreme environments with characteristics relevant to those inherent in space travel and habitation will play a crucial role in preparing humans for egress from planet Earth.

Given the disparate nature of these various environments, Suedfeld has proposed five key principles that may be useful guides in assessing the relevance of various extreme environments as viable analogues for space or providing the basis for cross-comparisons:

Principle 1: Researchers should think in terms of experiences within environments rather than of environmental characteristics;
Principle 2: Researchers should study differences and similarities between experiences, which are not the same as those between environments;
Principle 3: Analogies should be based on similarities of experience, not necessarily of environment;
Principle 4: Research should look at systematic links between personality factors and experience; and
Principle 5: Experience is continuous and integrated (Suedfeld 1991).

2.3 Critical Psychosocial Issues for Space

The research on teams has, to date, focused on and identified needs for further research under four broad categories. The intent here is not to recite the spectrum of findings across analogues within these areas, but to articulate how analogue environments can address these areas.

- *Selection issues* deal with the evaluation of existing ability, trainability, and adaptability of prospective team members. It is not merely a matter of selecting-out pathological tendencies, but, as importantly, selecting-in desirable characteristics. How can analogue environments allow us to investigate the impact of various individual and group characteristics upon individual and group performance?
- The impact of *isolation and confinement* has been shown to be significantly impacted by various moderator variables, e.g., the difficulty of rescue. While an emergency on the International Space Station certainly poses difficulties regarding time to rescue, one can argue that the difficulties inherent in a Mars mission or even here on Earth from the Antarctic in midwinter, where weather conditions may absolutely make rescue impossible for long periods, carry a qualitatively different psychological impact. An emergency on a mission to Mars will preclude any chance of rescue and necessitate a high degree of autonomy for the crew in making decisions without any real-time mission support. The degree to which such factors magnify the negative effects of isolation and confinement is critical to assess.
- *Group interaction and group processes* are not a simple sum of the individuals that make up the group. Complex interactions can reinforce, undermine, or create new behaviors in the individuals involved. Identification of group fusion (factors that encourage group cohesion) and fission (factors that contribute to group conflict) variables are elementary to creating habitats and work schedules, composing groups, and a myriad of other factors that will enable groups to function effectively and ensure individual and group well-being. For instance, in

a study of Antarctic winter-over personnel, Palinkas found that personnel at Palmer (a small station) spent 60% of their waking hours alone and retreated to their bedrooms extensively for privacy. These behaviors could be considered fission factors as they promote withdrawal, social isolation, and distancing from one's teammates. On the other hand, if the use of privacy served to control the amount of contact and decreased tensions and group conflict, they would be considered fusion factors. He also found that intermittent communication was a major source of conflict and misunderstanding between crews and external support personnel, a clear source of fission influence. Examples of fusion factors for this group were effective leadership styles, which played a significant role in station and crew functioning, as well as the ability to move furniture and decorate both common and private areas, which facilitated adaptation and adjustment (Palinkas 1990).

- *Individual and crew performance* is perhaps the clearest, most frequently studied outcome. Yet there are challenges in defining what constitutes acceptable outcomes at both the individual and group levels. They are not always the same thing, as investigations into missions that failed to meet expectations have repeatedly confirmed. It is a mistake to try to assess and maximize performance without understanding group dynamics, the effects of isolation and confinement or the environment in general on inhabitants. Given that our selection criteria have been little more than ruling out pathology and matching task requirements with technical proficiency within individuals, it is of little surprise that our efforts to implement performance improvements have been only modestly successful and fraught with inconsistent results. It is necessary to take the next steps to identify which individual and group characteristics are maximally associated with adaptation and functioning in these high-challenge environments. This is, in fact, one of the core principles of NASA's Human Research Program Integrated Research Plan (NASA 2011).

2.4 Terrestrial Analogues for Space

There are surprising similarities and differences found across environments. G. M. Sandal et al. found that coping strategies during confinement on polar expeditions were different from those in hyperbaric chambers (Sandal et al. 1996). Whereas polar teams evidenced a delay interval with a marked drop in aggression until after the first quarter, with concomitant increase in homesickness, chamber teams displayed a steady gradual increase in coping over time. A number of researchers have noted that it is not the site that seems to matter, but rather it is the differences in the mission profiles, e.g., tasks (daily achievement of a distance goal versus station-keeping) or duration (short versus long).

In fact, studies addressing Suedfeld's Principle four investigating personality characteristics have produced supporting evidence for a focus on the experience as the defining factor rather than the environment per se. The most persistently investigated personality assessment for the last 15 years has been the NEO-PI by Costa and McCrae (1978, 1985, 1989, 1991). This instrument assesses five global dimensions of personality: neuroticism, extraversion, openness to experience, agreeableness, and conscientiousness. These dimensions have been found to be associated with the previous personality "right stuff/wrong stuff/no stuff" profiles identified by Helmreich et al. in longitudinal studies of American astronaut candidate performance (McFadden et al. 1994). Additionally, measures of achievement motivation, interpersonal orientation, Type A, stress, and coping have been frequently evaluated. Recent studies have found evidence that agreeableness and conscientiousness seem to better predict performance at the global level, along with specific facets of extraversion (Suedfeld and Steel 2000; Rose et al. 1994; McCrae and Allik 2002). Conscientiousness, extraversion, and agreeableness have been found to be related more strongly to constructive change-oriented communication and cooperative behavior than to task performance. Cognitive ability appears to be related more strongly to task performance than to constructive change-oriented communication or cooperative behavior. Results also demonstrate contrasting relationships for agreeableness (positive with cooperative behavior and negative with constructive change-oriented communication) (Barrick et al. 1998; Ferguson et al. 2003; LePine 2003; Judge and Ilies 2002). However, another personality cluster has been identified in studies of successful polar trekking groups that is distinctly different from the "right stuff" profile in which factors indicative of individuals who are loners seem to be supportive of adaptation, i.e., happier alone than dependent on others, highly autonomous, independent, uncomfortable about and relatively uninterested in accommodating others in a group, task-oriented and somewhat competitive (Rosnet, Le Scanff, Sagal 2000). Since we do not have enough data to reliably draw inferences about these individuals, it is mere speculation at this time that perhaps the intense task focus of a polar trek, in which each individual is highly autonomous and individually self-reliant during the long travel each day, situated in an environment that precludes group interaction except for fundamental coordination of locomotion across the terrain, *selects for* individuals that are distinctly different from those who would occupy a habitat or confined environment for long durations. In other words, only individuals with this inward, self-focused personality would find such challenges rewarding and be successful at these tasks. Similarly, an apparently adaptive personality profile has emerged from winter-overers that is characterized by low levels of neuroticism, desire for affection, boredom, and need for order, as well as a high tolerance for lack of achievement, which would fit well in an environment where isolation and confinement prevented accomplishments and the participants experienced frequent shortages and problems (Palinkas et al. 2000). Those that would best adapt would be those who could more quickly adjust their expectations to the immediate situation and tolerate such obstacles. If this hypothesis is substantiated, then we must carefully match the characteristics of the individual to the environment as well as the group in order to maximize successful adaptation and performance.

Psychological research to date seems to support two general findings: (1) there do seem to be consistencies in the personality profile of functional and dysfunctional teams, and (2) characteristics of the mission may define very different personality profiles as best fit. Insomuch as it is possible to select for hardier and better-fit personalities by filtering individuals and teams through environmental challenges, selecting analogues with highly salient and relevant characteristics that match space mission profiles (e.g., long versus short duration, station-keeping versus expedition profiles) will be important.

2.4.1 The Expeditionary Analogue

Expeditions, by definition, revolve around movement. Expeditionary analogues (e.g., oceanic, polar, desert, caving, mountaineering) include various exploratory goals that are characterized by moving from one place to another rather than inhibiting a locale. Historical exploratory expeditions typically involved long durations (i.e., months to years) characterized by significant known and unknown risks, broad goals, a high degree of situationally driven contingency decision-making, and expectations of autonomy and self-sufficiency. Modern expeditions, in contrast, are typically of short duration (i.e., 2 weeks to 3 months), utilize the advantages of technology to minimize risks (e.g., weather forecasts to take advantage of the best weather of a region and satellite communications to maintain contact), are more narrowly goal-oriented and task-focused, and involve members with specialized roles and skills. In both expeditionary scenarios, teams were/are formed around appropriate skill sets and availability and a notable lack of any attempt to screen individuals psychologically except for medical factors. Research on team functioning is often secondary to expedition goals, personal goals, schedules, and contingencies. The expedition may be intended to recreate experiences of earlier explorers, such as the Polynesian *Kon-Tiki* oceanic traverse; set records or discover new territory, e.g., discover a route to India or explore a cave system; achieve personal challenges, such as climbing mountains or skiing to the North Pole; conduct scientific research, e.g., by means of oceangoing research vessels or polar ice drilling teams; or conduct commercial exploration, such as mineral and oil exploration (Bishop et al. 1998; Heyerdahl 1950; Bernard and Killworth 1973, 1974; Mallis and Deroshia 2005).

Ben Finney, Professor Emeritus in Anthropology at the University of Hawai'i and noted for his work on applying anthropological perspectives to humankind's expansion into space, has argued that from the earliest voyages to have scientific goals, "cultural" differences between scientists and seamen have led to conflict and that this inherent conflict of cultures is similarly reflected in our space program's structural differentiation between pilots and astronaut-scientists (Finney 1991). Voyages of scientific discovery began in the late eighteenth century, an age, Finney points out, that many have argued foreshadowed the space race of the 1960s (Goetzmann 1986). The first exploratory voyage to include scientists as crew and mission goals with explicit scientific objectives instead of commercial goals that serendipitously collected science data was the 3-year-long English expedition of the

Endeavour to Tahiti, 1768–1771, led by Captain James Cook. The on-board scientists were tasked to observe the transit of Venus across the face of the Sun to provide data needed to calculate the distance between Earth and the Sun. The success of the *Endeavour*'s expedition led to a second expedition, which sailed with a number of scientists, two astronomers, and a naturalist, an expedition that, in contrast to the first expedition, was rife with contentious relationships between the seamen and the scientists. Subsequent voyages with scientists on board were similarly plagued by conflicts between those pursuing scientific goals and those tasked with the piloting and maintenance of the ship. Historically, the English naval command eventually imposed an unofficial moratorium on the inclusion of non-naval scientists on board and pursued a policy of assigning any scientific duties to members of the crew. Not until a 100 years after Cook, in 1872, would the Royal Navy's *Challenger*, a three-masted, square-rigged, wooden vessel with a steam engine, sail around the world with six marine scientists and a crew and captain who were totally dedicated to the research (Linklater 1972).

Such troubles were not limited to the English. The French followed a similar pattern, beginning in 1766 and continuing through 1800, when scientists sailed with numerous expeditions that were summarily characterized by conflict and contention between the crews and scientists (Dunmore 1965, 1969). Finney further notes that such complaints are found in journals of early Russian scientists, as well as American scientists on the 4-year-long United States Exploring Expedition that sailed from Norfolk in 1838 with a contingent of 12 scientists (von Chamisso 1856, Stanton 1975).

Modern development of specialized ships complete with laboratories and equipment dedicated to oceanographic research has been primarily organized and maintained by universities and oceanographic institutes. Yet even aboard these dedicated floating research vessels, conflict between the ship's crew and the scientists whom they serve has not been eliminated. A dissertation study conducted by a resident at the Scripps Institute of Oceanography during 1973 concluded that tension between the two groups was inevitable because they formed two essentially separate and distinct subcultures with different values and goals, as well as different educational backgrounds and class memberships (Bernard and Killworth, 1973, 1974).

Finney argues that the same subcultures have become evident in the space program with the development of the role of payload specialists, who are considered visiting scientists rather than part of the elite astronaut corps. Tensions between payload specialists in pursuit of the scientific goals and the crew in pursuit of mission completion have routinely been in evidence. Finney eloquently states:

> . . . [I]f space research were to be made as routine to the extent that ocean research now is, subcultural differences, and hence tensions, between scientist and those pilots, station-keepers, and others whose job it will be to enable researchers to carry out their tasks in space may become critical considerations. If so, space analogues of the mechanisms that have evolved to accommodate differences between scientists and seamen aboard oceanographic ships may have to be developed (Finney 1991, p. 100).

The number and variety of expeditions examined for relevance to space is ever increasing as both modern expeditions and analyses of historical expeditions are scrutinized. An example of how examination of the records from past expeditions contributes to the current state of knowledge and provides the impetus for future studies in space can be seen in a meta-study by M. Dudley-Rowley et al. that examines written records from a sample of space missions and polar expeditions for similarities and differences in conflicts and perceptions of subjective duration of the mission. Ten missions were compared across a number of dimensions (Dudley-Rowley et al. 2000). The meta-study included three space missions that represented both long- and short-duration mission profiles: Apollo 11 (1969) and Apollo 13 (1970), ranging from 6 to 8 days apiece, and Salyut 7 (1982), which lasted over 200 days. Four Antarctic expeditions were included: the western party field trip of the Terra Nova Expedition (1913, 48 days), an International Geophysical Year (IGY) traverse (1957–1958, 88 days), the Frozen Sea expedition (1982–84, 480 days), and the International Trans-Antarctica Expedition (1990, 224 days). Finally, three early Arctic expeditions were also included: the Lady Franklin Bay (1881–1884, 1,080 days), Wrangel Island (1921–23, 720 days), and Dominion Explorers' (1929, 72 days). Seven factors emerged that seemed to coincide with the subjectivization of time and the differentiation of situational reality for the crews from baseline:

1. Increasing distance from rescue in case of emergency (lessening chances of "returnability");
2. Increasing proximity to unknown or little-understood phenomena (which could include increasing distance from Earth);
3. Increasing reliance on a limited, contained environment (where a breach of environmental seals means death or where a fire inside could rapidly replace atmosphere with toxins);
4. Increasing difficulties in communicating with Ground or Base;
5. Increasing reliance on a group of companions who come to compose a microsociety as time, confinement, and distance leave the larger society behind, in a situation where innovative norms may emerge in response to the new sociophysical environment;
6. Increasing autonomy from Ground's or Base's technological aid or advice; and
7. Diminishing available resources needed for life and the enjoyment of life.

The missions and expeditions were ranked by prevalence of the seven factors that might correspond with the differentiation in the subjectivization of the passage of time and in the situational reality for the crews from baseline. From highest to lowest in compromising factors, the rankings fell in the following order: Lady Franklin Bay (7); Wrangel Island, Apollo 13 (6); Salyut 7 (5); Terra Nova, Apollo 11 (4); Dominion Explorers' (3); Frozen Sea (2); IGY (1); International Trans-Arctic Expedition (0). The Lady Franklin Bay Expedition suffered 18 deaths of its complement of 25, and the rest were starving when found. The Wrangel Island expedition suffered four deaths out of its crew of five. Apollo 13 was a catastrophe that was remarkable in its recovery of the crew intact. The Salyut 7 mission, the Terra Nova western field party, and the Apollo 11 mission all had high degrees of risk. The later polar expeditions rank below these missions.

Both the space missions and the earliest polar expeditions are above or hover just below the median (3.5). Although the authors correctly note that the sample is too small to draw conclusions, the presence of similar factors in space and early polar exploration that contributed to perceptions of mission/expedition duration or of how their situational reality deviates from baseline is important to note. These results suggest that as control over their environment decreases, team members' subjective experiences of time and the situation increasingly differ from their baselines. The strong parallel between early expeditions and modern space missions lends support for historical analogues as viable substitutes for space.

2.4.2 Chamber Studies

Early evaluations for astronaut selection drew upon a history of sensory deprivation research initially begun by the military throughout the 1950s and 1960s to address performance concerns about two-person crews confined to armored vehicles for long durations and continued most notably through the series of studies conducted by J. P. Zubek (Zubek 1969; Honingfeld 1965). Initially, it was believed that space would represent a significant loss of normal sensory stimulation due to isolation from people, reduction in physical stimulation, and restricted mobility. Thus, sensory deprivation chambers were argued to be good analogues for astronauts (Flaherty 1961). Selection procedures, therefore, included stints in dark, small, enclosed spaces for several hours to observe how potential astronauts handled the confinement and loss of perceptual cues. As Dr. Bernard Harris, the first African American to walk in space, recounts, "They put me in this little box where I couldn't move or see or hear anything. As I recall, I fell asleep after a while until the test ended" (Harris 1995).

The first systematic attempts to investigate psychological adaptation factors to isolation and confinement in simulated operational environments were conducted in the 1960s and early 1970s by putting volunteers in closed rooms for several days, subjecting them to sleep deprivation and/or various levels of task demands by having them complete repetitive research tasks to evaluate various aspects of performance decrements (Haythorn and Altman 1966; Altman 1991). Chamber research, as it was to become known, encompassed a variety of artificial, constructed environments whose raison d'être was control over all factors not specifically under study. Later, specially constructed confinement laboratories such as the facility at the Johns Hopkins University School of Medicine or simulators at Marshall Space Fight Center in Huntsville, Alabama; the McDonnell Douglas Corporation in Huntington Beach, California; or Ames Research Center at Moffett Field, California, housed small groups of three to six individuals in programmed environments for weeks to months of continuous residence to address a variety of space-science-related human biobehavioral issues related to group dynamics (e.g., cohesion, motivation, effects of joining and leaving established groups), performance and work productivity, communication patterns, team cooperation, and social habitability factors. Other types of chambers have been utilized to evaluate various countermeasures, especially for stress and anxiety provoked by physiological stressors expected to be present in the space environment. For instance, Chinese behavioral scientists are the latest to employ the use of centrifuge training to test the efficacy of guided imagery to reduce anxiety, heart rates and heart rate variability (Jing et al. 2011). Critics of this approach point out such environments constitute *confinement* but not necessarily isolation. The stress of isolation is far more complex than that of confinement and involves a psychological dimension of separateness that may not be inherent in artificial chamber studies.

The epitome example of chamber research may be the series of five hyperbaric-chamber studies, sponsored by the European Space Agency investigating psychosocial functioning, in which groups were confined for periods lasting from 28 to 520 days (Sandal et al. 1995; Sandal 2004). Full mission protocols specifying all medical, technical, and operational parameters approximating expected living conditions of astronauts on a space station were used. The studies were intended to evaluate the efficacy of various psychosocial monitoring and assessment techniques for implementation on real space missions, as well as to investigate persistent occurrences of communication and interaction breakdowns between on-orbit teams and Mission Control anecdotally reported from space (Kanas et al. 2000; Sandal et al. 1995; Gushin et al. 1996). A number of opportunities and advances came from these studies, e.g., evaluating the efficacy of communication training for space teams or the opportunity to examine factors involved in an unplanned meltdown between crews precipitated by differences in cultural attitudes and norms about genders, authority, and control (Sandal 2004; Manzey 1998). However, skepticism regarding the verisimilitude of studies in which discontented members can simply quit has continued to raise real concerns as to how generalizable the findings from chamber studies are to space missions.

2.4.3 *The Middle Ground: Capsule Habitats in Extreme Unusual Environments*

Occupying the middle ground between traditional expeditionary missions with moving trajectories and the artificiality of laboratory spaces designated as space station habitats are capsule habitats, sharing the controlled, defined enclosure of the laboratory situated within an extreme unusual environment (EUE) (Suedfeld and Steel 2000). Characterized by a controlled, highly technological habitat that provides protection and life support from an environment that is harsh, dangerous, and life-threatening, capsule habitats occupy a wide range of environments. Some are true operational bases with missions in which biobehavioral research is only secondary. Others run the gamut from fundamental "tuna can" habitats with spartan support capabilities situated in locations of varying access to a full-fidelity Antarctic base constructed solely for the purposes of biobehavioral space analogue research.

2.4.3.1 Submersible Habitats

Due to their high military relevance, the best-studied of capsule habitats are submarines. As an analogue for space, submarines share a number of common characteristics: pressurization concerns (hyperpressurization for submarines and loss of pressurization for space), catastrophic outcomes for loss of power (e.g., the inability to return to the surface for submarines and degraded orbits for space), dependence on atmosphere revitalization and decontamination, radiation effects, and severe space restrictions. Prenuclear submarine environments were limited in the duration of submersions (72 h), crew size (9 officers and 64 enlisted men), and deployment periods without restocking of fuel and supplies. Structurally, these short-duration mission parameters mimicked those of the early years of space, albeit with vastly larger crews. With the launch of the nuclear-powered *Nautilus* in 1954, the verisimilitude of the submersible environment as an analogue for long-duration space missions was vastly improved. With the nuclear submarine, mission durations were extended to 60–90 days, crews were increased to 16 officers and 148 enlisted men, and resupply could be delayed for months (Weybrew 1991). Generalizing from submarine research to space regarding psychological and human factors related to adjustment and well-being, researchers have identified several salient issues:

- Atmospheric revitalization and contamination control;
- Development and validation of procedures for the medical and psychological screening of recruits;
- Identification of techniques for initiating and sustaining individual motivation and group morale; and
- Identification of stressors, assessment of the severity of patterns of stress reactivity, and development of effective stress coping strategies (Weybrew, Helmreich, Howard, 1986; Weybrew 1991).

An extension of the submersible operational environment of a military submarine is the NASA Extreme Environment Mission Operations program (NEEMO) being conducted in the Aquarius underwater habitat situated off Key Largo, Florida—the only undersea research laboratory in the world. Owned by the U.S. National Oceanic and Atmospheric Administration (NOAA) and operated by the National Undersea Research Center (NURC) of the University of North Carolina at Wilmington on behalf of NOAA, Aquarius is the submerged analogue to NOAA oceanic research vessels. First deployed in 1988 in the U.S. Virgin Islands and relocated to Key Largo in 1992, the underwater facility has hosted more than 80 missions and 13 crews of astronauts and space researchers since 2001. Aquarius provides a capsule habitat uniquely situated within an environment that replicates many of the closed-loop constraints of the vacuum of space: a hostile, alien environment that requires total dependency on life support; poses significant restrictions to escape or access to immediate help; and is defined by limited, confined habitable space and physical isolation. The complexity of NEEMO missions further parallels space missions in their mission architecture, with similar requirements for extensive planning, training, control, and monitoring via an external mission control entity. However, it has only been the most recent NEEMO missions in which stress, fatigue, and cognitive fitness, as well as individual and intrapersonal mood and interaction, have been the focus of study.

2.4.3.2 Polar Stations

First and foremost, Antarctica springs to mind when polar space analogues are raised. While there are other polar bases in the Arctic and subarctic, the bulk of sustained psychological research has been conducted in Antarctica (Harrison et al. 1991). G. M. Sandal, G. Leon, and L. Palinkas conducted a recent, extensive review of the literature on psychosocial adaptation by polar work groups, expedition teams, Antarctic bases, simulation, and space crews (Sandal et al. 2006). There are 47 stations throughout the Antarctic and sub-Antarctic regions, operated by 20 different nations, with populations running from 14 to 1,100 men and women in the summer to 10–250 during the winter months. The base populations vary from mixed-gendered crews to male-only crews, from intact families (Chile) to unattached singletons, for assignments that last from a few months to 3 years.

In 1958, after the IGY (1956–57) produced the first permanent bases in Antarctica, C. S. Mullin, H. Connery, and F. Wouters conducted the first systematic psychological study of 85 men wintering over in Antarctica (Mullin et al. 1958). Their study was the first of many to identify the Antarctic fugue state later dubbed the "big-eye," characterized by pronounced absent-mindedness, wandering of attention, and deterioration in situational awareness that surfaced after only a few months in isolation. The majority of subsequent studies up through the 1980s focused on the physiological changes evidenced in winter-over adaptation. Those that did address psychosocial factors tended to focus on the negative or pathological problems of psychological adjustment to Antarctic isolation and confinement, with persistent findings of depression, hostility, sleep disturbance, and impaired cognition, which quickly came to be classified as the "winter-over syndrome" (Gunderson 1973, 1974; Strange and Klein 1974). Sprinkled among Antarctic research have been findings that also report positive, or salutogenic, aspects of the winter-over experience in which winter-overers have reported enhanced self-growth, positive impacts to careers, and opportunities for reflection and self-improvement (Mullin 1960; Taylor and Shurley 1971; Wilson 1965; Palinkas 1986; Oliver 1991; Suedfeld 2005).

One of Antarctica's most prolific researchers, Dr. Larry Palinkas has analyzed 1,100 Americans who wintered over between 1963 and 2003 over four decades of research in Antarctica and proposed four distinct characteristics to psychosocial adaptation to isolation, confinement, and the extreme environment:

1. Adaptation follows a seasonal or cyclical pattern that seems to be associated with the altered diurnal cycle and psychological segmentation of the mission.

2. Adaptation is highly situational. Because of unique features of the station's social and physical environment and the lack of resources typically used to cope, baseline psychological measures are not as good predictors of depressed mood and performance evaluations as are concurrent psychological measures.
3. Adaptation is social. The structure of the group directly impacts individual well-being. Crews with clique structures report significantly more depression, anxiety, anger, fatigue, and confusion than crews with core-periphery structures.
4. Adaptation can also be "salutogenic," i.e., having a positive effect for individuals seeking challenging experiences in extreme environments (Palinkas 2003a).

Palinkas found that a depressed mood was inversely associated with the severity of station physical environments—that is, the better the environment, the worse the depression—and that the winter-over experience was associated with reduced subsequent rates of hospital admissions (Palinkas 2003a). He and others have speculated that the experience of adapting to the isolation and confinement, in general, improved an individual's self-efficacy and self-reliance and engendered coping skills that they used in other areas of life to buffer subsequent stress and resultant illnesses (Palinkas 2003a; Suedfeld 2005).

2.4.3.3 Concordia

In 1992, France initiated plans for a new Antarctic station on the Antarctic Plateau and was later joined by Italy. In 1996, the first French-Italian team established a summer camp at Dome C to provide logistical support for the European Project for Ice Coring in Antarctica (EPICA) and begin the construction of the permanent research station. Concordia Station became operational in 2005; the first winter-over took place in February 2005 with a staff of 13. The station consists of three buildings, which are interlinked by enclosed walkways. Two large, cylindrical three-story buildings provide the station's main living and working quarters, while the third building houses technical equipment, like the electrical power plant and boiler room. The station can accommodate 16 people during the winter and 32 people during the summer season. The typical winter population consists of four technicians for the station maintenance, nine scientists or technicians for the science projects, a chief, a cook, and a medical doctor.

Dome C is one of the coldest places on Earth, with temperatures hardly rising above −25 °C in summer and falling below −80 °C in winter. Situated on top of the Antarctic plateau, the world's largest desert, it is extraordinarily dry and supports no animals or plants. The first summer campaign lasted 96 days, from 5 November 2005 until 8 February 2006, with 95 persons participating. The 2006 season included seven crewmembers with two medical experiments and the first two psychological experiments sponsored by the European Space Agency for which the crew acted as subjects during their stay. The two experiments investigated psychological adaptation to the environment and the process of developing group identity, issues that will also be important factors for humans traveling to Mars. For this research, the crew completed questionnaires at regular intervals throughout their stay. The ESA's Mistacoba experiment to profile how microbes spread and evolve in the station—an isolated and confined environment—over time started in the 2005 season, when the first crew started living at the station, and has also continued with subsequent crews. Starting from a newly built clean environment, those conducting the study took samples from fixed locations in the base as well as from crewmembers themselves (The Concordia Station 2007).

2.4.3.4 Haughton-Mars Project

One of the first of dedicated research hybrid facilities was the Haughton-Mars Project (HMP), initiated in 1996 when the National Research Council of the U.S. National Academy of Sciences and NASA Ames Research Center sponsored a postdoctoral proposal to study the Haughton Crater on Devon Island in the Canadian Arctic as a potential analogue for Mars. The program has expanded from a four-member team in 1997 to a permanent habitat that hosts 8-week arctic summer field seasons with 50–90 participants, multiple teams, and research projects that run from instrument testing and development to biomedical and psychological evaluation. HMP routinely supports participation by NASA; the Canadian Space Agency (CSA); the Russian Institute for Space Research (IKI); various research institutions and universities in the United States, Canada, and the United Kingdom; and the U.S. Marine Corps. It has been the subject of various documentaries made by such groups as the National Geographic Society and Discovery Channel Canada (The Mars Institute 2007).

2.4.3.5 Flashline Mars Arctic Research Station

In 2000, a second dedicated research facility was deployed on Devon Island, jointly sponsored by the Haughton-Mars Project and the Mars Society: the Flashline Mars Arctic Research Station (FMARS). Running concurrently with HMP, the FMARS facility was the first of four proposed analogue research facilities to be built by the Mars Society, supporting smaller six-person crews for typically 2–8-week seasons. In summer 2007, the first 4-month-long FMARS mission was successfully completed with a crew of seven and a full complement of research studies covering technology, human factors, medicine, psychology, and communications. Results from this first attempt to extend, heretofore, short duration missions in a habitat developed specifically for research (instead of operational environments such as bases in Antarctica) revealed a number of issues (e.g., multicultural conflicts, coping inefficiencies) that underscore the need for longer studies situated in real environments that pose true isolation and risk (Bishop et al. 2010).

2.4.3.6 Mars Desert Research Station

The second Mars Society station, the Mars Desert Research Station (MDRS), came online in December 2001 and is situated in the Utah desert in the American Southwest. Because of its ease of access, the American station is considered well suited as a test bed for equipment that will later be sent to more remote and unforgiving locations. For the same reason, the American station has been the focus of short-duration isolation and confinement studies since its inception. A wide range of psychological studies investigating crew factors in short-duration missions has been in place since 2002. However, beyond preliminary descriptive results presented at conferences, the small sample size of crews has necessitated waiting until enough teams have rotated through the facility to allow meta-analyses (Bishop et al. 2004, 2006). Several international teams have also used the MDRS for studies investigating comparisons between homogeneous-gendered teams, comparisons between mission teams and backup crews, and international cultural factors, among others (Bishop et al. 2005, Bishop 2006).

The Mars Society plans additional facilities in Iceland and Australia that will capitalize on geological features that present opportunities to practice Mars exobiology field work. The Mars Society's Mars Analog Research Station Project envisions three prime goals to be served by these habitats (The Mars Society 2007):

- The stations will serve as effective test beds for field operations studies in preparation for human missions to Mars. They will facilitate the development and testing of key habitat design features, field exploration strategies, tools, technologies, and crew selection protocols that will enable and help optimize the productive exploration of Mars by humans. In order to achieve this goal, each station must be a realistic and adaptable habitat.
- The stations will serve as useful field research facilities at selected Mars analogue sites on Earth and will help further understanding of the geology, biology, and environmental conditions on Earth and on Mars. In order to achieve this objective, each station must provide safe shelter and be an effective field laboratory.
- The stations will generate public support for sending humans to Mars. They will inform and inspire audiences around the world. As the Mars Society's flagship program, the Mars Analog Research project will serve as the foundation of a series of bold steps that will pave the way to the eventual human exploration of Mars.

2.5 Conclusion

The use of analogues for space is an emergent field whose very short track record examining team dynamics and psychosocial factors impacting individual and group functioning vigorously supports the real value of these environments and generalizability to space environments. Unlike laboratory studies, where the threat of real danger is usually absent, teams operating within real extreme environments have unknown situational and environmental challenges to face. Even in circumstances in which death or injury occurs, there will always be questions regarding the ability to avoid negative outcomes. While post-mission analyses of behavior and performance add insight into contributing factors, it is seriously doubtful whether we will ever be able to accurately predict the entire range and complexity of interaction between the human-environment factors and the human-human factors. Risk is inherent in human exploration. Even so, the value of analogue experiences cannot be underestimated, regardless of whether they help us grapple with defining our levels of adequate preparation in the face of ideally predefined levels of "acceptable risk" or even "acceptable losses" (a concept familiar to those who perform military risk assessments).

One key methodological and validity issue is the added value of utilizing consistent measures across various analogues, allowing more accurate comparisons of individuals and teams across environments, including space. The necessity to validate multicultural questionnaires and methodologies that are relevant, reliable, and valid for international teams is of paramount importance as our reliance on these multinational teams will only increase in the future. To that extent, the various research endeavors in analogue environments have contributed significantly to validating such assessment instruments in a variety of teams.

Findings from analogues have clearly identified three major intervention points to affect group functioning outcomes:

- Selection: the development of reliable and valid methods of choosing the best fit at both the individual and the group levels.
- Training: improving the fitness of the group by prepping skills needed for interpersonal group dynamics as well as high-functioning self-monitoring and appropriate adaptation.
- Support: taking the form of prevention first, then early, proactive intervention second. To be successful, research to date strongly suggests that the support must include the group, the family, and all external participants (e.g., Mission Control) as partners.

A large portion of the current research represents opportunities to examine team dynamics and factors that impact team function in real-world groups that have been brought together for particular purposes that have little to do with research, e.g., geological field teams. Similarly, examinations of historical sources of past expeditions will continue to inform and provide additional insight into factors that have contributed to the success or failure of previous efforts. However, we need larger, more systematic studies in which the composition of the team is one of the driving factors under investigation instead of simply an extraneous variable. Our greatest hope lies with the new research facilities now available and coming online dedicated to such research.

References

Altman, I.: An ecological approach to the functioning of isolated and confined groups. In: Rasmussen, J. (ed.) Man in Isolation and Confinement, pp. 241–270. Aldine, Chicago (1973)

Anderson, C.: Polar psychology – coping with it all. Nature **350**(6316), 290 (1991)

Baddeley, A.D.: Selecting attention and performance in dangerous environments. Br. J. Psychol. **63**, 537–546 (1972)

Barrick, M.R., Stewart, G.L., Neubert, M.J., Mount, M.K.: Relating member ability and personality to work-team processes and team effectiveness. J. Appl. Psychol. **83**, 377–391 (1998)

Bernard, H.R., Killworth, P.: On the social structure of an ocean going research vessel. Soc. Sci. Res. **2**, 145–184 (1973)

Bernard, H.R., Killworth, P.: Scientist at sea: a case study in communications at sea. Report BK-103-74, Code 452, Contract N00014-73-4-0417-0001, prepared for the Office of Naval Research. National Technical Information Service, Springfield: (1974)

Bishop, S.L.: Psychological and psychosocial health and well-being at pole station. In: Cockell, C.S. (ed.) Project Boreas: A Station for the Martian Geographic North Pole, pp. 160–171. British Interplanetary Society, London (2006a)

Bishop, S.L.: Assessing Group Dynamics in a Mars Simulation: AustroMars Crew 48. Paper presented a part of the Mars2030: Interdisciplinary Workshop on Mars Analogue Research and AustroMars Science Workshop, University of Salzburg, Salzburg 24–26 Sept 2006b

Bishop, S.L., Santy, P.A., Faulk, D.: Team dynamics analysis of the Huautla Cave diving expedition: a case study. Hum. Perform. Extreme. Environ. **1**(3), 37–41 (1998)

Bishop, S.L., Grobler, L.C., Schjøll, O.: Relationship of psychological and physiological parameters during an Arctic Ski expedition. Acta Astronaut. **49**, 261–270 (2001)

Bishop, S.L., Primeau, L.: Assessment of group dynamics, psychological and physiological parameters during polar winter-over. Paper presented as part of the Human Systems Conference, Nassau Bay, 20–22 June 2001

Bishop, S.L., Dawson, S., Rawat, N., Reynolds, K., Eggins, R.: Expedition one: assessing group dynamics in a desert Mars simulation. Paper presented as part of the 55th International Astronautical Congress, Vancouver, 4–7 Oct 2004

Bishop, S.L., Sundaresan, A., Pacros, A., Patricio, R., Annes, R.: A Comparison of Homogeneous Male and Female Teams in a Mars Simulation. Paper presented as part of the 56th International Astronautical Congress, Fukuoka, Japan (2005)

Bishop, S.L., Dawson, S., Rawat, N., Reynolds, K., Eggins, R., Bunzelek, K.: Assessing teams in Mars simulation habitats: Lessons learned from 2002–2004. In: Clarke, J.D. (ed.) Mars Analog Research. American Astronautical Society Science and Technology Series, vol. 111, pp. 177–196. Univelt, San Diego (2006)

Bishop, S.L., Kobrick, R., Battler, M., Binsted, K.: FMARS 2007: Stress and coping in an arctic Mars simulation. Acta Astronaut. **66**(9–10), 1353 (2010). doi:10.1016/j.actaastro.2009.11.008

The Boeing Company.: Statistical summary of commercial jet aircraft accidents: worldwide operations, 1959–1993. In: Boeing Airplane Safety Engineering Report B-210B. Boeing Commercial Airplane Group, Seattle (1994)

The Concordia Station.: http://www.concordiastation.org/ Accessed 18 June 2007

Costa, Jr. P.T., McCrae, R.R.: NEO Five-Factor Inventory, Psychological Assessment Resources, Inc., Lutz (1978, 1985, 1989, 1991)

Dudley-Rowley, M., Whitney, S., Bishop, S., Caldwell, B., Nolan, P.D.: Crew size, composition and time: implications for habitat and workplace design in extreme environments. Paper presented at the SAE 30th international conference on environmental systems, 10–13 (2000)

Dunmore, J.: French Explorers of the Pacific, vol. 1. Claredon Press, Oxford (1965)

Dunmore, J.: French Explorers of the Pacific, vol. 2. Claredon Press, Oxford (1969)

European Space Agency, ESA Research News.: http://www.esa.int/esaHS/SEMBZA8A9HE_research_0.html#subhead1. Accessed 18 June 2007

Ferguson, L., James, D., O'Hehir, F., Sanders, A.: Pilot study of the roles of personality, references, and personal statements in relation to performance over the five years of a medical degree. Br. Med. J. **326**(7386) (2003): 326–429
Finney, B.: Scientists and seamen. In: Harrison, A.A., Clearwater, Y.A., McKay, C.P. (eds.) From Antarctica to Outer Space: Life in Isolation and Confinement, pp. 89–102. Springer, New York (1991)
Flaherty, B.E. (ed.): Psychophysiological Aspects of Space Flight. Columbia University Press, New York (1961)
Gilluly, R.H.: Tektite: unique observations of men under stress. Sci. News **94**, 400 (1970)
Goetzmann, W.H.: New Lands, New Men. Viking, New York (1986)
Greely, A.: Three Years of Arctic Service: An Account of the Lady Franklin Bay Expedition of 1881–1884, and the Attainment of the Farthest North. Scribner, New York (1886)
Gunderson, E.K.E.: Individual behavior in confined or isolated groups. In: Rasmussen, J. (ed.) Man in Isolation and Confinement, pp. 145–166. Aldine, Chicago (1973)
Gunderson, E.K.E.: Psychological studies in Antarctica. In: Gunderson, E.K.E. (ed.) Human Adaptability to Antarctic Conditions, pp. 115–131. American Geophysical Union, Washington, DC (1974)
Gushin, V.I., Kolintchenko, V.A., Efimov, V.A., Davies, C.: Psychological evaluation and support during EXEMSI. In: Bonting, S. (ed.) Advances in Space Biology and Medicine, pp. 283–295. JAI Press, London (1996)
Gushin, V.I., Zaprisa, T.B., Kolintchenko, V.A., Efimov, A., Smirnova, T.M., Vinokhodova, A.G., Kanas, N.: Content analysis of the crew communication with external communicants under prolonged isolation. Aviat. Space Environ. Med. **12**, 1093–1098 (1997)
Harris, B.: Personal communication, thesis committee member (1995–1996)
Harrison, A.A., Clearwater, Y.A., McKay, C.P.: From Antarctica to Outer Space: Life in Isolation and Confinement. Springer, New York (1991)
Haythorn, W.W.: The miniworld of isolation: laboratory studies. In: Rasmussen, J. (ed.) Man in Isolation and Confinement, pp. 219–240. Aldine, Chicago (1973)
Haythorn, W., Altman, I.: Personality factors in isolated environments. In: Trumbull, M. (ed.) Psychological Stress: Issues in Research. Appleton-Century-Crofts, New York (1966)
Heyerdahl, T.: Kon-Tiki. Rand McNally, Chicago (1950)
Honingfeld, R.: Group behavior in confinement: review and annotated bibliography, Report AD0640161, prepared for the Human Engineering Lab, p. 117. Aberdeen Proving Ground, Maryland (1965)
Jing, X., Liu, F., Wu, B., Miao, D.: Guided imagery, anxiety, heart rate, and heart rate variability during centrifuge training. Aviat. Space Environ. Med. **82**, 92–96 (2011)
Judge, T.A., Ilies, R.: Relationship of personality to performance motivation: a meta-analytic review. J. Appl. Psychol. **87**(4), 797–807 (2002)
Kanas, N.: Psychosocial factors affecting simulated and actual space missions. Aviat. Space Environ. Med. **56**, 806–811 (1985)
Kanas, N., Salnitskiy, V., Grund, E.M., et al.: Social and cultural issues during Shuttle/Mir Space Missions. Acta Astronaut. **47**, 647–655 (2000)
Kozlowski, S.W.J., Ilgen, D.R.: Enhancing the effectiveness of work groups and teams (Monograph). Psycholog. Sci. Pub. Interest. **7**, 77–124 (2006)
Lebedev, V.: Diary of a Cosmonaut: 211 Days in Space. Phytoresource Research, College Station (1988)
LePine, J.A.: Team adaptation and postchange performance: effects of team composition in terms of members' cognitive ability and personality. J. Appl. Psychol. **88**(1), 27–39 (2003)
Linklater, E.: The Voyage of the Challenger. John Murray, London (1972)
Lovell, J., Kluger, J.: Apollo 13 [Lost Moon: The Perilous Voyage of Apollo 13]. Pocket Books, New York (1994)
Mallis, M.M., Deroshia, C.W.: Circadian rhythms, sleep, and performance in space. Aviat. Space Environ. Med. **76**, B94–B107 (2005)

Manzey, D.: Space Psychology: Textbook for Basic Psychological Training of Astronauts AM-BMT-DLR-98-009. ESA/EAC, Cologne (1998)

The Mars Institute.: NASA Haughton-Mars Project History. available at http://www.marsonearth.org/about/history.html Accessed 14 June 2007

The Mars Society.: Mars desert research station project goals, available at http://www.marssociety.org/MDRS/mdrs01b.asp Accessed 14 June 2007

McCarthy, G.W.: Operational relevance of aeromedical laboratory research. Abstract no. 24. Paper presented as part of the Aerospace Medical Association's 69th Annual Scientific Meeting, Seattle, 17–21 May 1988, p 57

McCrae, R.R., Allik, J.: The Five-Factor Model of Personality Across Cultures. Kluwer, Dordrecht (2002)

McFadden, T.J., Helmreich, R., Rose, R.M., Fogg, L.F.: Predicting astronaut effectiveness: a multivariate approach. Aviat. Space Environ. Med. **65**, 904–909 (1994)

Mears, J.D., Cleary, P.J.: Anxiety as a factor in underwater performance. Ergonomics **23**(6), 549–557 (1980)

Mullin, C.S.: Some psychological aspects of isolated Antarctic living. Am. J. Psychiat. **111**, 323–325 (1960)

Mullin, C.S., Connery, H., Wouters, F.: A psychological-psychiatric study of an IGY station in Antarctica. Report prepared for the U.S. Navy, Bureau of Medicine and Surgery, Neuropsychiatric Division (1958)

NASA Human Research Program Integrated Research Plan.: HRP-47065 Rev C, http://humanresearchroadmap.nasa.gov/evidence/IRP%20Rev%20C.pdf (2011)

National Commission on Space: Pioneering the Space Frontier. Bantam Books, New York (1986)

National Science Board: The Role of the National Science Foundation in Polar Regions. National Academy of Sciences, Washington, DC (1987)

Oliver, D.: Psychological effects of isolation and confinement of a winter-over group at McMurdo station, Antarctica. In: Harrison, A.A., Clearwater, Y.A., McKay, C.P. (eds.) From Antarctica to Outer Space: Life in Isolation and Confinement, pp. 217–228. Springer, New York (1991)

Palinkas, L.A.: Health and performance of Antarctic winter-Over personnel: a follow-up study. Aviat. Space Environ. Med. **57**, 954–959 (1986)

Palinkas, L.A.: Sociocultural influences on psychosocial adjustment in Antarctica. Med. Anthropol. **10**, 235–246 (1989)

Palinkas, L.A.: Psychosocial effects of adjustment in Antarctica: lessons for long-duration spaceflight. J. Spacecraft. **27**(5), 471–477 (1990)

Palinkas, L.A.: Effects of physical and social environments on the health and well being of Antarctic Winter-Over personnel. Environment **23**, 782–799 (1991)

Palinkas, L.A.: On the ICE: Individual and group adaptation in Antarctica. (2003a), available at http://www.sscnet.ucla.edu/anthro/bec/papers/Palinkas_On_The_Ice.pdf Accessed 12 June 2007

Palinkas, L.: The psychology of isolated and confined environments: understanding human behavior in Antarctica. Am. Psychol. **58**(5), 353–363 (2003b)

Palinkas, L.A., Gunderson, E.K.E., Burr, R.: Social, psychological, and environmental influences on health and well-being of Antarctic Winter-over personnel. Antarc. J. US. **24**, 207–209 (1989)

Palinkas, L., Gunderson, E.K.E., Holland, A.W.: Predictors of behavior and performance in extreme environments: the Antarctic space analogue program. Aviat. Space Environ. Med. **71**, 619–625 (2000)

Pearce, R.: Marooned in the Arctic: Diary of the Dominion Explorers' Expedition to the Arctic, August to December 1929. Northern Miner, Winnipeg (1930)

Raymond, M.W., Moser, R.: Aviators at risk. Aviat. Space Environ. Med. **66**(1), 35–39 (1995)

Ricketson, D.S., Brown, W.R., Graham, K.N.: 3W Approach to the investigation, analysis, and prevention of human-error aircraft accidents. Aviat. Space Environ. Med. **51**, 1036–1042 (1980)

Rose, R.M., Helmreich, R.L., Fogg, L.F., McFadden, T.: Psychological predictors of astronaut effectiveness. Aviat. Space Environ. Med. **64**, 910–915 (1994)

Rosnet, E., Le Scanff, C., Sagal, M.: How self-image and personality affect performance in an isolated environment. Environ. Behav. **32**, 18–31 (2000)

Sandal, G.M.: Culture and crew tension during an international space station simulation: results From SFINCSS'99. Aviat. Space Environ. Med. **75**, C44–C51 (2004)

Sandal, G.M., Vaernes, R., Ursin, H.: Interpersonal relations during simulated space missions. Aviat. Space Environ. Med. **66**, 617–624 (1995)

Sandal, G.M., Vaernes, R., Bergan, P.T., Warncke, M., Ursin, H.: Psychological reactions during polar expeditions and isolation in hyperbaric chambers. Aviat. Space Environ. Med. **67**(3), 227–234 (1996)

Sandal, G.M., Endresen, I.M., Vaernes, R., Ursin, H.: Personality and coping strategies during submarine missions. Mil. Psychol **11**, 381–404 (1999)

Sandal, G.M., Leon, G.R., Palinkas, L.: Human challenges in polar and space environments. Rev. Environ. Sci. Biotechnol. **5**(2–3) (2006). doi:10.1007/s11157-006-9000-8

Sexner, J.L.: An experience in submarine psychiatry. Am. J. Psychiatr. **1**, 25–30 (1968)

Sexton, B., Marsch, S., Helmreich, R., Betzendoerfer, D., Kocher, T., Scheidegger, D.: Jumpseating in the operating room. J. Hum. Perform. Extreme. Environ. **1**(2), 36 (1996)

Space Science Board: A Strategy for Space Biology and Medical Science. National Academy Press, Washington, DC (1987)

Stanton, W.: The Great United States Exploring Expedition of 1838–1842. University of California Press, Berkeley (1975)

Stefansson, V.: The Adventure of Wrangel Island. MacMillan, New York (1925)

Strange, R., Klein, W.: Emotional and social adjustment of recent U.S. winter-over parties in isolated Antarctic station. In: Edholm, O.G., Gunderson, E.K.E (eds.) Polar Human Biology: The Proceedings of the SCAR/IUPS/IUBS Symposium on Human Biology and Medicine in the Antarctic, p. 410. Year Book Medical Publications, Chicago (1974)

Stuster, J.: Bold Endeavors. Naval Institute Press, Annapolis (1996)

Suedfeld, P.: Groups in isolation and confinement: environments and experiences. In: Harrison, A.A., Clearwater, Y.A., McKay, C.P. (eds.) From Antarctica to Outer Space: Life in Isolation and Confinement, pp. 135–146. Springer, New York (1991)

Suedfeld, P.: What can abnormal environments tell us about normal people? Polar stations as natural psychological laboratories. J. Environ. Psychol. **18**, 95–102 (1998)

Suedfeld, P.: Invulnerability, coping, salutogenesis, integration: four phases of space psychology. Aviat. Space Environ. Med. **76**, B61–B73 (2005)

Suedfeld, P., Steel, G.D.: The environmental psychology of capsule habitats. Annu. Rev. Psychol. **51**, 227–253 (2000)

Taylor, A.J.W., Shurley, J.T.: Some Antarctic troglodytes. Int. Rev. Appl. Psychol. **20**, 143–148 (1971)

Ursin, H.: Psychobiological studies of individuals in small isolated groups in the Antarctic and space analogue. Environ. Behav. **6**, 766–781 (1991)

von Chamisso, A.: Reise um die Welt mit der Romanoffischen Entdeckungs Expedition in den Jahren 1815–1818. Weidmann, Berlin (1856)

Weiner, B.L., Kanki, B.O., Helmreich, R.L.: Cockpit Resource Management. Academic, New York (1993)

Weybrew, B.B.: Three decades of nuclear submarine research: implications for space and Antarctic research. In: Harrison, A.A., Clearwater, Y.A., McKay, C.P. (eds.) From Antarctica to Outer Space: Life in Isolation and Confinement, pp. 103–114. Springer, New York (1991)

Weybrew, B.B., Helmreich, R.L., Howard, N.: Psychobiological and psychosocial issues in space station planning and design: inferences from analogous environments and conditions unpublished report prepared for NASA (1986)

Wiegmann, D.A., Shappel, S.A.: Human factors analysis of postaccident data: applying theoretical taxonomies of human error. Int. J. Aviat. Psychol. **7**, 67–81 (1997)

Williamson, J.A., Webb, R.K., Sellen, A., Runciman, W.B., van der Walt, J.H.: Human failure: an analysis of 2000 incident report. Anesth. Intensive. Care. **21**, 678–683 (1993)
Wilson, O.: Human adaptation to life in Antarctica. In: Van Meigheim, J., van Oue, P., Schell, J. (eds.) Biogeography and Ecology in Antarctica. Monographiae Biologicae, vol. 15, pp. 690–672. W. Junk, The Hague (1965)
Wilson, G., Skelly, J., Purvis, B.: Reactions to emergency situations in actual and simulated flight. Presented as a paper at the Aerospace Medical Panel Symposium, The Hague (1989)
Yacovone, D.W.: Mishap trends and cause factors in naval aviation: a review of Naval Safety Center Data, 1986–1990. Aviat. Space Environ. Med. **64**, 392–395 (1993)
Zubek, J.P.: Sensory Deprivation: Fifteen Years of Research. Appleton-Century-Crofts, New York (1969)

Chapter 3
Patterns in Crew-Initiated Photography of Earth from the ISS: Is Earth Observation a Salutogenic Experience?

Julie A. Robinson, Kelley J. Slack, Valerie Olson, Michael H. Trenchard, Kimberly J. Willis, Pamela J. Baskin, and Jennifer E. Boyd

Abstract To provide for crewmember well-being on future exploration missions, understanding coping strategies that International Space Station (ISS) crewmembers adopt to mitigate the inherent stress of long-duration confinement is important. A recent retrospective survey of flown astronauts found that the most commonly reported psychologically enriching aspects of spaceflight had to do with their perceptions of Earth. ISS crewmembers photograph Earth both volitionally and in response to requests from Crew Earth Observations (CEO) scientists. Automatically recorded data from the camera can be used to test hypotheses about factors correlated with self-initiated crewmember photography. The present study used these objective in-flight data to investigate the nature of voluntary photographic activity. We examined the distribution of photographs with respect to time, crew, and subject matter. We determined whether the frequency fluctuated in conjunction with major mission events such as vehicle dockings and extravehicular activities (EVAs, or spacewalks), relative to the norm for the relevant crew. We also examined the influence of geographic and temporal patterns on frequency of Earth photography activities. We tested the hypotheses that there would be peak photography intensity over locations of personal interest, as well as on weekends.

Of nearly 200,000 photographs taken on eight ISS expeditions, 84.5% were crew-initiated. Once a crewmember went to the window for a CEO request,

J.A. Robinson (✉) • M.H. Trenchard • K.J. Willis
NASA Johnson Space Center, Houston, TX, USA
e-mail: julie.a.robinson@nasa.gov

K.J. Slack • P.J. Baskin
Wyle Laboratories, Houston, TX, USA

V. Olson
Rice University, Houston, TX, USA

J.E. Boyd
San Francisco Veterans Affairs Medical Center and University of California, San Francisco, San Francisco, CA, USA

D.A. Vakoch (ed.), *On Orbit and Beyond*, Space Technology Library 29,
DOI 10.1007/978-3-642-30583-2_3, © Springer-Verlag Berlin Heidelberg 2013

he or she was more likely to take photographs for his or her own interest. Fewer self-initiated images were taken during and immediately preceding major station events. Crewmembers were more likely to take self-initiated images during periods when they had more free time. Analysis indicated some phasing in patterns of photography during the course of a mission, although it did not suggest that psychological functioning was lower during the third quarter of confinement (i.e., no third-quarter effect was found). Earth photography is a self-initiated positive activity of possible importance for salutogenesis (increase in well-being) of astronauts on long-duration missions. Scientific requests for photography through CEO play an important role in facilitating crew-initiated photography. Consideration should be given to developing substitute activities for crewmembers in future exploration missions where there will not be the opportunity to look at Earth, such as on long-duration transits to Mars.

3.1 Background

3.1.1 Earth Observation Throughout Human Spaceflight

John Glenn, the first U.S. astronaut in orbit, talked NASA into letting him carry a camera on *Friendship 7* on 20 February 1962 (Apt et al. 1996). On reaching orbit, Glenn told capsule communicator Alan Shepard over the radio, "Oh, that view is tremendous." Glenn proceeded to describe each of the three sunrises and sunsets he saw during the flight, and he continues to recount that experience in interviews today (Ethier et al. 1997). A number of the astronauts who have followed have verbally recounted emotional experiences related to seeing and photographing Earth, and several astronauts have documented in written form their responses to views of Earth linked to their photography activities while in space. Space Shuttle astronaut Kathryn D. Sullivan wrote in an article documented with her Earth photography, "It's hard to explain how amazing and magical this experience is. First of all, there's the astounding beauty and diversity of the planet itself, scrolling across your view at what appears to be a smooth, stately pace ... I'm happy to report that no amount of prior study or training can fully prepare anybody for the awe and wonder this inspires" (Sullivan et al. 1991). Observations of familiar places on Earth can also have strong emotional connections. NASA-*Mir* astronaut Jerry Linenger recorded photographing his hometown in Michigan in his crew notebook, "Great View—Michigan + Great Lakes cloudfree—ready to go home, now!" (Lulla et al. 2000, p. 252).

From Apollo to the current ISS, scientists have assisted astronauts with crew-initiated and science-specific photography of Earth. All the imagery is archived in a searchable online database maintained by the descendent of the previous programs on the International Space Station, CEO, which provided statistics summarized here. Over 2,500 photographs of Earth were taken by Mercury and Gemini astronauts. Apollo astronauts photographed both Earth and Moon views, with over 11,000 photographs taken, and have been credited with initiating the interest in Earth observations from space (Lowman 1999). Handheld photography of Earth by astronauts on *Skylab* accompanied the extensive imagery obtained by an automated multispectral camera system (NASA 1974; Wilmarth, Kaltenbach, and Lenoir 1977). Over the three *Skylab* missions, crewmembers took around 2,400 images of Earth, and the automated camera systems an additional 38,000 photographs with specialized films.

Building from this experience and the growing interest in Earth observations from space, a program called the Space Shuttle Earth Observations Project (SSEOP) was established in 1982 to support the acquisition and scientific use of Earth photography from Space Shuttle flights. Located at the center of astronaut training, Johnson Space Center, SSEOP scientists were assigned to each Shuttle crew. Astronauts were trained in geology, geography, meteorology, oceanography, and environmental change for a total of approximately 12 instructional hours prior to flight. Also before flight, about 20–30 sites were chosen for the crew to photograph while on orbit. The mission-specific sites were chosen from a list of previously identified environmentally dynamic terrestrial areas visible from the Space Shuttle. Each crew was given a preflight manual consisting of their unique sites that included photographs and scientific information. The decision on when to take photographs was at the astronauts' discretion. A list of targets was sent to the Shuttle crew on a daily basis during the flight. The main camera used for Earth observation was the 70-mm Hasselblad with the 50-, 100-, 110-, and 250-mm lenses commonly used, and both color and infrared film was made available per crew preference (Robinson et al. 2002). After each flight, the Earth-viewing film was cataloged and entered into a database. Paper catalogs were also mailed to a subscriber list of interested scientists and educational and government users. To date, Shuttle crewmembers have captured over 287,000 images of Earth.

From March 1996 through June 1998, the scientists of SSEOP supported Earth photography by crewmembers spending longer durations in space as part of the NASA-*Mir* program. U.S. investigators collaborated with the Institute of Geography, Russian Academy of Sciences, in developing Earth observation objectives for astronauts on board *Mir* (Lulla et al. 2000). The documentation of dynamic environmental changes on Earth's surface was a primary objective for both SSEOP and the Russian Institute of Geography. Another objective was to develop scientific approaches and procedures that could later be applied to the same kinds of dynamic observations from the ISS. With the advent of Shuttle-*Mir* and the ISS, the focus of SSEOP changed from short-term observation to long-term observation.

The cameras used on Shuttle-*Mir* were the same as on the Shuttle, with the 70-mm Hasselblad (film) as the main camera, but the Nikon F3 35-mm camera was also available. A joint list of sites was chosen by U.S. and Russian scientists for Shuttle-*Mir*. Earth observation target sites were sent to the Shuttle-*Mir* crews weekly. Training was modified from the typical Shuttle briefings to enable the Shuttle-*Mir* crews to document unanticipated dynamic events as well as targets of opportunity that would be encountered more often on long-term missions. Another benefit of long-term Earth observing missions was the ability to document seasonal change and long-term climatic effects. Approximately 22,500 photographs were taken during the seven Shuttle-*Mir* missions.

Crew Earth Observations began as a formal ISS research activity ("payload") on the first mission, Expedition 1, in October 2000. Training for ISS crews evolved from experiences gained in the Shuttle and Shuttle-*Mir* programs. Rather than discipline-specific training, ISS crews were trained on science topics such as coral reefs, global urban systems, deltas, and glaciers. The emphasis was more on observing Earth as a system than on documenting independent events. An overall science plan tied together the target sites and crew training and is still used and updated by increment for ISS crews today. Due to the extensive training ISS astronauts receive regarding all aspects of their missions, CEO training is limited to 4 h. Typically this training occurs during the early part of the training cycle. Since an ISS mission is longer than a Shuttle mission, the number of targets per increment varies from approximately 140 to 160 sites, and they are updated with the change of each ISS increment.

The digital camera, a Kodak 460 DSC, was introduced on STS-73; however, the Hasselblad film camera remained the favorite of the Shuttle crews, most likely because of their experience with that camera. Improvements in digital technology coincided with the change in focus of the Shuttle program to the assembly of the International Space Station. Following the Space Shuttle *Columbia* accident in 2003, NASA's support of Earth observations by crewmembers has been focused on the ISS. Although SSEOP was dissolved, individual Shuttle crewmembers on missions to the ISS could still use the on-board cameras to take images of Earth, but without scientific support.

3.1.2 Earth Observation in Human Spaceflight Today

The digital camera was favored by ISS crews over the film cameras because it allowed them to review their imagery while on orbit. The immediate review of their imagery enabled the crews to view and improve their photographic techniques. Digital images could also be down-linked to the CEO scientists for review, and the scientists in turn could provide feedback to the crew. The issue of film versus digital cameras was settled in 2003 when mission length was extended to about 6 months. The extension of crew time on orbit made film more susceptible to radiation "fogging." While digital cameras are not immune to radiation, they are better able to cope with longer exposures to the space environment, and eliminating the need to return film to Earth was also an important improvement.

With the use of the 400-mm lens and 2× extender available for the digital camera, ISS crews have been able to document dynamic events at a higher resolution than was possible from the Shuttle with the 250-mm lens (Robinson and Evans 2002). The 400- and 800- mm lens options are clearly the favorites of ISS crews. An additional benefit of the camera is the automatic logging of the time as well as the date the image was acquired, along with other camera settings. Currently, the Kodak 760 DSC is used for CEO; however, this camera was upgraded with the higher resolution Nikon D2× in the latter part of 2008.

In addition to watching Earth, ISS crewmembers photograph Earth through the windows of the ISS and are able to share those images with the world. The CEO activity provides a venue to transmit requests for photographs of areas of scientific or public interest to the astronauts each day and to distribute the acquired photographs to scientists and the public. Crewmembers take photographs of the targets during their free or unscheduled time; Earth photography is never a scheduled crew activity. A list of candidate targets is sent to them on a daily basis, and crewmembers can make attempts to photograph those targets, choose to take no images, or, on their own initiative, photograph Earth at any time. These self-initiated images would seem to be of special importance to crewmembers since the taking of these images is purely volitional. Whether requested by scientists or self-initiated, images of Earth taken from the ISS are identified and distributed via the *Gateway to Astronaut Photography of Earth* Web site (NASA 2012).

3.1.3 Earth Observation and Behavioral Health in Human Spaceflight

While NASA has always engaged in space exploration research, *The Vision for Space Exploration* and subsequent definitions of specific exploration mission architectures have required a much more focused use of the ISS (NASA 2004). In particular, the ISS is to be used for research on human health on long-duration space missions, as well as for technology development and testing (NASA 2006). Behavioral health and performance has been identified as a discipline with additional research needs requiring the ISS (Ball and Evans 2001; NASA 2005). Maximizing psychological well-being and performance of the crew, while in a confined space with interpersonal interactions limited to a small number of people, is important for the success of ongoing ISS missions. Knowledge about behavioral health gained from ISS missions is also important for the success of future missions to a lunar base and provides key data for a 4- to 6-month Mars transit. A particular concern is maintaining crew psychological well-being for the duration of a round-trip mission to Mars that could last as long as 3 years (NASA 2005).

Positive (or "salutogenic") experiences while in space may promote psychological well-being by enhancing personal growth and may be important for offsetting the challenges of living and working in a confined and isolated environment (Suedfeld 2001; Suedfeld and Weiszback 2004). In a survey of flown astronauts aimed at identifying the positive or salutogenic effects of spaceflight, Eva Ihle and colleagues identified positive changes in perceptions of Earth as the most important change experienced by astronauts (Ihle et al. 2006).

If viewing Earth is an important component of positive experiences in spaceflight, then having Earth "out of view" may be an important challenge for crews going to Mars because it could increase the sense of isolation (Kanas and Manzey 2003). To the extent that observing Earth is a positive experience for ISS crewmembers, replacement activities or new psychological countermeasures may be needed to ensure the well-being of crewmembers on a Mars mission.

3.1.4 Objectives

In this chapter, we mine the dataset of Earth observation photography to see whether additional information could be gleaned about the importance to crewmembers of the positive experience of viewing Earth. Our first objective was to quantify the extent to which photography of Earth was self-initiated. A second objective was to identify patterns in photography, or conditions under which crewmembers were more likely to take self-initiated images. From this we hoped to gain quantitative (although correlative) insight into whether Earth observation activities are important to long-duration crewmembers on the ISS and use this to infer whether Earth observation activities might play a role in maintaining the psychological well-being of at least some of these crewmembers.

3.1.5 Hypotheses

Prior to analyzing the photographic incidence data, we generated the following hypotheses:

3.1.5.1 Hypothesis 1: Fewer Self-Initiated Images Are Expected to Be Taken During Periods of, and Preparation for, Extraordinary Activities

Daily activities on the Station can be very crudely dichotomized into regular daily activities and extraordinary activities. Extraordinary activities include EVAs as well as docking and undocking (i.e., of Space Shuttle, Soyuz, and Progress spacecraft). Further, these extraordinary activities require substantial focus and preparation leading up to the event. These extraordinary activities generally consume more time than regular daily activities, leaving less time for volitional activities such as taking images. In the mission timelines, extensive EVA and docking preparation ramps up prior to an event, with restrictions on the ability to schedule other, noncritical activities beginning 1 week prior to the EVA, so we considered 1 week prior as our preparation period.

3.1.5.2 Hypothesis 2: More Self-Initiated Images Are Expected to Be Taken During Weekends or Other Light-Duty Times

Typically, crewmembers have fewer set tasks to accomplish on weekends, so they have increased periods of time in which they can choose their activities. Given the volitional nature of self-initiated images coupled with the enjoyment crews have stated that they receive from viewing Earth, we expected crewmembers to take more Earth photographs during periods of decreased workload.

3.1.5.3 Hypothesis 3: More Self-Initiated Images Are Expected to Be Taken of Geographic Areas of Personal Interest to Crewmembers

Past crews have placed great importance on viewing Earth, (Ihle et al. 2006) and most Shuttle and ISS crewmembers have requested support in photographing their hometowns and other places of personal interest. If such interest provides an indirect linkage of crewmembers in space to the people and place they have left behind, the photographing of places that hold special meaning for crewmembers, such as their childhood home or their alma mater, might be expected to be of particular relevance.

3.1.5.4 Hypothesis 4a: Phasing Occurs Such That Differing Numbers of Self-Initiated Images Will Be Taken Over the Course of a Mission. Hypothesis 4b: During the Third Quarter of the Mission, Increased Numbers of Self-Initiated Images Will Be Taken

Previous research, both in space and in analogue environments such as the Antarctic, has found mixed results regarding the existence of either phasing or a third-quarter effect (Bechtel and Berning 1991; Connors, Harrison, and Akins 1985; Kanas, Weiss, and Marmar 1996; Sandal 2000; Stuster, Bachelard, and Suedfeld 2000). The term *phasing* suggests that isolated individuals experience a cycle of ups and downs in psychological well-being during their time in confinement. While the term phasing is more general, the term *third-quarter effect* specifically refers to a period of lowered psychological well-being during the third quarter of an extended confinement. Thus, we looked for several possible temporal patterns in the incidence of self-initiated photography.

3.2 Methods

3.2.1 Participants

Images taken by up to 19 ISS crewmembers, beginning with ISS Expedition 4 (December 2001, when the full capability of the digital camera began to be used) and continuing through Expedition 11 (October 2005), were included in this study. Ten were astronauts with NASA, and nine were Russian cosmonauts. The expeditions consisted of three crewmembers through Expedition 6, when the number of crewmembers on the Station dropped to two, one Russian and one American. Gender of the crew for Expeditions 4 through 11 was predominantly male with only one female astronaut. It is not known whether every individual on board the ISS actually used the camera, nor which individuals took which images.

3.2.2 Data and Analyses

Digital photographs are taken on orbit and downlinked to the ground during the course of the mission. These are separated by content (Earth, hardware, people). All Earth images become part of the Database of Astronaut Photography of Earth, which was used for these analyses and is available online (NASA 2012).

We analyzed the Earth photography patterns using the digital data recorded on the back of the digital cameras used on the ISS. The cameras automatically record the date and time when the photograph was taken, as well as specific photographic parameters. The data do not identify the individuals using the camera, as any crewmember may pick up any camera to take pictures, and individuals often stop briefly at a window to take pictures throughout the day. Crews are cross-trained in the use of the imagery equipment. Some crews share the responsibility of taking images of Earth; in other crews, one member might have more interest and thus be the primary photographer. Regardless, crewmembers report photographing areas known to be of interest to fellow crewmembers.

Additional datasets compiled for use in analyses were (1) lists of areas of known geographic interest to crews based on publicly released biographical information, (2) orbital track parameters to relate images taken to the log of scientific requests sent to the crew, and (3) records of on-orbit activities to determine the incidence of EVAs, the docking of visiting vehicles, and days of light duty/holidays. We used the orbit tracks and message logs to identify which photographs were in response to CEO requests and which were self-initiated by the crew. Occasionally, battery changes and camera resets were conducted on orbit without resetting the date and time on the camera. Because of this, not all camera time stamps were accurate. We screened those data for inaccuracies (such as an incorrect year for a specific expedition), and these records were eliminated from the analyses.

For each day, we determined the number of images of Earth that were self-initiated, were of areas of known geographic interest to any member of that crew, were in response to a scientific request, and used the 800-mm (high-magnification) lens setup. The use of the 800-mm lens was tracked because it represents a significant skill that requires much effort to achieve the best results, and the resulting images provide the most detail (up to 6-m spatial resolution). The crewmembers must practice tracking the motion of Earth beneath the ISS using the camera equipped with the 800-mm lens and learn how to focus properly through the lens (Robinson and Evans 2002). Although this was not one of our original hypotheses, we realized that use of the 800-mm lens could be an indicator of crew interest in Earth photography as a challenging, self-motivated hobby.

In general terms, the analyses looked for relationships between self-initiated image-taking and when the images were taken, as well as between self-initiated image-taking and the geographic location of those images. For the benefit of statistically minded readers, hypotheses 1 and 2 were addressed by examining zero order correlations and using general linear models in a statistical analysis package (GLIMMIX [generalized linear mixed models] procedure in SAS).

This procedure fits generalized linear mixed models to the data and allows for normally distributed (Gaussian) random effects (SAS 2006). Hypothesis 3 was tested using a related procedure that could incorporate categorical data into the model (GENMOD procedure for generalized linear models in SAS). Hypothesis 4a was tested using regression, while general linear model repeated measures analysis was used for hypothesis 4b.

3.3 Results

From December 2001 (Expedition 4) through October 2005 (Expedition 11), crewmembers took 144,180 images that had accurate time and date data automatically recorded by the camera. Of time-stamped photographs, 84.5% were crew-initiated and not in response to CEO requests.

3.3.1 Comparison of Variables

These comparisons were made by examining the degree to which the variables are related (correlation), the average for each variable (mean), and the degree to which the values for a variable differ from its average (standard deviation). See Table 3.1 for all measures included in the study. For subsequent analyses, we considered only self-initiated images and excluded images taken in response to CEO requests.

The correlations presented in Table 3.1 provide a preliminary examination of the data rather than a formal test of the hypotheses. When conducting statistical analyses, correlations typically are examined first, and then a priori hypotheses are tested using more robust statistical approaches. Based on the correlations in Table 3.1, the following inferences can be made.

A crewmember with a camera in hand was more likely to take self-initiated photos in addition to the requested images (self-initiated images were correlated with requested images—$r = .36$, $p < .01$). Also, 800 mm was the focal length more frequently chosen when crewmembers took images of their own choice, even though taking images at 800 mm was more challenging (self-initiated images were correlated with images taken at an 800-mm focal length—$r = .41$, $p < .01$). Further, crewmembers also were more likely to take self-initiated images of geographic areas of Earth that were of personal interest to them ($r = .25$, $p < .01$). A crew containing a member, for example, whose childhood home was in a small town in Illinois, would be more likely to take images of that area than of areas not holding personal significance for any member of that crew.

Crews were more likely to take self-initiated images on weekends (self-initiated images correlated with weekends—$r = .07$, $p < .01$) However, contrary to expectations, neither activity nor holiday was related to self-initiated images (due to space limitations, the variables of weekends, activities, and holidays are not

Table 3.1 Means, standard deviations, and correlations across all missions. Each parameter is measured on a daily basis across all expeditions combined

		Mean	Std. Dev.	1	2	3	4	5	6
Daily number of:									
1	Total images taken	102.3	119.1	—					
2	Self-initiated images taken	86.4	107.5	.98[b]	—				
3	Images of geographic interest	1.6	5.1	.25[b]	.25[b]	—			
4	Requested images taken	15.9	25.3	.54[b]	.36[b]	.10[b]	—		
5	Images taken with 800-mm	17.8	34.4	.41[b]	.41[b]	.15[b]	.19[b]	—	
Proportion of days:									
6	Higher availability to take images	.3	.4	.06[a]	.07[b]	−.01	−.03	.07[b]	—

[a]Correlation is significant at the .05 level (2-tailed)
[b]Correlation is significant at the .01 level (2-tailed)

included in Table 3.1). It could be that crewmembers did not necessarily have more time available on holidays or that self-initiated images are more tightly linked to scientist-requested approaches to the window. To address the possibility that holidays and planned activities were not indicative of whether crewmembers had time available to take self-initiated images, the variables of holiday, weekend, and activity were combined to create a measure of general availability to take images. Using this new measure of general availability, we found that crewmembers were more likely to take self-initiated images when they had time available (general availability was correlated with self-initiated images—$r = .07, p < .01$).

3.3.2 *Tests of Hypotheses*

3.3.2.1 Hypothesis 1: Activity

Hypothesis 1 stated that fewer self-initiated images would be taken 1 week prior to and during extraordinary mission events, such as EVAs and dockings. The data analysis supported this hypothesis. Crewmembers were less likely to take self-initiated images while preparing for and during these mission events than during regular daily station activities. To state in statistical terms, conditional upon the degree of variability that could not be described by the model (the random effect of the intercept) and controlling for time (days) on the ISS, the type of activity on the Station predicted whether self-initiated images were taken ($t = -2.50$, $p < .01$). Further, crewmembers were more likely to take self-initiated images on days of regular Station activities as the mission progressed. In other words, the longer crewmembers had been on orbit, the more likely they were to take self-initiated images (the effect of activity also was different over time—$t = -4.65$, $p < .01$) (see Fig. 3.1).

Extraordinary mission events, along with the period spent in orbit, influenced whether or not images were taken each day but did not influence the number of images taken on a given day. To use other words, although general activity

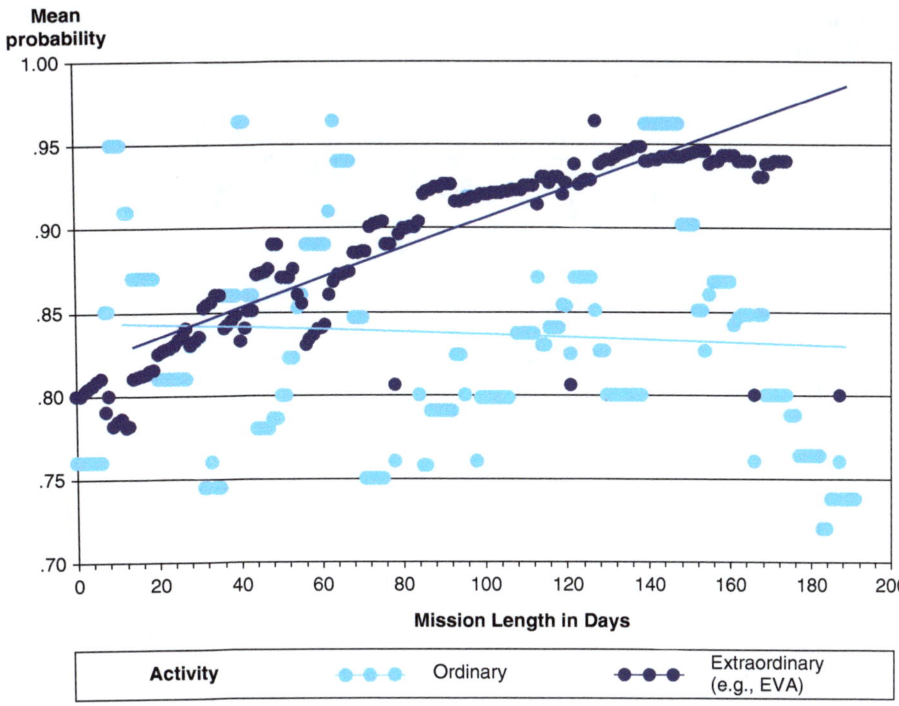

Fig. 3.1 Probability that self-initiated images were taken, using activity as a predictor across mission

predicted whether or not any photographic activity occurred, it did not predict the *number* of self-initiated images taken during regular or extraordinary mission events ($t = -.77$, not significant [ns]).

3.3.2.2 Hypothesis 2: Weekends

According to hypothesis 2, more self-initiated images would be taken on weekends. This hypothesis was not supported ($t = .65$, ns), perhaps because life on the Station does not always allow weekends off. This finding, in apparent conflict with the preliminary finding that self-initiated images taken were correlated with weekends, is due to hypothesis 2's being tested with more robust statistical methods than those of correlation alone.

In response to this supposition, the composite variable of days available, or general availability, was used to reflect workload on the ISS more accurately. This composite took into account whether activity on the Station was extraordinary, whether it was a weekend, and whether a particular day was considered an off or partial-duty day or a regular-duty day. General availability was thus a more realistic representation of days with crewmember free time.

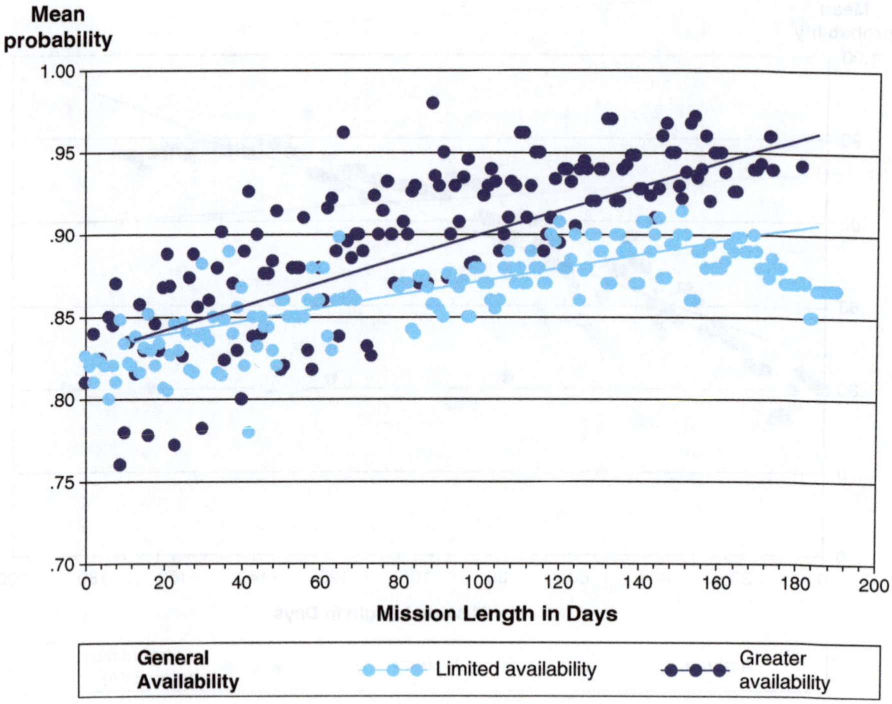

Fig. 3.2 Probability that self-initiated images were taken, using general availability as a predictor across mission

This post hoc hypothesis was supported. Regarding the *taking* of images, more self-initiated images were taken when crewmember schedules allowed (see Fig. 3.2). Crewmembers were less likely to take self-initiated images on a Saturday, for example, if they were also preparing for an EVA. In more statistical terms, general availability was associated with whether self-initiated images were taken ($t = 4.37, p < .01$), conditional upon the intercept and after controlling for time on the Station. Additionally, later in the mission, crewmembers became even more likely to take self-initiated images when their schedules allowed, indicating that general availability predicted differentially as the mission progressed ($t = 2.26$, $p = .02$).

Regarding the *number* of images taken, the longer crewmembers had available time, the more images they took. In more statistical terms, general availability predicted the number of self-initiated images taken ($t = 2.10$, $p = .02$) after controlling for the intercept and time on the Station. Further, crewmembers took just as many images when they had time available at the beginning of the mission as at the end of the mission. In other words, general availability did not differentially predict over the length of the mission ($t = -1.22$, ns).

3.3.2.3 Hypothesis 3: Geographic Interest

Hypothesis 3 stated that crewmembers would take more self-initiated images of areas that were of personal interest to them.[1] This hypothesis was not supported by the data (χ^2 (df = 1) = 1.14, ns). Crewmembers were no more likely to take self-initiated images of geographic areas of personal interest to them than they were to take any other self-initiated images. The fidelity of the measure of areas of personal interest is questionable, though. The geographic areas of interest were determined by our survey methods rather than by direct reports from crewmembers themselves.

3.3.2.4 Hypotheses 4a and 4b: Phasing/Third-Quarter Effects

Hypothesis 4a stated that phasing would occur such that self-initiated images would be taken differentially across the mission. This hypothesis was supported in that the number of self-initiated images taken is not consistent across the duration of the mission. Statistically, the quadratic term of the regression equation used to assess curvilinearity in the data was significant ($R^2 = .02, p < .01$). The temporal effects identified in the tests of hypothesis 1 and, to a lesser extent, hypothesis 2 lend further support to hypothesis 4a. Conditional upon the type of activity, the longer the crewmembers were on the Station, the more likely they were to take self-initiated images. In more statistical terms, when using type of activity as a predictor, time on the Station was a significant predictor of whether self-initiated images would be taken ($t = 3.16, p < .01$).

Hypothesis 4b stated that greater numbers of self-initiated images would be taken during the third quarter of the mission. This hypothesis was not supported; Fig. 3.3 effectively illustrates the lack of a third-quarter effect. The statistical method used to test hypothesis 4b was general linear model repeated measures.

3.4 Discussion

In this study, we made novel use of data available on the patterns of Earth photography by crews aboard the International Space Station. Although the data are observational, they allow additional insights into the role that observing Earth plays in the daily lives of crews in orbit. Perhaps the most important observation is the importance that photography of Earth has for at least some crewmembers, as evidenced by the degree to which it is self-initiated (84.5% of 144,180 photographs)

[1] Due to limitations in the GLIMMIX procedure, a model fitting geographic interest could not converge. As an alternative, the GENMOD procedure was used. Given that GLIMMIX uses the GENMOD procedure to fit generalized linear models when random effects are absent, this change in statistical procedure is not significant.

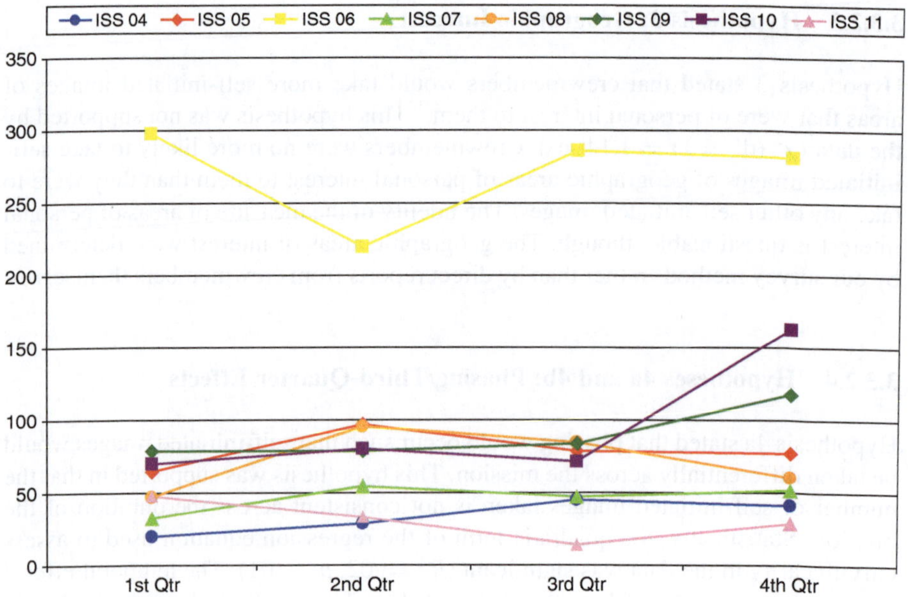

Fig. 3.3 Quarterly estimated marginal mean number of self-initiated images taken by mission

and conducted as a leisure activity (for an example, see Fig. 3.4). As of mid-2006, active photography of Earth has continued, with a total over 250,000 images taken through Expedition 13.

Earth photography also offers several venues for personal accomplishments for ISS crewmembers. For example, the use of motion tracking with the 800-mm lens combination to achieve high-quality, high-magnification images of Earth is a challenge that some crewmembers have relished. Anecdotally, some crewmembers do not try to use the combination; others make it a personal challenge, and several crewmembers have become expert at its use, taking many thousands of such images during their mission (for an example, see Fig. 3.5). We cannot distinguish whether the most important element of this activity is taking and sharing these sometimes astounding images of Earth, the challenge of learning/perfecting a new skill, or a combination (for an example, see Fig. 3.6).

The correlation between scientific requests and self-initiated photography might reflect the practicality of a crewmember who continues to photograph Earth once he/she has a camera in hand. However, it is also suggestive of the importance of the scientific basis and public use of photographs in making the activity worthwhile for the crews. The scientific support from Crew Earth Observations enables self-initiated photography of Earth by providing opportunities and reminders to go to the window with a camera. It is likely that the image identification and Web distribution completed by CEO adds value to the self-initiated photography by ensuring that it can be shared with the world—but confirming this hypothesis would require a more structured survey of crewmembers. The database structure of CEO

Fig. 3.4 Example of a crew-initiated image of São Paulo, Brazil, at night. While staying on the ISS, astronaut Don Pettit assembled a homemade tracking system to photograph long-exposure images under low light conditions (Image number ISS006-E-44689, 12 April 2003)

Fig. 3.5 Example of a crew-initiated image using the 800-mm lens combination. This view shows a portion of an image of the Golden Gate Bridge, San Francisco, California, taken by astronaut Jeff Williams from the ISS during Expedition 13. Expedition 13 held the record for the largest number of Earth images taken during an expedition (74,266 Earth images taken) until it was surpassed when Jeff Williams was again on the ISS (this time with Soichi Noguchi and T. J. Creamer) during Expedition 22 (88,779 images), November 2009-March 2010 (Image number ISS013-E-65111, 6 August 2006)

also adds personal value to crewmembers, making it easier for them to search for images they have taken for their own use.

From our analyses for hypotheses 1 and 2, self-initiated images were less likely to be taken when workload prevented it—and since 84.5% of images taken are

Fig. 3.6 View of Mount McKinley (Denali), Alaska, taken from the ISS using a powerful 800-mm lens to photograph this peak while the spacecraft was over the Gulf of Alaska, 800 miles to the south of the mountain. Cosmonaut Sergei Krikalev has assembled a collection of these views of major world mountain peaks during his stays on the ISS during Expeditions 1 and 11. (Image number ISS011-E-11806, 14 August 2005)

self-initiated, Earth photography is clearly a leisure activity. However, crews are more likely to take self-initiated images as the mission progresses—perhaps due to acclimation and familiarity with life and duties on the Station or a growing realization that their time in space, and thus their ability to photograph Earth from space, is limited. This trend over the duration of the mission was the only mission phasing observed. A more careful examination of Fig. 3.3 suggests that the phasing effect might be due more to individual differences pertaining to specific missions or perhaps to an increasing competency with the photographic equipment. It is not clear to what extent this phasing effect might reflect differences in mission profiles or characteristics of the particular crewmembers assigned to the particular missions.

3.4.1 Future Research and Applications

In spite of the importance of behavioral health and performance for the success of human spaceflight missions, relatively few studies have been done on the ISS to date (Evans et al. 2009). This analysis of data collected for other purposes serves as an example of mining data collected as part of ISS operations to increase knowledge. Such analyses can inform surveys for future data collection and influence future behavioral studies on the ISS. Given these observations, future studies should consider crew motivations in photographing Earth. Psychological component testing could reveal whether taking more self-initiated images is associated with increased crew psychological well-being.

The importance of viewing Earth as reported in crew surveys, crewmember anecdotes to CEO personnel (and the published quotations in the introduction), and the patterns in photography of Earth reported in our analysis all point to a positive psychological role on the part of these activities. Conclusions from this type of correlative data mining are only suggestive of the importance of Earth observation for crewmembers. Future studies of behavioral health should consider quantitative assessment of salutogenic effects of leisure activities such as Earth observations.

As we begin to plan for interplanetary missions, it is important to consider what types of activities could be substituted. Perhaps the crewmembers best suited to a Mars transit are those individuals who can get a boost to psychological well-being from scientific observations and astronomical imaging. Replacements for the challenge of mastering 800-mm photography could also be identified. As humans head beyond low-Earth orbit, crewmembers looking at Earth will only see a pale blue dot, and then, someday in the far future, they will be too far away to view Earth at all.

Acknowledgments We would like to thank Edna Fiedler and Frank Carpenter for their encouragement to pursue these analyses. Chuck Green advised us on the statistical analyses and helped with access to SAS procedures. We also thank Cindy Evans, Edna Fiedler, and Nick Kanas for their reviews and suggestions for improvement.

References

Apt, J., Wilkinson, J., Helfert, M.: Orbit: NASA Astronauts Photograph the Earth. National Geographic Society, Washington, DC (1996)

Ball, J.R., Evans, C.H., Jr. (eds.): Safe Passage: Astronaut Care for Exploration Missions. Committee on Creating a Vision for Space Medicine During Travel Beyond Earth Orbit, Board on Health Sciences Policy, National Institute of Medicine, National Academy Press, Washington, DC (2001)

Bechtel, R.B., Berning, A.: The third-quarter phenomenon: do people experience discomfort after stress has passed? In: Harrison, A.A., Clearwater, Y.A., McKay, C.P. (eds.) From Antarctica to Outer Space: Life in Isolation and Confinement, pp. 261–266. Springer, New York (1991)

Connors, M.M., Harrison, A.A., Akins, F.R.: Living Aloft: Human Requirements for Extended Spaceflight, NASA SP-483, Washington, DC (1985)

Ethier, B.: John Glenn: First American to Orbit the Earth, American History (1997). Available at http://www.historynet.com/john-glenn-first-american-to-orbit-the-earth.htm. Accessed 13 Mar 2012

Evans, C.A., Robinson, J.A., Tate-Brown, J., Thumm, T., Crespo-Richey, J., Baumann, D., Rhatigan, J.: International Space Station Science Research Accomplishments During the Assembly Years: An Analysis of Results from 2000–2008, pp. 110–159, NASA Johnson Space Center, NASA/TP-2009-213146–REVISION A, Houston (2009)

Ihle, E.C., Ritsher, J.B., Kanas, N.: Positive psychological outcomes of spaceflight: an empirical study. Aviat. Space Environ. Med. **77**, 93–101 (2006)

Kanas, N., Weiss, D.S., Marmar, C.R.: Crew member interactions during a MIR space station simulation. Aviat. Space. Environ. Med. **67**, 969–975 (1996)

Kanas, N., Manzey, D.: Space Psychology and Psychiatry. Kluwer Academic, Dordrecht, The Netherlands (2003)

Lowman Jr., P.D.: Landsat and Apollo: the forgotten legacy. Photogramm. Eng. Remote Sens. **65**, 1143–1147 (1999)

Lulla, K.P., Dessinov, L.V., Evans, C.A., Dickerson, P.W., Robinson, J.A.: Dynamic Earth Environments: Remote Sensing Observations from Shuttle–Mir Missions. John Wiley, New York (2000)

NASA: Skylab Earth Resources Data Catalog, JSC-09016. Johnson Space Center, Houston (1974)
NASA: Bioastronautics Roadmap: A Risk Reduction Strategy for Human Space Exploration, NASA Johnson Space Center SP-2004-6113, Houston (2005)
NASA: Gateway to Astronaut photography of Earth Web site (2012), Available at http://eol.jsc.nasa.gov. Accessed 13 Mar 2012
NASA: The NASA Research and Utilization Plan for the International Space Station (ISS), A report to the committee on science of the United States house of representatives and the committee on commerce, science, and transportation of the United States Senate, NASA Headquarters, pp. 1–20, NASA Headquarters, Washington, DC (2006)
NASA: The Vision for Space Exploration, NASA NP-2004-01-334-HQ, Washington, DC (2004)
Robinson, J.A., Evans, C.A.: Space station allows remote sensing of Earth to within six meters. Eos. Trans. Am. Geophys. Union **8**(3), 185 (2002)
Robinson, J.A., Amsbury, D.L., Liddle, D.A., Evans, C.A.: Astronaut-acquired orbital photographs as digital data for remote sensing: spatial resolution. Int. J. Remote Sens. **23**, 4403–4438 (2002)
Sandal, G.M.: Coping in Antarctica: is it possible to generalize results across settings? Aviat. Space. Environ. Med. **71**(9, Supplement), A37–A43 (2000)
SAS: The GLIMMIX Procedure, SAS Institute Inc., Cary (2006), Available at http://support.sas.com/rnd/app/papers/glimmix.pdf. Accessed 13 Mar 2012
Stuster, J.W., Bachelard, C., Suedfeld, P.: The relative importance of behavioral issues during long-duration ICE missions. Aviat. Space. Environ. Med. **71**(9, Supplement), A17–A25 (2000)
Suedfeld, P.: Applying positive psychology in the study of extreme environments. J. Hum. Perform. Extrem. Environ. **6**, 21–25 (2001)
Suedfeld, P., Weiszbeck, T.: The impact of outer space on inner space. Aviat. Space. Environ. Med. **75**(7, Supplement), C6–C9 (2004)
Sullivan, K.D.: An Astronaut's view of earth, Update (newsletter of the National Geographic Society's Geography Education Program) (Fall 1991) **1**, 12–14, full article available at http://eol.jsc.nasa.gov/newsletter/uft/uft1.htm. Accessed 13 Mar 2012. Quoted passage available at http://eol.jsc.nasa.gov/newsletter/uft/uft2.htm. Accessed 13 Mar 2012
Wilmarth, V.R., Kaltenbach, J.L., Lenoir, W.B. (eds.): Skylab Explores the Earth. NASA SP-380, Washington (1977)

Chapter 4
The Roles of NASA, U.S. Astronauts, and Their Families in Long-Duration Missions

Phyllis J. Johnson

Abstract Johnson Space Center oral histories, pre-flight interviews, and journals of residents of Skylab, Mir, and the International Space Station (ISS) were analyzed qualitatively to identify the U. S. astronauts' views of the roles of NASA, the astronauts and their families in helping the astronauts to create a daily life that mirrors some aspects of their life on Earth. Findings are that NASA's role is to establish scheduling and facilitate communication avenues. The astronaut's role is to personalize leisure, to connect looking out the window with specific aspects of life on Earth, to find ways to make daily routines fun, and to celebrate traditions and historical space events. The family's role is to participate in sufficient two-way communication and to send care packages with treats and reminders of home. These strategies, which provided a semblance of their life on Earth, may have helped the astronauts to remain connected to life on Earth and to deal better with time away from home and family.

4.1 Introduction

Long-duration space missions are characterized by isolation, confinement and physical separation from family and Earth. Considerable attention has been given to identifying psychological countermeasures to maintain astronauts' productivity, motivation, and psychological well-being in the space environment. Such countermeasures are also designed to alleviate potential boredom and monotony, especially on future long-duration missions where there may be extensive periods of non-work activities (Harrison 2001; Kanas and Manzey 2003; Stuster 1996).

Adaptation of Johnson (2010), reprinted with permission from Elsevier.

P.J. Johnson (✉)
University of British Columbia, Vancouver, BC, Canada
e-mail: phyllis.johnson@ubc.ca

D.A. Vakoch (ed.), *On Orbit and Beyond*, Space Technology Library 29,
DOI 10.1007/978-3-642-30583-2_4, © Springer-Verlag Berlin Heidelberg 2013

Some of the countermeasures relevant to this chapter (e.g., private audio-video conferences with family and friends, uplinks of news and e-mails from home, and individualized leisure materials) have been incorporated in NASA's protocol for in-flight support of ISS residents (Kanas and Manzey 2003).

In addition to developing countermeasures for possible psychological and behavioral problems, it is important to recognize and enhance the positive aspects of being in space (Suedfeld 2005). One approach to fostering psychological well-being of astronauts is for them to create a daily life in space that approximates life on Earth. This approach, which is the focus of this study, has been assessed previously using diaries and journals of Arctic whalers and explorers. The Arctic sojourners coped with isolation and danger by recreating a home-like environment, with traditions, special foods, and celebration of family occasions, all of which provided them with continuity and a feeling of being connected with family back home (Johnson and Suedfeld 1996). Drawing upon existing archival oral histories, the roles of NASA, U.S. astronauts and their families in creating a home-like environment in space are identified.

The purpose of oral histories is to provide first-hand narrative accounts of the participants' experiences in particular historical events. Qualitative analyses of these accounts are used to identify themes across the experiences of the specific group, with the themes potentially serving as a basis for developing explanatory theories (Chaitin 2008; Berg 1998; Ritchie 1995). For this study, the transcribed interviews with American astronaut participants provided information about how each experienced being an astronaut during their particular historical phase of the space program. Themes, exemplifying specific ways the astronauts created a home-like environment in space, were identified from the first-hand accounts within each long-duration mission and across the three missions. The resulting explanatory theory could serve as a basis for developing a quantitative study to measure the concepts, thus allowing for generalizing the results to the astronaut corps.

4.2 Methods

Data sources were Johnson Space Center oral history interviews conducted with NASA astronaut residents of Skylab (8 men; no interview available for Conrad) and Shuttle-Mir (5 men, 1 woman; Linenger's published autobiography was also analyzed because he has not been interviewed for the oral history project), Johnson Space Center archives of pre-flight interviews and mission journals by NASA residents on the International Space Station (18 men and 3 women, Expeditions 1-17), and selected archived interviews with ISS residents (a CNN interview with Expedition 1 and Japan Space Agency interviews with two ISS residents). All of these sources are open to the public and provide a narrative database. The astronauts are asked to participate in the Johnson Space Center oral history project if they wish to do so. Some of the astronauts decide to keep a narrative account of their activities during their time on the International Space Station, and with their permission, these have become part of the available public narrative database.

These archival sources were analyzed qualitatively to identify the role of NASA, the U.S. astronauts and their families during the long-duration missions. While there are other roles played by each of these groups in making the space environment more like home, the only aspects that can be recorded in this analysis are those the astronauts described in the archival materials. The interviewers conducting the Johnson Space Center oral histories asked questions about the astronauts' experiences, but did not focus specifically on ways the astronaut thought NASA, the family or the astronauts themselves created a life in space that resembled life on Earth.

The qualitative analysis was guided by previous research on the role of home and family in helping Arctic whalers and explorers cope with stress (Johnson and Suedfeld 1996). In that study, the thematic categories included activities approximating life at home (communication, educational and recreational activities, celebrations of special occasions, and food and drink) and reflections of home (missing family and seeing the ship as home). For this study, similar categories were used: the archival materials were read for information about how the astronauts spent their non-work hours (use of free time), indications of how they made routine activities such as exercise and eating more interesting, celebrations of traditions and holidays, descriptions of the station as home, and thoughts of home and family on Earth.

The astronauts' views about work scheduling, crew arrivals, communication options, and materials received from family and friends were noted and subsequently classified as the roles NASA and the astronauts' families carried out to help the astronauts maintain a semblance of their life on Earth.

4.3 Findings

The astronauts' perception of NASA's role is summarized briefly; however, quotes are not provided. Instead, the focus is mainly on the astronaut's role and the family's role, with supporting quotes identified for these two roles. Implicit in the astronaut's and the family's role is NASA's role in scheduling and in facilitating opportunities for the astronauts and their families to carry out activities that mirror some aspects of home life.

4.3.1 NASA's Role

NASA's role included (1) scheduling important and worthwhile activities to perform; allowances made for settling in time and for the crew's judgment about coordinating the activities; allowing sufficient time for maintenance of the station and for personalized leisure activities, which are necessary to maintain productivity; and timing the arrival of supplies and new crew, which provides an anticipated break in routine and (2) regular communication related to work, publicity, and education, all of which provided connection to Earth and reduced the perception of isolation. The scheduling mirrored the daily life of high-level professionals on Earth.

4.3.2 *Astronaut's Role*

The astronaut's role is best described as what he or she has done to make the space environment more like home to the point that it is viewed as home (e.g., how they use their time off, alone or with the crew, and how they personalize the experience; ways of turning routines of eating, exercise, and experiments into fun; carrying out space traditions such as handover ceremonies on the ISS; celebrations for special holidays, birthdays, and space events [e.g., job well done, recognition of record set or an historic space flight]; and thoughts of home to close the gap between space and home).

4.3.2.1 Use of Free Time

Skylab

The astronauts spent much of their free time looking out the windows. As expressed by Weitz, they "never got tired of looking out the window" (Weitz 2000, p. 27) and by Gibson, "The Earth is so fascinating. There's something different coming over the horizon all the time, and the more you look at it, the more you see" (Gibson 2000, p. 70). According to Garriott, "anyone who runs out of something to do must have had a failure in their imagination block up here somewhere. Because if there's nothing else, you can look out the window, which would fascinate me for weeks on end" (Garriott 2000, p. 15).

Shuttle-Mir

Assuming there was not a need to catch up on work or on the station's maintenance, there was some free time on weekends and evenings. During this time, they all mentioned the value of looking out the windows, and described the activities they did alone such as reading or with the rest of the crew, usually movies. A number of leisure materials were already available on Mir, especially books, music, and videos.

A popular activity during free time was looking out the window, identifying various sites, and noting the location of their home or other personally relevant places on Earth, and taking photos of what they saw outside of the windows. Not all felt that looking out a window was an activity that would continue to engage them. For example, Thagard (1998, p. 31) said "But what I found was, after a couple of weeks you look, say, 'Oh, yeah. Been there. Seen that.' It's always going to be a neat thing to look out the window, because it is kind of like looking at a map, and with binoculars you can see some detail, although not tremendous detail. But again, Earth's only so big, it's not infinite, so after a while there was a certain sameness even to some of that."

Most took along personal leisure activities. Thagard (1998) had received a *New York Times* crossword puzzle from his wife. However, because his crewmates were so busy and he was not, he said he did not feel comfortable completing the puzzle even though that was an activity he would have thoroughly enjoyed. Shannon Lucid said "As long as I have some books, I'll be happy" (Lucid 1998, p. 7). Her daughter, an English major, selected the books following her mother's directive: "Just go pick some out. Go to Half-Price Bookstore and just get some things that you think I haven't read." "The only requirement is they had to have a lot of words per page [laughter]. I mean, we had to make sure we had the most–they had to be sort of dense. [S]he picked out a lot of English writers that she thought I should have read. And there was a reason why I had not read them. [Laughter]. You know, you make do with what you have" (Lucid 1998, p. 8).

Thomas noted the importance of recreation serving to recharge the person to be more productive and that the options for such activities are limited on a space station. Therefore, "you've got to have personal recreation there, something that you are interested in and which you can escape in, because you can't physically escape the environment, but psychologically you can, so you need something to remove yourself from the environment and get regenerated so that you can have a productive flight" (Thomas 1998, p. 4). His personal activity was drawing sketches of what he captured in a freeze-framed video shot from a window. He described his sketching as "a very rewarding activity because you can get so involved in it, it distracts you completely from everything else you've been thinking about, and you're thinking creatively, which I like to do, and that turned out to be an immense pleasure during the flight. I would do a sketch on a Saturday, or do some sketching on a Saturday, and I'd find suddenly that hours had passed, and at the end of it I felt just refreshed and ready to go, plus it has given me a personal record of the trip, which is perhaps a little more personal than just a whole series of photographs" (Thomas 1998, p. 5). Thomas said "The crewperson does have to have a personal recreation device at his disposal which is something he or she really derives a personal benefit from. NASA can provide tapes and videos and CDs and things like that, but the crewperson needs to think about what they really need for themselves for that time, what hobby is it they can take with them that will give them the recreational needs that they have, because it's really important that you have good recreation for a flight like this to make your off time productive and to get you away, psychologically remove you from the environment, so that you can have productive times during your work, just like here on Earth" (Thomas 1998, p. 6).

Crew time together was around meals, tea breaks and the communication times with ground control. Lucid noted that the communication passes (Com) worked well "because they gave us little breaks all through the day when we were together, and we could have a few minutes of socializing" (Lucid 1998, p. 13). Foale (1998) also noted the importance of these communication sessions. Blaha (1998) said that the crew was so busy that there was little time to do activities together. Linenger and his crewmate ("Sasha" Yurievich Kaleri) shared pictures and talked about their young sons (Linenger 2000).

ISS

In their pre-flight interviews, astronauts mentioned items they were planning to take along, such as CDs and books. Several mentioned that there were a number of books, CDs, and movies on the station so they did not need to take much along. In their journals or in-flight interviews, most mentioned that they spent time looking out the window and taking photos, indicating that time goes fast and there is no time to be bored (Anderson 2007; Bursch 2002; Chamitoff 2008; Pettit 2003; Tani 2007; Williams, J 2006). Ed Lu provided a description of what follows the work day: "Following that [work day] is dinner (always a fun time!), and then we have a few hours of free time before bedtime to do what we choose. I spend this time working on some science experiments of my own (I'll describe those in a later installment), sending and reading e-mails from home, and taking photographs out the window. There is also a small electronic piano up here that I like to tinker around on. The other night was amateur barber night here, as we were getting a little shaggy and it was time for a haircut. These hours go by really quickly. Just like at home it seems there are never enough hours in the day to get all of my personal projects done! Finally it is time for bed" (Lu 2003, Letter 9).

Looking out the window was the top activity given for use of leisure time, with each one personalizing the experience. For example, Phillips, interested in geography and the history of exploration, focused on noting the route of explorers when he was doing earth observations (Phillips 2005a, Letter from LEO Number 7), Chiao wanted to photograph the Great Wall of China (Japan Aerospace Exploration 2007b), Foale wanted "to learn about the countries over which I am flying" (Japan Aerospace Exploration 2007a), and others highlighted personal geographic areas such as their home. Sunita Williams summarized the importance of windows in long-duration missions: "I think windows are really important for the psychological health of crew members on long duration missions. It is nice to look outside, even if it is dark" (Williams 2006, Science is Fun).

NASA added to the opportunities for leisure in several ways: made it possible for the astronauts to watch a football game as it was being played, and to chat with people at a Houston hangout. Sunita Williams described the football game and their excitement about having the opportunity to watch it: "What a game!!! How about those Patriots! Unbelievable. We were able to watch about 60 % of the game up here thanks to some dedicated folks in Houston who hooked us up. Even NASA has the technology to ship a football game to space! Amazing, except for the last 15 s or so when San Diego was going for the kick, we lost video and audio and were dying up here wondering what happened. Next we heard someone was hugging someone and we couldn't tell who was being happily or unhappily hugged. Finally the video came back after the game had been over a good minute and we saw the score. Talk about being on the edge of your seat. *Imagine the 20-min delay if you were on Mars* and wanted to know if the Patriots had won!!!" (Williams 2006, Taking out the Trash).

As described by Sunita Williams, "There is a group of folks who work for NASA to help make sure we are happy onboard up here—they are called psychological support. One of the cool things that they can do is set up communication between us folks in Houston through Mission Control and a phone line. They did this at a local hang-out. There was a microphone and speaker set up so that folks could come by and chat with us. At the same time we could pretty much hear everyone down there talking. We miss all of you in Houston!" (Williams 2006, Party Week).

The Shuttle's arrival was highly anticipated, with preparations such as cleaning house quite similar to how one prepares for house guests at home. As stated by Clay Anderson, "Prepping for the arrival of the shuttle crew is not a trivial task. I find that it is reminiscent of the same type of work you might do when you have guests scheduled to arrive in your home! They are my former crewmates, but more importantly, they are my friends! Time to put out the guest towels!" (Anderson 2007, Chap. 5), and by Dan Bursch, "I am very much looking forward to the arrival of Atlantis and her crew. They promise to bring new care packages from home, fresh 'smells' of the Earth and old friends" (Bursch 2002, STS-110/8a Launch: April 4, 2002). With the additional Shuttle crew on the Station, it became crowded. Although the residents talked extensively about the enjoyment of the visit, working, eating, and playing together, a few also mentioned escaping periodically to a less crowded area on the Station. However, they were sorry to see the visitors go as expressed by Clayton Anderson, "And then, in what seemed like the blink of an eye, it was time to say good-bye. Their departure was too quick and the Station seemed so empty" (Anderson 2007, Chap. 1).

4.3.2.2 Making Daily Activities Fun

Skylab

Among the games they were supposed to evaluate were a Velcro dart board and a deck of cards. Weitz said "[W]ho wants cards when you can look out the window?" (Weitz 2000, p. 33). However, they did try to use the dart board. They also made up games. For example, they used a blue rubber ball, stood by the trash airlock and tried to see who could bounce the ball off the airlock and into the command module (100 ft away). They ran around the food lockers, did gymnastics, simple jogging, and push-ups with one to three crew members on the astronaut's back. They also were creative in designing and filming experiments, which they thought might be useful in the schools.

A sense of humor between crew members and with ground control was evident in the oral history interviews with Skylab astronauts. For example, Al Bean's crew made up three dummies from spare clothing, attached the next crew's names to them, and placed the dummies in the work room on the exercise bike, at the control panel, and in the storage locker. These dummies were the first thing the crew saw when they entered the work room. Gibson (2000, p. 57) said "Your mind tells you it's a person, so it was kind of eerie for a little while, but we liked the joke." According to Garriott (2000, p. 51), such jokes "give a good feeling among the crew."

Garriott, with the help of the Capsule Communicator (CapCom), made it appear that his wife was actually on board Skylab. He took a pre-recorded tape of her voice with him and played it, with the CapCom completing the conversation as if he were talking in real time with Garriott's wife on board Skylab. When she identified herself, the CapCom said "What are you doing up there?" And she answered: "Well, we just came up to bring the boys a fresh meal, or a hot cooked meal. They haven't had one for quite a while. We thought they might enjoy that." And then proceeded to talk a bit more, finally saying "Well, I see the boys are floating in my direction. I've got to get off the line. I'm not supposed to be talking to you." How this happened remained a mystery to Mission Control until Garriott explained it at the 25th anniversary of the mission (Garriott 2000).

The next two examples show that ground control also had a sense of humor. Ground control played a trick by setting up a Polaroid camera with *Playboy* centerfolds. When Garriott took a photo of a crew member what came out of the camera was a *Playboy* centerfold, not the picture of the crew member (Garriott 2000). In the second example, the crew had been calling the Saturn (rocket) "Old Humpty Dumpty" because it had cracked fins that were delaying the launch. When it was almost time to launch, ground control sent the crew a message "Good luck and God speed from all the king's horses and all the king's men." (Gibson 2000, pp. 52, 53).

Shuttle-Mir

Linenger described a live media interview they had on the way to Mir. John Grunsfeld, one of the crew, knew, from his undergraduate days at MIT, two men who later had a radio program called *Car Talk*. So John arranged with the producers of that show to call in with a problem. "John began by complaining to the show's hosts about a government-issued vehicle that was really giving him a hard time. According to John, this government-issued excuse for a vehicle had horrendous gas mileage . . . a real gas-guzzler. Furthermore, this particular vehicle would run extremely rough for the first 2 min—lots of shimmying and shaking—but then would smooth out and ride beautifully. But at after 8 min into the trip, the darn thing would die completely! No, John replied to a concerned query by one of the hosts, 'the acceleration was good, very good'. After more concerned bantering on the part of the show's hosts, John began tipping his hand. 'No, the vehicle hits pretty good speeds. About 17,500 miles per hour'. At this point, the astonished, and now clearly suspicious, hosts asked, 'Who is this?' Still controlled but smiling, John responded that he was an astronaut onboard the Space Shuttle *Atlantis*" (Linenger 2000, p. 77).

The astronauts on Mir did not describe any particular ways they made the routines of food, exercise, and work more fun. Lucid emphasized the importance of having variety in the types of experiments and the need to have productive and worthwhile work to do on a mission. When they described their days, they talked of work, specific experiments, time spent in maintenance and repair of the station, and having some time together during the evening meal and with the communication passes.

ISS

Several of the astronauts found ways to make exercising more interesting. Shepherd (CNN 2000) experimented with watching a movie while he exercised and Suni

Williams set up her exercise routine as if she were running a marathon, and listened to "A Short History of Nearly Everything" while running her marathon (Williams 2006, As the World Turns and the ISS Orbits). Clayton Anderson said, "I look forward to exercising because I sweat and that makes me feel like I am doing something as if I were on Earth" (Anderson 2007, Chap. 2). Ed Lu had a wager with Peggy Whitson that he would do better than she did on bone density measures; whoever would win would not let the other one forget (Lu 2003, Letter 7).

Enjoying food and meals together was included in the journals written by the ISS astronauts. They recognized the importance of receiving food from home: "One truly bright spot in our culinary prospects is food that is sent up from our families or friends in so-called 'bonus' containers or care packages. The food is neither dehydrated nor thermo-stabilized; it's just as you might see it in the store" (Lopez-Alegria 2006, What's for Dinner?). Some of the astronauts described how they experimented with food combinations to make their meals more interesting, especially as their tastes changed over the mission. Suni Williams was inventive in creating food treats from supplies on hand. She wrote in her journal, "I was remembering how much I love chocolate-covered pretzels and thought, hmmm this could be a good idea. I have frosting left over from cookie decorating, so I thought, what a great mixture! Then, LA [Lopez-Alegria] saw what I was doing and came up with the idea to melt some chocolate. He had some really great Swiss dark chocolate that we put in a drink bag and melted in the oven. I dipped the pretzels in the chocolate on one end and frosting in the other end. It was a great combination—reminded me of home eating frosting with my sister..." (Williams 2006, Happy St. Patrick's Day).

4.3.2.3 Space Traditions

Skylab

None of the interviews included examples of traditions that might have been followed. Each crew entered an unoccupied station and had to spend considerable time in getting it up and operating as needed for the time there.

Shuttle-Mir

Handover ceremonies were described by Thagard, the first NASA astronaut resident on Mir. The Russian and U. S. flags were sent through first, and as the guest, Thagard (1998) was the first one to go into the station where he was welcomed with the traditional Russian bread and salt ceremony. Linenger (2000) described welcoming the incoming crew in a similar fashion, and noted that they reciprocated with a special treat of fresh fruits.

ISS

Traditions for change of command, welcome and farewell ceremonies were established by Bill Shepherd, the first commander of ISS. These traditions reflect naval traditions, as he was a Navy SEAL. There is a ship's log, a ship's bell, and a tradition of a poem for each New Year (Williams 2006, Merry Russian Christmas) and mission patches are displayed. Whitson had to find out quickly what the protocol on bell ringing was for an arriving Shuttle, which is "ring twice for arriving/departing commanders and twice for arriving/departing vehicles" (Whitson 2003, Letters Home #12). Culbertson described the importance of having a handover ceremony and when it would occur: "The actual handover of command of the station will occur more than likely on the last docked day. And we'll make it a formal ceremony so the ground and everybody else in the world knows that we have actually …done a change of command.… If you don't make it a somewhat formal situation, you have kind of a fuzzy time when you're not sure who's who. And I want to make sure that that's a very clean and well-understood, period of time in the life of the station, that, you know, one moment you have one Commander and his or her crew, the next moment you have another Commander and the crew. And that the ground has no questions about who that is at any given time" (Culbertson 2001).

4.3.2.4 Celebrations/Holidays

Skylab

Carr (2000) described their Christmas tree built from food can liners, with decals for decorations and a foil star with a trailer to remind them of the comet Kohoutek. They also did a televised show in which they talked about their experiences in space. Ground control then told them where to find their Christmas presents, and gave them a beta cloth Christmas tree.

Shuttle-Mir

Thagard (1998) and Blaha (1998) mentioned celebrations that were personal milestones, major holidays, or recognition of a job well done. Thagard celebrated his 52nd birthday on Mir and received a gift of Russian cognac (Thagard 1998). Celebrations of a job well done occurred after spacewalks. These celebrations typically included a great meal, conversation, and music. Wolf said they celebrated Christmas and Chanukah: "Couldn't light the candles, but it was a nice time. We slowed down a bit in the work, actually took some time off. Special communications periods. But I kind of felt like I do at Christmas on Earth. I'm ready to get the world moving again and back to normal activity" (Wolf 1998, p. 11).

ISS

Birthdays of the crew and their friends on earth were celebrated. Peggy Whitson (2003, Letters Home #4) described a 3-way video conference with friends in

Houston and Star City to celebrate a friend's 40th birthday. She also told how the crew arranged for the Progress (Russian unmanned cargo ship) to bring up items (happy birthday banner, kazoos, singing candle, party hats and cupcakes) for Sergey's birthday (Whitson 2003, Letters Home #9). For another cosmonaut's birthday, Suni Williams and the crew ate special foods (rye bread, herring, and smoked pork) (Williams 2006, As the World Turns and ISS Orbits).

In her journal, Suni Williams mentioned specific holidays, including Martin Luther King Day, St. Patrick's Day, Easter, Mother's Day, Christmas, and New Year. The ISS crew valued having holidays off, although they did not get off work for all of them, with the exception of Christmas (Williams 2006). Several astronauts described how they celebrated holidays. Shepherd said "We have got some goodies stashed away to do some of that. We'll have to get a little inventive with the menu. Sergei (Krikalev) says 'we will probably open two cans instead of one.' (laughter) We have some other things we have stashed away that we will break out" (CNN 2000). Leroy Chiao said "We actually will have some time off during the holidays, during Thanksgiving and Christmas and then New Year's and also the Orthodox Russian Christmas Day, which is January 7th, and, on board we understand, we do have some things from previous crews. And also I understand some surprises are being packed in some of our supplies that are going to come up, and so I'm not sure exactly what they are, but I think we're going to be getting something" (Chiao 2004). Suni Williams noted: "Not sure if you all knew this, but the New Year is really a big holiday in Russia, and then comes Christmas. Their Christmas is January 7th as per the Russian Orthodox Church Calendar. So their holiday season has just gotten under way. We are going to watch a classic Russian New Year's movie this evening called 'Irony of Fate'" (Williams 2006, Happy New Year). Peggy Whitson described how they celebrated the 4th of July: "In order to make the holiday more fun for the folks who had to work in Mission Control, I arranged to have food and red/white/blue banners for the Mission Control folks. For the first shift, as soon as I got up, I played wake-up music for them (shuttle tradition is that the ground plays wake-up music for the shuttle crews... not a tradition we have for station). I picked 'Born in the USA' since I thought it appropriate for the day. Later, when the second shift came to replace the first crew, I asked for live video down. I had used markers to paint red/blue stripes and stars on our faces. Valery and Sergey were good sports and let me paint on them too!" (Whitson 2003, Letters Home #4).

Celebrations of space activities included specific space history landmarks that occurred during their time on the station, personal landmarks of the crew such as breaking a previous record of time in space, and celebrations of recognizing a job well done. For example, Suni Williams noted that they had a feast on Cosmonautics Day, April 12th, which "is a very special day in the space biz. It is the day the first man launched into space, Yuri Gagarin, as well as the first flight of the Space Shuttle Columbia by John Young and Bob Crippen. It is a huge holiday in Russia and in the U.S., 'Yuri's Night' is a great party in the Johnson Space Center area, Ames area and probably others" (Williams 2006, Holidays with Friends). Carl Walz said that during their space walk, he and his crewmate reflected on the 40th anniversary of John Glenn in Friendship 7. After their EVA they talked with Senator Glenn (Walz 2002).

Celebrations recognized work well done on the Station, such as EVAs and redocking. Such events were usually celebrated with a special meal. Suni Williams described one of these celebrations: "We had herring in burgundy sauce (tomato sauce really), rye bread, pickles, borscht and mashed potatoes with mushrooms and onions. Misha and I split a 'fresh' garlic clove and followed it up with some hot tea. A little odd all these tastes, but it was fun just hanging around the table chatting about the view from outside and what we had experienced [on the EVAs]. Dinner went late, but we had 1/2 day on Friday, so we slept in!" (Williams 2006, Playing Outside).

Personal landmarks, such as a crew member breaking a previous record of time in space, were also celebrated. As described by Suni Williams: "We've had a lot to celebrate up here this week. Some pretty neat happenings on the good ship Alpha. First of all, last Sunday was a fun day as it marked my 100th day in space and 6 months for Misha and LA [Lopez-Alegria]. So of course we had a good dinner. We had one of LA's Spanish delicacies, a mixture of deer, pheasant, fava beans and a mild beef type gravy. It was a great Sunday meal. To mark this occasion in a more official manner, everyone in the control room at Houston Mission Control had a flat Gorby [Suni's dog] on their console. I really appreciate that support—folks know I miss him a lot, but thank God they weren't real Jack Russell Terriers. I don't think I can handle more than one. In addition to our dinner, we had a party for all the folks in Houston who have supported us these last 100 (Suni) + 200 (LA) + 200 (Misha) days on Friday. That was really fun for us believe it or not" (Williams 2006, Party Week). As described by Peggy Whitson: "Yesterday we celebrated 100 days on orbit! We arranged, with some extremely helpful folks on the ground, for a party in Houston, as well as food for the folks that had to work in Mission Control in Houston, Huntsville and Moscow. We had pre-recorded a video message to all the folks, and it was played on the TV at the party. Later we made phone calls to the folks in Star City and Houston, just to chat with our friends during the celebration. Everyone seemed to be having a good time" (Whitson 2003, Letters Home #10). John Phillips said "[W]e have a milestone to note. Sergei Krikalev has now spent over 2 years in space, and is the second person in history to pass this mark" (Phillips 2005a, Letter 3).

4.3.2.5 Thoughts of Home in Space and on Earth

While in space, the astronauts began to think of the station as a home away from home. They talked about the station as "home" while in their thoughts maintaining a strong connection to their family on earth. Communication with, as well as thoughts about, family may have helped to close the gap between their home on earth and in space.

Skylab

Gibson (2000, p. 75) described how Skylab seemed like home to the crew "The space station felt so stable. It felt so much like a home in the sky that we felt completely at home there, very comfortable."

Shuttle-Mir

Until the Shuttle crew left, Thomas said that the station was not yet home because he had not been able to settle in to his new home with them there "because until you get into a day-to-day routine of living and have a comfortable routine that balances your personal needs of recreation with the programmatic needs of work, you're not really living here, and you can't begin to enjoy the environment and establish a stable lifestyle until you do that, until you develop your routine" (Thomas 1998, p. 3). Wolf said that when the Shuttle launched to come get him he started "thinking about Earth thoughts and driving to the Stop and Go to get coffee, and, you know, all the Earth thoughts. I let them flood back in. But until then, I didn't let them. This metal house and processed air was my world, and that was what I was living in" (Wolf 1998, p. 11). Foale described his arrival at Mir: "Going into the Mir living area, I was pleasantly surprised at the cheerfulness of the atmosphere there. It was kind of a warm, welcoming, cozy place, in spite of the masses of cables and equipment and wires that are on the walls. Nonetheless, it looked like a home" (Foale 1998, p. 9) Linenger described how it felt to return after having been in the Soyuz and redocking: "Looking into the station I could see a lone ray of light shining through the port window and outlining the dining table. We had left some food out for dinner. It was the only time during my stay in space that *Mir* looked warm, inviting, and *spacious*. It reminded me of opening the door to a summer cottage that had been boarded up for the winter, looking inside, and seeing familiar surroundings. It had felt good to go for a Sunday ride and to escape Mir's confining clutches. But at the same time, it was good to be back again. Mir no longer seemed just a mechanical object designed to keep us alive, but home" (Linenger 2000, p. 95).

Although they felt the station was home, they also wanted reminders of their home on Earth. For example, Blaha timed looking out the window to see the city where his wife was visiting, or where his son was living, which were ways he thought of his family during the mission (Blaha 1998, p. 14). Linenger did the same, taking photos of home, and saying hello when looking at where his hometown was. Linenger also thought of his deceased father, saying "I felt his presence strongly, perhaps because I was up there in the heavens, nearer to him. I would hold a silent conversation with him, and tell him that I missed him" (Linenger 2000, p. 179).

ISS

Whitson, who had resided on ISS before, described her arrival the second time: "[I]t felt even more like coming home than I expected it would. Viewing the truss and arrays with the curve of blue and white below us, our home looks like a gigantic sailing ship. It's GREAT to be back!" (Whitson 2008, Launch). Clayton Anderson wrote, "With hatches opened we arrived in my new home and I felt a tinge of 'I have been here before'; undoubtedly from my training in Houston and Star City, Russia" (Anderson 2007, Chap. 1). Chamitoff said "So maybe as a result of putting my hands on just about everything and moving it around, or maybe just because I've been here for 88 days now, this place has really become, let's say, a second home to me" (Chamitoff 2008). Suni Williams contrasted how she felt when she arrived with after being on the Station for a while: "I almost can't remember how unnatural and difficult it was to live in space when we first got here . . . space is now my home and very comfortable" (Williams 2006, Holidays with Friends).

Thoughts of home or family were evident in their journals. In carrying out her daily activities, Sunita Williams likened them to something about her mother or father, e.g., while measuring if her calves had become smaller, she said she had her Dad's calves and they couldn't get any smaller; and while growing soybeans, she said "I have my mom's green thumb up here!" (Williams 2006, Happy Martin Luther King Day; Taking out the Trash). Lu described the food as "It isn't quite like Mom's cooking, but it isn't bad!" (Lu 2003, Letter 3). Some mentioned special dates in their family, e.g., Sunita Williams said it was her husband's birthday dinner as she described the dinner she was eating (Williams 2006, Taking out the Trash), and Anderson noted that the day of the launch was his father's birthday (Anderson 2007, Chap. 1).

Several looked out the window to see or photograph their home's location, noting their family was watching, or where they had lived at various times in their lives. Lu described an annual air show, the Experimental Aircraft Association fly-in at Oshkosh, and the photo he took of it. He wanted the photo because he was missing the event for the first time in 7 years (Lu 2003, Letters 8 and 10).

4.3.3 *Family's Role*

The astronauts' view of the role of family centers on the ways in which family members supported the astronauts in their temporary home (e.g., sufficient two-way communication and care packages). With such support, the astronaut could be an integral part of the family back home.

4.3.3.1 Communication with Family

Skylab

On the first manned Skylab mission, the astronauts' private conversations with family were to be monitored. The argument was that this project was government funded and the public had a right to know what was going on. There were also concerns that the astronauts might tell family members if there was a serious problem but would hide this from the press or mission control. The way the astronauts decided to handle the requirement of no private communication was to agree that they would not have any communication with their families during the 28-day mission. The rationale was that as Navy wives, their wives were used to their absences and it would not be a problem for the wives to tough it out until the astronauts were back home (Kerwin 2000, p. 32). On subsequent missions a compromise was found: the astronauts had private conversations with one NASA person listening in to be sure that nothing potentially problematic was overlooked, but the transcripts remained private (Kerwin 2000).

Lousma thought that NASA allowing private communication with his family twice a week was a very helpful situation. This meant that the astronauts could feel they were involved in the daily life of their families. These times for communication were planned for a time that maximized the time available for talking with family (Lousma 2001, p. 32).

Shuttle-Mir

The opportunity to talk with family during the Shuttle-Mir missions was viewed positively by all involved, but the amount of time spent in this way varied given the couple's prior communication patterns, the other demands on their time, and the choices they had to make for using their time.

Thagard and his wife had an established pattern of talking once a week whenever he was out of town. Therefore, the weekly communication they had while he was on Mir was consistent with their usual pattern. He compared this pattern to when he was in Vietnam for a year, and their opportunities to communicate were even more limited, consisting of letters and a brief visit in Japan (Thagard 1998).

Blaha said he could have talked more with his wife Brenda, but he chose to do other things that he would not have the opportunity to see again, or he had pressing work to do: "You can go over there and talk to Brenda, John, right now, on the radio, or you can come over into this room and, guess what, watch a miracle roll by. You can see Texas coming, and Florida over there, and oh, there's Boston out on the horizon, and oh, Chicago and the Great Lakes, and four and a half minutes later it's rolled by. Or I can go in that room and talk to Brenda on the radio" (Blaha 1998, p. 12).

Foale (1998) did not talk about communication with his wife in the oral history interview. Kathryn Linenger had stayed in Russia because the Linengers wanted the weekly visual communication that was possible with the satellite communication system. However, it was not working and they had to rely on a 5-min weekly conversation. He said: "As trivial as a weekly 5 min conversation might sound, I can tell you that to an astronaut-cosmonaut cut off from the world, those 5 min are precious. Just hearing my wife's voice saying that our son was fine, that she was proud of what I was doing, and that everyone back home sent their love, would make my week" (Linenger 2000, p. 195).

Lucid (1998) emphasized that the reason that being away from her family had worked while she was training in Russia was because she had been able to have constant communication with her family through e-mail as the phone systems were not reliable. She insisted that this be possible during her time on Mir, leaving it to NASA to figure out how best to ensure this, and she made sure her family had computers so they were able to receive messages. The system worked, so that she received and sent regular e-mails. The bi-weekly phone conversation usually worked, but the video conferencing was less reliable. However, all of the methods together kept her connected with her family. In answering about the psychological importance of the communication for the crew members, she said "Oh, I think the most important was having the e-mail connection, because in that way you had daily messages, so you felt like you were still plugged in on a daily basis to your family's life. After that, it was the telephone conversations, because they came through pretty good. Actually, the best conversations were those that you were just one on one. When they tried to conference and had a lot of people, then all that happened was people just say, 'Hello,' and, 'How are you?' and that was it. You never got any information passed. Then at the bottom was the video conferences, as far as satisfaction of keeping in contact with your family" (Lucid 1998, p. 7). Lucid's daughter and son-in-law got their ham operator's license so they could talk with her, and she felt that this method was also effective for keeping up with her family's activities (Lucid 1998).

Wolf emphasized the role his flight surgeon played in being a gatekeeper for communication to him during the mission, and in answering some of the mail for him. He stressed the importance of the flight surgeon knowing his family and him well in order to carry out this role effectively. Of all the sources of communication, Wolf liked e-mail the best because he could feel what the person was saying better than in the other methods. He also thought they were candid in their e-mails, even though he knew the e-mails were not totally private (Wolf 1998).

Thomas also preferred e-mail communication to a voice contact. "You know, it's like getting a letter in the mailbox, something you can sort of hold in your hand and you can reread and prepare a response to. You can relive it" (Thomas 1998, p. 8).

Linenger wrote nightly letters to his son, John, even though it would be years before he was old enough to read them. Linenger said: "I wanted him to know that, even though I was in space and far from him, I was thinking about him, that I cared and worried about him. . ." (Linenger 2000, p. 208).

ISS

Weekend video visits and phone chats helped get the astronauts through the separation. Bursch, Anderson, S. Williams and Fincke mentioned the importance of having e-mail and phones to keep in contact with their families (Bursch 2002; Anderson 2007; Williams 2006; Fincke 2004). Anderson wrote about how much the regular communication meant to him: "It meant so much to me to be able to see your faces as well as hear your voices. It brought me 'home' for just a while and helped carry me through the days" (Anderson 2007, Chap. 13). Bursch said "It was very special to be able to call family and friends, but what became most entertaining were peoples' reactions when we said we were calling from space! I didn't expect to get the chance to talk to so many visiting relatives of my friends on Earth!" (Bursch 2002, Just 'Us' on Station).

Others described their excitement of hearing from family or described specific conversations. For example, Sunita Williams said "[G]ot to talk to MOM! How cool was that! What a wonderful surprise to hear her lovely voice on the space-to-ground loops and on the Earth airwaves!" (Williams 2006, As the World Turns and ISS Orbits). Williams also described how fat her dog was getting: "His little body is taking up the entire screen nowadays!" (Williams 2006, Happy Mother's Day). After experiencing delays due to the weather on earth, Whitson called her father and "told him that farming was not the only profession that depended on the weather!" (Whitson 2003, Letters Home #7). Whitson had thought her rehydrated hamburger was special, so she described it to her mother who did not seem to think it sounded great (Whitson 2003, Letters Home #10). Anderson said "I joke with my kids that this is probably a record length of time for them to have not been scolded by their father!" (Anderson 2007, Chap. 6).

4.3.3.2 Care Packages

Skylab

Carr's family gave him a gold icthus (Christian fish symbol) for his Christmas present (Carr 2000). The other Skylab astronauts did not mention gifts or other items taken along with them to remind them of family and life on Earth.

Shuttle-Mir

Foale mentioned videos and things from his family, as well as chocolate (Foale 1998, p. 2), and Thomas talked about the Progress saying "[P]eople would put letters and photographs and things from home in there for me plus treats, like we'd get some fresh fruit, some fresh vegetables, and got a nice big bag of M&Ms, a nice big bag of Oreo cookies, things like that" (Thomas 1998, p. 8). Blaha likened receiving the crew packages to "Christmas and your birthday all rolled together when you were 5 years old" (Blaha 1998, p. 26). "The nice thing is, all crew members get a five-kilogram care package that's put together by families, friends, that sort of thing" (Thagard 1998, p. 29). Linenger described the care package he received, which contained pictures of his son, ultrasound films of his child to be, birthday and Valentine cards, hundreds of notes from family and friends, and letters from his wife. He said "I read and reread those letters [from my wife] everyday until the flight ended" (Linenger 2000, p. 157).

ISS

Bursch explained his excitement about the arrival of the Progress: "I thought that because we had e-mail and the 'phone,' that I wouldn't think the Progress was such a big deal...but it was! I underestimated how much I would anticipate the arrival of fresh fruit and care packages...something from Earth...something from home...that my friends and family had touched not too long ago!!" (Bursch 2002, Cargo Ship Arrives). Sunita Williams described its arrival: "Not only did we get food, and of course CANDY (lots of it), we also got some cards and gifts from folks back home. It was great, sort of like a little Christmas. Everyone took a little time to themselves to sit back, read and enjoy the news (though old) from home. I even got some launch letters and notes, particularly sparkly notes (thanks, Twinkle) from the prelaunch party!" (Williams 2006, It's All About Food).

4.4 Discussion

During the long-duration missions so far—Skylab, Mir, and the International Space Station—US astronauts adjusted to the constraints of the space environment and created a daily life that mirrored some aspects of their life at home. NASA, the astronauts, and their families all had roles to play in creating this daily life that connected the astronauts to life on Earth.

4.4.1 Astronaut's Role

Throughout the history of long-duration space missions, astronauts have coped with isolation and a potentially stressful environment by creating a microcosm of home life, including the celebration of special events as is customary on Earth. Establishing and carrying out traditions, as well as celebrating holidays and personal landmarks, helped make the new environment seem more like home. Mission control and the astronauts cooperated in the planning and celebrating of some events (e.g., Christmas, 4th of July, birthday for Ed Lu). In others, the crew prepared a special meal to recognize birthdays and personal space markers such as a job well done or a record for an astronaut's time in space. After 50 years of human spaceflight, it is not surprising that historic space anniversaries are commemorated. Traditions have been established and their importance is recognized by the crews. As longer missions occur and there is less opportunity to bring things along for the celebrations, the spacefarers may become inventive, as explorers in the past did, and create plays, games, newsletters, etc. (Stuster 1996; Johnson and Suedfeld 1996).

The astronauts found ways to make routines enjoyable, and used their leisure time in personally satisfying ways. Looking out the window was the favorite leisure activity—a point that others have made (Harrison 2001; Kanas and Manzey 2003; Stuster 1996). However, the astronauts showed considerable variability in what they chose to focus upon in their viewing, indicating the importance of personalizing even this common activity. Other ways in which the value of personalizing leisure was emphasized was in how they used their spare time and the items they brought with them to the station. The astronauts appear to be following recommendations for recreational activities, i.e., personalizing them (Stuster 1996) and creating ways to carry out their routines enjoyably so they remain motivated to complete them (Harrison 2001).

The use of practical jokes to make daily activities more fun was evident in the Skylab missions, but not on the Mir or ISS missions. Some forms of humor may not translate well with international crews, or with crews including both men and women.

The ISS astronauts in particular highlighted ways that NASA made it possible for them to enjoy leisure activities they would likely be pursuing if they were on Earth (e.g., watching a football game as it is being played). Such uplinks are part of the psychological countermeasures that NASA has implemented for ISS (Kanas and Manzey 2003). This is an example of how technological advances have made it possible to transport Earth life to space, which was not as available with the earlier long-duration missions—and may again become less possible on a Mars mission. Based on the astronauts including these examples in their journals, these activities have become an important part of their lives in space, providing an immediate connection to life on Earth.

4.4.2 Family's Role

The main contributions of family members were engaging in two-way communication and sending care packages of favorite foods, gifts, photos and letters. With such support, the astronaut in space could still be an integral part of the family back home. As evidenced by the comments of the astronauts, NASA's role in facilitating communication with family has improved over time. Another key point is that in spite of all the high-tech communication, having a chance to read letters and get care packages is highly important in the daily lives of astronauts in space.

Creating a home-like experience in space may have helped the astronauts deal better with the time away from home and family. These findings are similar to the patterns that Johnson and Suedfeld (1996) found in their analysis of the writings of Arctic whalers and explorers. The space station, like the whaling and explorer ships, became a home which they grew fond of and were ambivalent about leaving. Creating a home-like environment that mirrors life on Earth will continue to be important on future exploration missions, and the experiences of space station astronauts, as well as early explorers, as suggested by Suedfeld (2009), may be the most pertinent sources of information for designing countermeasures.

The next step in this research is to analyze similar qualitative data from cosmonauts who participated in the Shuttle-Mir and the ISS long-duration missions, to note cross-cultural applicability of how life on Earth is recreated in the space capsule environment. Then, a quantitative measure can be developed to assess the importance of creating a space life that resembles life on Earth. Scores on this quantitative measure could be related to standard measures of psychological well-being to note the effect of this particular countermeasure on astronauts' and cosmonauts' well-being during long-duration missions.

Acknowledgements The preparation of this chapter was made possible by Contract No. 9F007-033006/01/ST with the Canadian Space Agency, Peter Suedfeld, Principal Investigator. Thanks to Jelena Brcic and Katya Legkaia for assistance with the research.

References

Anderson, C.: ISS Expedition 15: Journals. http://www.nasa.gov/mission_pages/station/expeditions/expedition15/journals_clayton_anderson.html (2007). Accessed 13 Mar 2012

Berg, B.: Historiography and Oral Traditions, Qualitative Research Methods for the Social Sciences, pp. 198–211. Allyn and Bacon, Boston (1998)

Blaha, J.E.: NASA Johnson Space Center Oral History Project, pp. 1–30. http://www.jsc.nasa.gov/history/oral_histories/participants.htm (1998). Accessed 13 Mar 2012

Bursch, W.D.: Expedition Four: 120 Day Report. http://spaceflight.nasa.gov/station/crew/exp4/120days.html (2002). Accessed 13 Mar 2012

Carr, G.P.: NASA Johnson Space Center Oral History Project, pp. 1–64. http://www.jsc.nasa.gov/history/oral_histories/participants.htm (2000). Accessed 13 Mar 2012

Chaitin, J.: Oral history, In: Givon, L.M. (ed.), Sage Encyclopedia of Qualitative Research Methods, pp. 584–585. Sage, Thousand Oaks, doi: 10.4135/9781412963909.n301

Chamitoff, G.: Greg Chamitoff's Journal. http://www.nasa.gov/mission_pages/station/expeditions/expedition17/journal_greg_chamitoff_3.html (2008). Accessed 13 Mar 2012

Chiao, L.: The International Space Station Expedition 10 Crew Interviews (Pre-flight Interview). http://www.spaceflight.nasa.gov/station/crew/exp10/intchiao.html (2004). Accessed 13 Mar 2012

CNN: Space station crew interview (Transcript of Expedition 1 exclusive interview by Miles O'Brien). http://archives.cnn.com/2000/TECH/space/12/18/miles.interview.transcript/index.html (2000). Accessed 13 Mar 2012

Culbertson, F.: The International Space Station Expedition 3 Crew Interviews (Pre-flight Interview). http://www.spaceflight.nasa.gov/station/crew/exp3/intculbertson.html (2001). Accessed 13 Mar 2012

Fincke, M.: The International Space Station Expedition 9 Crew Interviews (Pre-flight Interview). http://www.spaceflight.nasa.gov/station/crew/exp9/intfincke.html (2004). Accessed 13 Mar 2012

Foale, C.M.: NASA Johnson Space Center Oral History Project, pp. 1–20. http://www.jsc.nasa.gov/history/oral_histories/participants.htm (1998). Accessed 13 Mar 2012

Garriott, O.K.: NASA Johnson Space Center Oral History Project, pp. 1–80. http://www.jsc.nasa.gov/history/oral_histories/participants.htm (2000). Accessed 13 Mar 2012

Gibson, E.G.: NASA Johnson Space Center Oral History Project, pp. 1–91. http://www.jsc.nasa.gov/history/oral_histories/participants.htm (2000). Accessed 13 Mar 2012

Harrison, A.A.: Spacefaring: The Human Dimension. University of California Press, Berkeley (2001)

Japan Aerospace Exploration Agency: Interview with Expedition 8 Commander Michael Foale: To the Moon and Mars from the International Space Station. http://www.jaxa.jp/article/interview/vol12/index_e.html (2007a). Accessed 13 Mar 2012

Japan Aerospace Exploration Agency: Leroy Chiao- ISS Mission Accomplished. http://www.jaxa.jp/article/interview/vol20/index_e.html (2007b). Accessed 13 Mar 2012

Johnson, P.J.: The roles of NASA, U.S. astronauts and their families in long-duration missions. Acta. Astronautica. **67**, 561–571 (2010)

Johnson, P.J., Suedfeld, P.: Coping with stress through the microcosms of home and family among Arctic whalers and explorers. Hist Fam **1**, 41–62 (1996)

Kanas, N., Manzey, D.: Space Psychology and Psychiatry. Kluwer, Boston (2003)

Kerwin, J.P.: NASA Johnson Space Center Oral History Project, pp. 1–58. http://www.jsc.nasa.gov/history/oral_histories/participants.htm (2000). Accessed 13 Mar 2012

Linenger, J.: Off the Planet: Surviving Five Perilous Months aboard the Space Station, Mir. McGraw-Hill, New York (2000)

Lopez-Alegria, M.: Space Station Journals http://www.nasa.gov/mission_pages/station/expeditions/expedition14/journals_lopez-alegria.html (2006). Accessed 13 Mar 2012

Lousma, J.R.: NASA Johnson Space Center Oral History Project, pp. 1–63. http://www.jsc.nasa.gov/history/oral_histories/participants.htm (2001). Accessed 13 Mar 2012

Lu, E.: Greetings, Earthlings: Ed's Musings from Space, Letters 1–15. http://spaceflight.nasa.gov/station/crew/exp7/luletters (2003). Accessed 13 Mar 2012

Lucid, S.: NASA Johnson Space Center Oral History Project, pp. 1–24. http://www.jsc.nasa.gov/history/oral_histories/participants.htm (1998). Accessed 13 Mar 2012

Pettit, D.: Don Pettit Space Chronicles. http://www.spaceflight.nasa.gov/station/crew/exp6/spacechronicles9.html (2003). Accessed 13 Mar 2012

Phillips, J.: Expedition 11 Letter from LEO Number 7. http://www.nasa.gov/mission_pages/station/expeditions/expedition11/journals_phillips_ch7.html (2005a). Accessed 13 Mar 2012

Phillips, J.: Expedition 11 Letter from LEO Number 3. http://www.nasa.gov/mission_pages/station/expeditions/expedition11/journals_phillips_ch3.html (2005b). Accessed 13 Mar 2012

Ritchie, D.A.: Doing Oral History. Twayne, New York (1995)

Stuster, J.: Bold Endeavors: Lessons from Polar and Space Exploration. Naval Institute Press, Annapolis (1996)

Suedfeld, P.: Invulnerability, coping, salutogenesis, integration: four phases of space psychology. Aviat. Space Environ. Med. **76**, B61–B66 (2005)

Suedfeld, P.: Historical space psychology: early terrestrial explorations as mars analogues. Planet. Space. Sci. **58**(4), 639–645 (2009)

Tani, D.: The International Space Station Expedition 16 Crew Interviews (Pre-flight Interview). http://www.nasa.gov/mission_pages/station/expeditions/expedition16/exp16_interview_tani.html (2007). Accessed 13 Mar 2012

Thagard, N.E.: NASA Johnson Space Center Oral History Project, pp.1–56. http://www.jsc.nasa.gov/history/oral_histories/participants.htm (1998). Accessed 13 Mar 2012

Thomas, A.S.W.: NASA Johnson Space Center Oral History Project, pp. 1–16. http://www.jsc.nasa.gov/history/oral_histories/participants.htm (1998). Accessed 13 Mar 2012

Walz, C.E.: Expedition Four: Spacewalking, ISS Style. http://spaceflight.nasa.gov/station/crew/exp4/eva/spacewalking.html (2002). Accessed 13 Mar 2012

Weitz, P.J.: NASA Johnson Space Center Oral History Project, pp. 1–67. http://www.jsc.nasa.gov/history/oral_histories/participants.htm (2000). Accessed 13 Mar 2012

Whitson, P.: Expedition Five: Letters Home by Peggy Whitson. http://spaceflight.nasa.gov/station/crew/exp5/lettershome.html (2003). Accessed 13 Mar 2012

Whitson, P.: Expedition 16: Journals (Launch). http://www.nasa.gov/mission_pages/station/expeditions/expedition16/journals_peggy_whitson.html (2008). Accessed 13 Mar 2012

Williams, J.: Jeff Williams' Journals. http://www.nasa.gov/mission_pages/station/expeditions/expedition13/journals_williams.html (2006). Accessed 13 Mar 2012

Williams, S.: ISS Mission Journals. http://www.nasa.gov/mission_pages/station/expeditions/expedition15/journals_sunita_williams.html (2006). Accessed 13 Mar 2012

Wolf, D.A.: NASA Johnson Space Center Oral History Project, pp. 1–19. http://www.jsc.nasa.gov/history/oral_histories/participants.htm (1998). Accessed 13 Mar 2012

Part II
Interpersonal Dimensions of Space Exploration

Chapter 5
Human Interactions On-orbit

Nick Kanas

Abstract Anecdotal reports from space and studies from space analogue missions on Earth have suggested four areas of importance that have relevance for human interactions during on-orbit missions: time effects, displacement, leadership roles, and cultural issues (both national and organizational). In a communications questionnaire survey given to astronauts and cosmonauts who had flown in space, it was found that fluency in a common language during the mission was also important, and a number of factors that improved or hindered intra-crew and crew-ground communication were revealed. Following a pilot study of human interactions during the HUman BEhaviour Study (HUBES), my colleagues and I conducted two major studies involving on-orbit missions to the Russian Mir and International Space Stations. We did not find evidence supporting time effects (in particular during the second half or the third quarter of the missions), but we did find evidence for the displacement of crew tension and unpleasant emotions to the ground and support for the importance of task and support leadership roles for group cohesion. We also found cultural differences in work pressure and tension that may have been related to national and/or organizational factors. In a separate study, we found that space is a positive experience for astronauts and cosmonauts, especially with respect to gaining an appreciation for the Earth and its beauty. These studies suggest a number of countermeasures that can be employed during future space mission in terms of crew selection, pre-launch training, mission monitoring and support, and post-return readaptation.

N. Kanas (✉)
University of California, San Francisco, CA, USA
e-mail: nick.kanas@ucsf.edu

D.A. Vakoch (ed.), *On Orbit and Beyond*, Space Technology Library 29,
DOI 10.1007/978-3-642-30583-2_5,

5.1 Human Interactions in Isolated and Confined Environments

Anecdotal reports from space and studies conducted in space analogue environments on Earth (e.g., Antarctic, submarines, space simulation habitats) have isolated a number of psychological, psychiatric, and interpersonal issues that can affect the safety and well-being of crewmembers working on-orbit, such as in the International Space Station (ISS) (Kanas and Feddersen 1971; Kanas 1985, 1987, 1990, 2004; Connors et al. 1985; Harrison et al. 1991; Sandal et al. 1995; Palinkas et al. 2000; Sandal 2000; Stuster et al. 2000; Kanas and Manzey 2008). These issues have particular relevance for long-duration missions lasting 6 weeks or more, where the stressors of working weightless in a dangerous isolated and confined environment begin to take their toll. This chapter will focus on one aspect of this on-orbit experience: its impact on the interpersonal interactions of the crewmembers, both with each other and with mission control. Four issues can be identified.

The first relates to the influence of time on the crew. Some crewmembers in space or space analogue environments have reported having psychological and interpersonal difficulties after the halfway point of their mission. These difficulties have included increased homesickness and depression, irritability, isolation from others, territorial behavior, and drops in group cohesion. It is not so much the number of days that have transpired but the perception of "halfway" that has the most psychological relevance. One explanation for this occurrence is that the crewmembers experience a sense of relief that the mission is half over, but this is followed by an emotional letdown when they realize that there is another half yet to go before they return home. Bechtel and Berning have called this the "third quarter phenomenon" (Bechtel and Berning 1991). This phenomenon has received empirical support in some studies (Palinkas et al. 2000; Sandal 2000; Stuster et al. 2000) but not in others (Kanas et al. 1996; Steel and Suedfeld 1991; Wood et al. 1999).

A second issue concerns crew-ground communication and how it relates to interpersonal tension and other negative dysphoric emotions on-board. The findings from some analogue studies on Earth and reports from returning space crew debriefings and diaries suggest that crewmembers may vent their frustrations on people in mission control who are supporting their activities from Earth. In group dynamic terms, they are "displacing" negative affect from the on-board in-group to a convenient out-group that is more distant and less able to retaliate. Bill Feddersen and I identified the relevance of this phenomenon for space travel in 1971: "a crew often handles interpersonal friction by displacing their hostility to outside personnel" (Kanas and Feddersen 1971, p. 45). Many of us experience displacement when we cannot express angry feelings at our boss for fear of retaliation, so we instead go home and yell at our spouse or kick at the cat, safer targets who are innocent victims of our anger. Although offering temporary release of emotion, displacement does little to help crewmembers resolve on-board tension in the long run, and it can lead to miscommunication and conflicts with the ground that can prove disastrous during times of danger.

A third issue deals with the different roles of the team leader and how he or she relates to the rest of the group. Research dating back to the 1960s that involved Antarctic teams found that the leader had two major roles in such an isolated and confined group (Kanas and Feddersen 1971; Kanas 1990; Gunderson and Nelson 1963; Nelson 1964). The first relates to motivating the team members to work and making sure that all of the mission objectives are reached: the task role. The second deals with supporting the group members and paying attention to their morale: the support role. These aspects of leadership take on more or less importance at different times during the mission. For example, during emergencies the task role is crucial, whereas during monotonous periods (such as the long winter-over when people are confined indoors with little to do) the support role becomes more relevant. Ideally, the commander of a space mission is comfortable with both roles and knows how and when to use them.

A final issue relates to crew heterogeneity, especially in terms of cultural differences. More and more, on-orbit space missions (such as to the ISS) involve people of both sexes, diverse professional and experiential backgrounds, and different nationalities. Such heterogeneity can be beneficial, since it provides novelty and stimulation when people begin to tire of the routine and look for something new to talk about. However, diversity can also be stressful, especially when people do not understand its subtleties. Take, for example, cultural norms. People from Mediterranean countries typically are animated and comfortable being close to one another when interacting, whereas people from Northern European countries are more reserved behaviorally and have less tolerance for someone gesticulating close to them, which they may perceive as boorish behavior or as a sign of aggression. If not understood as culturally determined, such differences in attitudes and behaviors can cause problems in the confines of a space vehicle (Kanas 2009). Heterogeneity in space may even relate to organizational culture. Space programs vary in their degree of dependence on written procedures and equipment redundancy versus expert advice and "duct tape" engineering to resolve problems. An astronaut or cosmonaut familiar with one system may have difficulty adapting to another during a long-duration space mission. Ease in speaking a common mission language also is important in dealing with crew heterogeneity, not only to enhance efficient communication of ideas during work activities and emergencies, but also to improve bonding with fellow crewmates through a better understanding of the connotations of their speech and the meaning behind their comments and jokes. For more on general psychosocial and cultural issues that affect astronauts and cosmonauts, see the chapter in this book entitled "Psychology and Culture during Long-duration Space Missions" (Chap. 9).

5.2 A Study of Intra-crew and Crew-Ground Communication

Aspects involving intra-crew and crew-ground communication were examined in a questionnaire study involving 54 astronauts and cosmonauts who had flown in space that was conducted by Kelly and Kanas (1992, 1993, 1994). All of the respondents acknowledged that it was important for space crewmembers to be fluent in a common language, and 63% rated this as being "very important". American and Russian space travelers were less tolerant of dialect difference than their international counterparts from other countries. The respondents rated Shared Experience, Excitement of Space Flight, Close Quarters, and Isolation from Earth as factors that significantly helped them communicate with each other, and they perceived the first two factors as significantly helping them communicate with mission control. Spacecraft Ambient Noise, Space Sickness, and Facial Swelling were rated as significantly impairing intra-crew communication, with the first two factors seen as significantly impairing crew-ground communication.

Subjects acknowledged the value of contact with loved ones on Earth as having a significantly positive influence on mission performance—cosmonaut ratings were higher than those of astronauts, and long-duration space travelers scored higher than those spending less than 20 days in space. Several subjects mentioned the need to have private space-ground audio-visual (A-V) links available for crewmembers to talk with their family and friends. Cosmonauts and long-duration space travelers missed the absence of contact with people on Earth significantly more than astronauts and short-duration space travelers. Communication with Earth was seen as an important way to fill leisure time, with both A-V and written media endorsed (since most respondents flew in the pre-internet era, written media largely consisted of family letters transported into space by resupply rockets). In terms of issues that crewmembers preferred to hear about, international and national events and historical topics were rated as being most important.

5.3 Psychosocial Research On-orbit

5.3.1 HUman BEhaviour Study (HUBES)

My colleagues and I have conducted two large international on-orbit studies of psychological and interpersonal issues that have affected crewmembers and mission control personnel during a series of missions to the Russian Mir Space Station and to the ISS. These missions were preceded by a space simulation study that allowed us to frame our hypotheses and test our methods (Kanas et al. 1996). This simulation, termed HUBES (for HUman BEhaviour Study), was a joint European and Russian venture that involved three crewmembers who were isolated in the Mir simulator in Moscow for 135 days. Our part of the project focused on the human interaction issues discussed in Sect. 5.1, above. However, since the crewmembers were all male Russian physicians in their 30s who spoke the same language, we could not study cultural and other factors of crew heterogeneity.

Weekly, subjects completed a research questionnaire that included items from three well-known, valid, and reliable psychosocial measures: the Profile of Mood States (POMS), the Group Environment Scale (GES), and the Work Environment

Scale (WES). Based on scores from subscales of these measures, we found significantly more total mood disturbance and more tension during the first half of the mission than during the second, and cohesion scores dropped at a significantly greater rate during the last third of the seclusion. Although indicating time effects, these findings did not support our predicted second half decline or the presence of a third quarter phenomenon. However, these results may have been influenced by a mid-mission resupply event that brought the crew favorite foods, letters from home, and replacement equipment that visibly improved morale in the third quarter. Overall, there was less tension during the quieter seclusion period than during the busy pre-isolation training period. As predicted, there was evidence for the displacement of tension and unpleasant emotions to the outside, as measured by the significantly negative correlations between some of our measures of on-board dysphoria and a measure of support from outside monitoring personnel. There were also significant correlations between measures of both the task and the support role of the leader and a measure of crew cohesion. These findings suggested that issues involving time, displacement, and leadership roles were important and could be studied in isolated and confined groups. They also suggested that our measures could be useful in examining these areas during future on-orbit experiments.

5.3.2 Studies On-board the Mir and ISS

Our on-orbit activities began in the mid-1990s during the Shuttle/Mir Program, which was viewed as the first phase of the project to construct the ISS. It was a partnership program between the United States and Russia, whereby crewmembers and mission control personnel from both countries would gain valuable experience working together during long-duration on-orbit missions to the Russian Mir Space Station in preparation for their joint activities involving the ISS. Our Shuttle/Mir study was followed by a second study involving the ISS itself. In both studies, missions lasted 4–7 months, consisted of two or three permanent crewmembers from the U.S. and Russia, and examined a number of hypotheses relating to intra-crew and crew-ground human interactions. The Mir study sample involved a total of 5 American astronauts, 8 Russian cosmonauts, and 42 American and 16 Russian mission control personnel (Kanas and Manzey 2008; Kanas et al. 2000a, b, 2001a, b, c, d, 2002). The ISS study sample involved 8 American astronauts, 9 Russian cosmonauts, and 108 American and 20 Russian mission control personnel (Kanas and Manzey 2008; Kanas et al. 2006, 2007a, b; Boyd et al. 2009). As in our HUBES experiment, all subjects in these two on-orbit studies completed a weekly research questionnaire that included items from the POMS, GES, and WES.

The main findings were not significantly different between the Mir and ISS missions, so the main results will be presented and discussed below in terms of four broad categories: time effects, displacement, leadership roles, and cultural issues (due to the paucity of female crewmembers, we could not evaluate sex differences statistically). Two exceptions are the findings from a Culture and Language Questionnaire that was only part of the ISS study and the results from a Critical Incident Log that asked subjects to list and rate the impact of important events that occurred each week; these will be presented in separate sections.

5.3.2.1 Time Effects

We found no significant changes in subscale scores measuring mood or interpersonal environment over time, in either the Mir or ISS study. For example, there was no general increase in tension or other dysphoric emotions or decrease in group cohesion after the halfway point of the missions, nor was there any evidence to support the typical presence of a third quarter phenomenon. In fact, there were no general differences in subscale scores between any of the quarters of the missions. This is not to say that an individual astronaut or cosmonaut did not have a third quarter drop in morale—a few did. But such drops were offset by other crewmembers who showed improvement or no change in emotional status in the third quarter. Taken together, the subjects were remarkably stable emotionally throughout the missions. These results echo those of others who did not find evidence for a general third quarter phenomenon (Kanas et al. 1996; Steel and Suedfeld 1991; Wood et al. 1999).

How can we explain these results in the face of those studies in other isolated and confined environments that found a significant third quarter effect (Palinkas et al. 2000; Sandal 2000; Stuster et al. 2000)? Perhaps the stringent selection criteria and intensive training that astronauts and cosmonauts underwent prior to launch resulted in crewmembers who could deal with stress in an emotionally calm and positive manner, reacting with less lability than, say, Antarctic workers or polar explorers. Or perhaps the excitement of being in the special on-orbit environment stimulated the crewmembers in positive ways that prevented emotional or interpersonal problems from developing. These ideas are supported by the finding that ISS crewmember mood was more positive during the missions than before launch (Boyd et al. 2009; Ritsher 2005).

Another explanation for the absence of time effects may be due to the fact that the crewmembers were supported by space psychologists and flight surgeons in mission control who utilized a variety of countermeasures to help them deal with stress and monotony as their mission progressed. For example, whenever it was felt that a crewmember was becoming homesick or despondent, he or she was encouraged to increase communication with family and friends on the ground via A-V links or e-mail in order to boost morale. Also, surprise presents, favorite foods, and letters from home were sent up during resupply missions. In contrast, this intensity of outside communication and support has been difficult to maintain in some space analogue settings on Earth, such as during the winter-over period in the Antarctic, where communication and resupply opportunities are limited by radio static and the grounding of airplanes due to the harsh weather conditions.

A final explanation deals with experimental methodology. We used measures of mood and group social climate that all subjects had to complete, and from their responses we tested predictions resulting from a hypothesized third quarter phenomenon (e.g., increased scores on dysphoric mood subscales; decreased score in the cohesion subscale). By averaging the subscale scores from all crewmember or mission control subjects, we could see if the third quarter emerged as an outlier from the other quarters of the mission in ways that were consistent with the presence of a general third quarter phenomenon. In contrast, in anecdotal reports from space or in studies using open-ended measures or diary entries from isolated and confined subjects, more weight typically has been given to a mentioned negative response than to the absence of a response (a "squeaky wheel" effect). So for example, if one out of three crewmembers verbalizes problems related to a possible third quarter effect and the others are silent on the matter, then this occurrence might be given credence as a common problem affecting the entire crew. This is compounded where vocal subjects are involved who provide most of the feedback (see Sect. 5.3.2.6, below). Unless proper statistical controls are used to give equal weight to the reactions of all crewmembers, there is a danger that an effect like the third quarter phenomenon will enter the astronaut "lore" as a common issue, when in fact it is an important issue but likely a comparatively rare event.

There was one additional interesting finding from the Mir study: American crewmembers scored higher on subscales measuring order and organization, task orientation, and self-discovery during the first few weeks of being on-orbit. This suggests a novelty effect, probably reflecting the rapid learning curve due to the lack of experience that Americans had at the time with participating in multinational space station missions involving a Russian space vehicle. This effect was not found during the ISS missions, as both Americans and Russians had by then gained more familiarity with being on-orbit together in a more neutral facility.

5.3.2.2 Displacement

In both studies, there was strong evidence for the occurrence of displacement in both crewmembers and mission control personnel. During HUBES, we identified six tension and mood subscales that we felt were indicative of external and interpersonal stress affecting our subjects, and we predicted that on-orbit these subscale scores would correlate negatively with scores from a measure of perceived support from outside personnel in mission control. We reasoned that if crewmember subjects were experiencing high internal dysphoria or interpersonal tension, they would displace their mood state onto the outsiders and perceive them as not being very supportive. As predicted, all six correlations were significant and in the predicted negative direction.

Although providing temporary relief, in the long run this displacement strategy can lead to interpersonal problems with mission control personnel, and it does not address the source of the original conflict on-board. It would have been better for crewmembers to identify the causes of intra-psychic and intra-group stress and to have applied strategies for coping with these causes directly. In this way, the stressors would not be allowed to fester, thus improving crewmember morale and well-being and reducing the chance of crew-ground communication difficulties.

It is possible that crewmembers in space were in fact being less supported by people in mission control during the times that they were experiencing tension amongst themselves. However, there was little objective evidence for this in the daily mission activity reports or in the post-mission debriefings. Consequently, the relationship between measures of tension and negative affect in space and perceived lack of support from the ground was most likely due to the displacement construct than to real events that occurred during the missions.

5.3.2.3 Leadership Roles

In both studies, the mission control subjects significantly related subscales scores that measured the task and support roles of their leader to the scores of a group cohesion subscale, as we predicted (recall that studies in the Antarctic have found that both leadership roles were related to improved morale and cohesion (Kanas and Feddersen 1971; Kanas 1990; Gunderson and Nelson 1963; Nelson 1964)). Although crewmember scores showed a similar significant relationship between the support role of their mission commander and crew cohesion, they did not result in a significant relationship between the task role of the commander and cohesion.

This latter situation might have been an artifact of crew size. As mentioned earlier, the on-orbit crews that we studied consisted of only two or three people, and in such a crew each person has specialized job skills that make him or her a unique leader in those areas that are related to these skills. For example, the commander might focus on piloting and navigation, whereas the engineer might specialize in equipment maintenance and repair. So when a piece of equipment breaks down, the commander might very well defer to the skill and judgment of the engineer, even though the former is the responsible task leader. There is not the same type of leadership hierarchy with two or three people that exists in a larger group of six, seven, or even more crewmembers. In addition, in such a small isolated group in space, one's social network is limited, and it is important to be cordial and supportive of your fellow crewmembers in order to avoid being even more isolated. Thus, cohesion in such a small group would be expected to be more sensitive to support from the leader than to his or her specific task-related activities.

5.3.2.4 Cultural Issues

In looking at the results related to possible cultural differences between American and Russians subjects, the findings from the two studies revealed that compared with the Russians, the Americans reported significantly more work pressure in both studies and less tension during ISS missions (also during Mir missions, but the difference was not statistically significant). These findings may have been influenced by national culture differences. For example, the Americans might have felt more pressure to perform than the Russians due to on-the-job expectations rooted in typical American attitudes about competition and achievement. However, it is also possible that these differences mirrored space agency organizational cultures. It has been suggested that the American space program relies more on written procedures and the maintenance of redundant systems to deal with potential problems, whereas the Russian program relies more on expert opinion from the ground to address issues as they arise (Boyd et al. 2009; Ritsher 2005). Consequently, American astronauts might have experienced more work pressure than their Russian colleagues as they tried to deal with various agency-directed procedural activities in a timely manner during both Mir and ISS missions. The relative tension felt by the Russian crew and ground subjects during the largely U.S.-managed ISS missions may have reflected their lack of familiarity with American operational tasks and procedures.

In both studies, the crewmembers reported significantly lower levels of negative emotions on the mood subscales than mission control personnel. This result may be explained by psychological differences between people who become astronauts and cosmonauts and people who become mission control employees. By nature and training, the former may characteristically experience less emotional distress in response to on-the-job stress than the latter. In addition, being on-orbit is the exciting and positive fulfillment of a long-standing dream of most astronauts. Finally, due to the active involvement of flight surgeons and specially-trained psychologists on the ground, crewmembers receive more psychological support during the missions than people in mission control. Notably, both crewmembers and mission control subjects scored significantly lower on most dysphoric subscales than normative samples of people on Earth working in non-space related fields. This might be due to the kinds of people who end up working for space programs, or perhaps involvement with space missions is more exciting and fulfilling than most other jobs.

5.3.2.5 Culture and Language Questionnaire

This measure was included only in the ISS study and dealt with some of the same issues that were explored in the Kelly/Kanas survey mentioned above in Sect. 5.2. Although a few of the same questions were included, most were different and expanded further into areas of culture and language that were specifically designed for participants in a multinational space program (Boyd et al. 2009).

The ISS crewmembers scored higher in cultural sophistication than mission control personnel, which might be expected since they trained in various locations and as a result interacted more with people from other countries. American mission control personnel scored lower than their Russian counterparts, which was a bit surprising given the diversified nature of U.S. culture and the relative financial ability of Americans to travel. Perhaps this finding reflected the physical proximity of Russia to other Asian and European countries, especially to centers of the multinational European Space Agency, with whom Russia has worked on a number of projects.

Americans scored significantly higher than Russians on the importance of all crewmembers and mission control personnel speaking the same dialect of a common language during space missions. Russians involved with international space activities are probably exposed more to English than Americans are to Russian, since most conferences and meetings involving ISS partners are conducted in the English language (as are U.S. movies and television programs that are widely transmitted abroad). This familiarity might have accounted for the greater language flexibility endorsed by the Russian subjects.

5.3.2.6 Critical Incident Log

The responses to the Critical Incident Log were variable in both studies, with many subjects not contributing and with just a few vocal subjects accounting for the majority of the responses. Consequently, statistical analyses were not possible, and the results could only be reported descriptively. In addition, the Log was altered in the ISS study to include positive as well as negative incidents, so a direct comparison of the two settings was not possible. Nevertheless, a few interesting trends were found that might inform future experimental or operational activities.

In the Shuttle/Mir study, all subject groups listed negative events, such as a fire and several equipment breakdowns that occurred on the Mir, as important sources of critical incidents (Kanas and Manzey 2008; Kanas et al. 2001d). American crewmembers and mission control personnel also cited interpersonal conflicts and misunderstandings as being important, perhaps as a result of working in a program that was operationally managed by people from another culture. Russian mission control personnel mentioned resource and salary deficits as important, which reflected real issues related to political changes in Russia and the fact that many of them experienced delays in receiving their salaries.

In the ISS study, nearly half of the responses from the crewmembers mentioned neutral or positive events, such as birthday celebrations and holiday parties (Kanas and Manzey 2008). Most of the negative events were due to interpersonal conflicts or psychological problems. The majority of the critical incidents that reflected operational issues were expected (e.g., dockings, EVAs), and there was a paucity of responses describing emergencies or accidents in comparison to the Shuttle/Mir experience. This result was likely due to the newer and improved environment on the ISS, in comparison with that on the aging Mir space station. The increased number of total responses might have been due to the higher frequency of positive experiences, or it may have reflected an increased tendency of ISS subjects to respond to the Log in comparison with their Shuttle/Mir counterparts.

5.4 Positive Aspects of Being in Space

The concept of salutogenesis refers to the health-promoting, growth-enhancing effects of a challenging situation. Some people gain strength and wisdom from successfully coping with stressful conditions. Larry Palinkas (1991) and Peter Suedfeld (1998) have discussed the salutogenic reactions some people have to the adverse conditions found in polar environments (e.g., increased fortitude, perseverance, independence, self-reliance, ingenuity, comradeship). Suedfeld has further argued that we should pay more attention to positive psychology and salutogenesis in planning for future space missions (Suedfeld 2001), and he and his colleagues found positive attitude changes reported in the memoirs of 125 space travelers as a result of being in space (Suedfeld et al. 2010).

In an attempt to examine the positive aspects of space flight, my colleagues and I conducted a questionnaire study involving astronauts and cosmonauts who had participated in at least one space mission (Ihle et al. 2006). Subjects were recruited anonymously from two sources: the Association of Space Explorers (an organization of people who have flown in space) and the NASA astronaut corps at the Johnson Space Center. The final sample consisted of 39 respondents, who then completed our Positive Effects of Being in Space questionnaire. This measure consists of 36 items that assess areas of personal growth likely to be influenced by space flight.

We found that every respondent reported at least some positive change as a result of flying in space. The items clustered into eight subscale categories: Perceptions of Earth, Perceptions of Space, New Possibilities, Appreciation of Life, Personal Strength, Changes in Daily Life, Relating to Others, and Spiritual Change. Of these, only one was significantly different from the others and produced a "moderate" level of change in the subjects: Perceptions of Earth. One item from this subscale, "I gained a stronger appreciation of the Earth's beauty", had the highest mean score: the average rating translated into a "great degree" of change. Perhaps this reflected an appreciation and longing for the familiar comfort of our home planet's natural environment versus that in a space habitat, which may have been perceived comparatively as confining and sterile.

For some of the questionnaire items, elements of attitude change seemed to translate into behavioral change after the respondent returned to Earth. For example, three of these items ("I realized how much I treasure the Earth," "I learned to appreciate the fragility of the Earth," and "I gained a stronger appreciation of the Earth's beauty") were significantly associated with the behavioral item "I increased my involvement in environmental causes" after returning.

Interestingly, 10 of the respondents indicated that they were reporting no change in at least one item because no further shift was possible (i.e., the described experience was already optimal for them and could not be enhanced by being in space). The item most frequently designated as unchangeable was "I became more excited about space exploration", followed by two items from the Spiritual Change subscale: "I have a better understanding of spiritual matters" and "I have a stronger religious faith."

5.5 Lessons Learned: Countermeasures

Based on the above findings, a number of countermeasures are suggested for future on-orbit missions, which can be categorized in terms of crew selection, pre-launch training, mission monitoring and support, and post-return readaptation. Crews should be selected to include people who are cognizant of and sensitive to psychological, interpersonal, and cultural (both national and organizational) issues. They need to be able to monitor their own feelings and engage in problem-solving activities around interpersonal difficulties. They should be able to work alone on a project when necessary, but also value teamwork and enjoy interacting with other people. Potential mission commanders should have demonstrated an ability to use both task and support leadership roles appropriately in previous work activities.

Both crewmembers and mission control personnel need to receive pre-launch psychosocial education training. Specific topic areas for this training include: the psychological, interpersonal, and cultural issues that can affect isolated and confined groups; the relationship between crewmembers and mission control personnel and the possible impact of displacement on this relationship; the appropriate use of leadership roles; and the importance of national and organizational norms on international activities. Not only familiarity but mastery of the main language(s) used in the mission should be a training goal for crewmembers and key mission control personnel; this might include visits to the countries of origin to gain language and cultural perspective. Training should utilize both didactic and experiential techniques, some of which would involve the crew and relevant mission control personnel working together in simulated mission scenarios.

During the mission, crewmembers and mission control personnel should take computer-based psychosocial education refresher courses to remind them of key issues discussed prior to launch. This refresher training could be done periodically throughout the mission, but at least at the start and again mid-way. Periodic "bull sessions" should be scheduled for crewmembers and mission control personnel to take stock of their interactions in order to recognize and deal with stress before it begins to fester. Crewmembers need to have leisure time and window space available to view or photograph the Earth. Private contact with family and friends on Earth should be facilitated using both A-V approaches (e.g., teleconferencing via a camera or computer, traditional audio links, internet protocol (or IP) phone) and e-mail approaches (e.g., traditional written messaging, social networking and tweeting). Crewmembers should know that their families at home are supported during the mission, both informally (e.g., peer-led support groups) and formally (e.g., individual and group counseling with a trained therapist).

Finally, post-return readaptation debriefings and supportive activities (e.g., time alone with families, individual and family counseling) need to be utilized to help the crewmembers readjust to life on Earth. In high-profile missions, dealing with issues of fame and glory should be discussed, and the privacy needs of both the astronauts and their family members need to be respected as they readapt to life together on the home planet.

For a discussion of psychosocial issues and countermeasures affecting astronauts and cosmonauts on expeditions beyond the Earth-Moon neighborhood, see the chapters in this book entitled "From Earth's Orbit to the Outer Planets and Beyond: Psychological Issues in Space" (Chap. 15) and "High versus Low Crewmember Autonomy in Space Simulation Environments" (Chap. 12).

References

Bechtel, R.B., Berning, A.: The third-quarter phenomenon: do people experience discomfort after stress has passed? In: Harrison, A.A., Clearwater, Y.A., McKay, C.P. (eds.) From Antarctica to Outer Space. Springer, New York (1991)

Boyd, J.E., Kanas, N.A., Salnitskiy, V.P., Gushin, V.I., Saylor, S.A., Weiss, D.S., Marmar, C.R.: Cultural differences in crewmembers and mission control personnel during two space station programs. Aviat. Space Environ. Med. **80**, 532–540 (2009)

Connors, M.M., Harrison, A.A., Akins, F.R.: Living Aloft: Human Requirements for Extended Spaceflight, NASA SP-483. National Aeronautics and Space Administration, Washington, DC (1985)

Gunderson, E.K.E., Nelson, P.D.: Adaptation of small groups to extreme environments. Aerospace Med **34**, 1111–1115 (1963)

Harrison, A.A., Clearwater, Y.A., McKay, C.P. (eds.): From Antarctica to Outer Space. Springer, New York (1991)

Ihle, E.C., Ritsher, J.B., Kanas, N.: Positive psychological outcomes of space flight: an empirical study. Aviat. Space Environ. Med. **77**, 93–101 (2006)

Kanas, N.: Psychosocial factors affecting simulated and actual space missions. Aviat. Space Environ. Med. **56**, 806–811 (1985)

Kanas, N.: Psychological and interpersonal issues in space. Am. J. Psychiatry **144**, 703–709 (1987)

Kanas, N.: Psychological, psychiatric, and interpersonal aspects of long-duration space missions. J Spacecraft Rockets **27**, 457–463 (1990)

Kanas, N.: Group interactions during space missions. Aviat. Space Environ. Med. **75**(7 Suppl.), C3–C5 (2004)

Kanas, N.: Psychology and Culture During Long-Duration Space Missions. International Academy of Astronautics, Paris (2009)

Kanas, N.A., Feddersen, W.E.: Behavioral, Psychiatric, and Sociological Problems of Long-duration Space Missions, NASA Technical Memorandum (TM X-58067). NASA Manned Spacecraft Center, Houston (1971)

Kanas, N., Manzey, D.: Space Psychology and Psychiatry, 2nd edn. Springer/Microcosm Press, Dordrecht/The Netherlands/ El Segundo (2008)

Kanas, N., Weiss, D.S., Marmar, C.R.: Crewmember interactions during a Mir space station simulation. Aviat. Space Environ. Med. **67**, 969–975 (1996)

Kanas, N., Salnitskiy, V., Grund, E.M., Gushin, V., Weiss, D.S., Kozerenko, O., Sled, A., Marmar, C.R.: Social and cultural issues during Shuttle/Mir space missions. Acta Astronaut. **47**, 647–655 (2000a)

Kanas, N., Salnitskiy, V., Grund, E.M., Gushin, V., Weiss, D.S., Kozerenko, O., Sled, A., Marmar, C.R.: Interpersonal and cultural issues involving crews and ground personnel during Shuttle/Mir space missions. Aviat. Space Environ. Med. **71**(9 Suppl.), A11–A16 (2000b)

Kanas, N., Salnitskiy, V., Weiss, D.S., Grund, E.M., Gushin, V., Kozerenko, O., Sled, A., Bostrom, A., Marmar, C.R.: Crewmember and ground personnel interactions over time during Shuttle/Mir space missions. Aviat. Space Environ. Med. **72**, 453–461 (2001a)

Kanas, N., Salnitskiy, V., Grund, E.M., Weiss, D.S., Gushin, V., Kozerenko, O., Sled, A., Marmar, C. R.: Human interactions during Shuttle/Mir space missions. Acta. Astronaut. **48**, 777–784 (2001b)

Kanas, N., Salnitskiy, V., Grund, E.M., Weiss, D.S., Gushin, V., Kozerenko, O., Sled, A., Marmar, C. R.: Human interactions in space: results from Shuttle/Mir. Acta Astronaut. **49**, 243–260 (2001c)

Kanas, N., Salnitskiy, V., Grund, E.M., Weiss, D.S., Gushin, V., Bostrom, A., Kozerenko, O., Sled, A., Marmar, C.R.: Psychosocial issues in space: results from Shuttle/Mir. Gravit. Space Biol. Bull. **14**(2), 35–45 (2001d)

Kanas, N., Salnitskiy, V., Grund, E.M., Gushin, V., Weiss, D.S., Kozerenko, O., Sled, A., Marmar, C.R.: Lessons learned from Shuttle/Mir: psychosocial countermeasures. Aviat. Space Environ. Med. **73**, 607–611 (2002)

Kanas, N.A., Salnitskiy, V.P., Ritsher, J.B., Gushin, V.I., Weiss, D.S., Saylor, S.A., Kozerenko, O.P., Marmar, C.R.: Human interactions in space: ISS vs. Shuttle/Mir. Acta Astronaut. **59**, 413–419 (2006)

Kanas, N.A., Salnitskiy, V.P., Ritsher, J.B., Gushin, V.I., Weiss, D.S., Saylor, S.A., Kozerenko, O.P., Marmar, C.R.: Psychosocial interactions during ISS missions. Acta Astronaut. **60**, 329–335 (2007a)

Kanas, N.A., Salnitskiy, V.P., Ritsher, J.B., Gushin, V.I., Weiss, D.S., Saylor, S.A., Kozerenko, O.P., Marmar, C.R.: Crewmember and mission control personnel interactions during international station missions. Aviat. Space Environ. Med. **78**, 601–607 (2007b)

Kelly, A.D., Kanas, N.: Crewmember communication in space: a survey of astronauts and cosmonauts. Aviat. Space Environ. Med. **63**, 721–726 (1992)

Kelly, A.D., Kanas, N.: Communication between space crews and ground personnel: a survey of astronauts and cosmonauts. Aviat. Space Environ. Med. **64**, 795–800 (1993)

Kelly, A.D., Kanas, N.: Leisure time activities in space: a survey of astronauts and cosmonauts. Acta Astronaut. **32**, 451–457 (1994)

Nelson, P.D.: Similarities and differences among leaders and followers. J. Soc. Psychol. **63**, 161–167 (1964)

Palinkas, L.A.: Group adaptation and individual adjustment in Antarctica: a summary of recent research. In: Harrison, A.A., Clearwater, Y.A., McKay, C.P. (eds.) From Antarctica to Outer Space. Springer, New York (1991)

Palinkas, L.A., Gunderson, E.K.E., Johnson, J.C., Holland, A.W.: Behavior and performance on long-duration spaceflights: evidence from analogue environments. Aviat. Space Environ. Med. **71**(9 Suppl.), A29–A36 (2000)

Ritsher, J.B.: Cultural factors and the international space station. Aviat. Space Environ. Med. **76** (Suppl 6), B135–B144 (2005)

Sandal, G.M.: Coping in Antarctica: is it possible to generalize results across settings? Aviat. Space Environ. Med. **71**(9 Suppl.), A37–A43 (2000)

Sandal, G.M., Vaernes, R., Ursin, H.: Interpersonal relations during simulated space missions. Aviat. Space Environ. Med. **66**, 617–624 (1995)

Steel, G.D., Suedfeld, P.: Temporal patterns of affect in an isolated group. Environ. Behav. **23**, 749–765 (1991)

Stuster, J., Bachelard, C., Suedfeld, P.: The relative importance of behavioral issues during long-duration ICE missions. Aviat. Space Environ. Med. **71**(9 Suppl.), A17–A25 (2000)

Suedfeld, P.: Homo invictus: the indomitable species. Can Psychol **38**, 164–173 (1998)

Suedfeld, P.: Applying positive psychology in the study of extreme environments. J Human Perform Extreme Environ **6**, 21–25 (2001)

Suedfeld, P., Legkaia, K., Brcic, J.: Changes in the hierarchy of value references associated with flying in space. J. Pers. **78**, 1–25 (2010)

Wood, J.A., Lugg, D.J., Hysong, S.J., Harm, D.L.: Psychological changes in hundred-day remote Antarctic field groups. Environ. Behav. **31**, 299–337 (1999)

Chapter 6
Managing Negative Interactions in Space Crews: The Role of Simulator Research

Harvey Wichman

Abstract This chapter argues that a watershed period has been reached in the history of spaceflight that requires a "paradigm shift" in the way spacecraft are designed and people are selected and trained for spaceflight. In the beginning, space programs had minimal spacecraft, and flights were of short duration. Heroic human specimens were then recruited and extensively trained to perform in these machines no matter how difficult or uncomfortable it was. Spacecraft technology is now sufficiently sophisticated to design spacecraft to be much more accommodating to human occupants. The historical timing of this shift in thinking is heralded by the coming together of sophisticated space technology, the rise of space tourism, and the desire for spaceflights of greater duration than brief sorties into Earth orbit.

Simulator technology has developed in step with spacecraft technology. However, simulators are used primarily in training. The chapter concludes with an illustration of how simulators can be used as behavioral research laboratories. A study conducted for the McDonnell Douglas Aerospace Corporation is presented; in it, a spaceflight simulator was used to explore both applied and theoretical questions with a diverse group of civilian passengers in a simulated 45-h orbital spaceflight, including the value of preflight instruction in effective interpersonal behavior and conflict resolution.

6.1 Introduction

In more than a half century since Yuri Gagarin became the first person in space and the first person to orbit Earth, several hundred cosmonauts and astronauts have successfully flown in space. Clearly, there is no longer any doubt that people can live and work successfully in space in Earth orbit. This ability has been

H. Wichman (✉)
Claremont McKenna College and Claremont Graduate University, Claremont, CA, USA
e-mail: hwichman@cmc.edu

D.A. Vakoch (ed.), *On Orbit and Beyond*, Space Technology Library 29,
DOI 10.1007/978-3-642-30583-2_6, © Springer-Verlag Berlin Heidelberg 2013

demonstrated in spacecraft as tiny as the Mercury capsules, in Space Shuttles, and in various (and much more spacious) U.S. and Soviet/Russian space stations. Spending up to half a year in space with a small group of others is no longer unusual. However, plans are afoot to return to the Moon and establish a permanent settlement there and then to proceed to Mars. Big challenges are on the horizon, and their likely success is predicated on three historical series of events: first, the long series of successes in Earth-orbital flights since the launch of Sputnik on 4 October 1957; second, the six successful excursions of Apollo astronauts on the Moon; and third, the successful robotic landings to date on Mars.

In addition to the challenges that lie ahead for the big government-sponsored Moon and Mars projects of the future, the challenging era of civilian space tourism is about to begin. Five persons, beginning with Dennis Tito in 2001, have purchased flights to the International Space Station on board Russian Soyuz spaceships and have had short stays of about a week in space. On 4 October 2004, Burt Rutan and his team at Scaled Composites, LLC, won the $10 million Ansari X Prize by successfully flying a privately developed and funded suborbital rocket ship capable of reaching space. The British company Virgin Galactic is now planning to fly commercial passengers to the edge of space in a larger suborbital spaceship being developed by the X Prize–winning Scaled Composites in Mojave, California. These will be short rides in space, lasting only minutes, but it is clear that the era of space tourism is at hand. Travel and touring form a powerful human motive, an observation corroborated by the fact that tourism is the world's largest industry. Spaceflight involving humans used to be exclusively the domain of the massive governmental programs in the space race between the United States and the Soviet Union. Now, however, other countries have smaller but significant space programs with human spacefarers. Robert Bigelow's private aerospace company in Las Vegas, Nevada, now has flying in Earth orbit a pair of proof-of-concept scale models of a proposed generic habitat that could become a space hotel or a private space factory or laboratory.

Both the Soviet/Russian and U.S. space programs have demonstrated that humans, men and women of different national and ethnic groups—can live and work together in Earth-orbiting habitats for modest periods that have quite precise beginning and ending times. But all of these successful experiences have taken place in the quasi-military social structures of the astronaut and cosmonaut space programs. In addition, both of these programs managed the expected problems of human physical and social frailty by very stringent selection procedures and lengthy training regimens. None of this is compatible with space tourism. Space tourists will compose a much wider spectrum of spacefarers who will have to be accommodated and whose training periods will have to be dramatically shorter. In addition, space tourists will not be government employees and heroes of national renown. Rather, they will be purchasing their flights to space and will expect services commensurate with the cost of the tour.

I am proposing that, as we approach the end of the first half century of spaceflight, the accelerating maturation of the space program has brought us to a historical threshold. To move successfully beyond that threshold will require a significant shift in the way engineers, designers, and flight managers approach their tasks. The time is past when getting to space was a dangerous experiment; technology was in an early stage of development, and it was sufficient (and perhaps necessary) to design minimal spacecraft that required highly selected and arduously trained test pilots in top physical condition to tolerate them as environments.

At this point in the history of spaceflight, we face two big questions: (1) even though the cosmonaut and astronaut social systems have adapted to the demands placed on them so far, will they be able to cope with the much greater challenges that lie ahead, and (2) is there any chance that space tourism, with a much more fluid social structure and a vastly broader spectrum of participants than in the current space program, will work at all? This chapter deals primarily with the second question, but some of the things that make civilian spaceflight possible will apply to facilitating astronaut and cosmonaut success with their new challenges of establishing a permanent Moon base and then going on to Mars.

Social scientists have been studying the behavior of humans in what have come to be called "extreme environments" even before the space program began. Extreme environments are those that are characterized by such features as isolation, risk, confinement, crowding, restricted privacy, and the inability to leave at will. Illustrations of such environments are submarines, Arctic and Antarctic research stations, undersea habitats, and high-altitude research stations. Once there was interest in how people might perform in tiny, crowded spacecraft, these studies of other extreme environments were looked at as analogues for spaceflight since all of the environments had salient social stressors in common. It seemed reasonable to assume that it did not much matter if one was deep in the ocean or up in space if one was confined in a metal canister with other people. A good summary of this work can be found in Albert A. Harrison's book, *Spacefaring: The Human Dimension* (Harrison 2001). Suffice it to say here that these were good analogues for what spaceflight would turn out to be like. The design of the interiors of spacecraft and the management of daily life on board benefited much from the information gleaned from experiences with other extreme environments. These experiences contributed to the development of selection criteria and training regimens in the early days of the space program. When reviewed by social scientists, these studies of real-life situations generated hypotheses and theories that were then subjected to rigorous tests in laboratories and added to the general fund of knowledge in social psychology. An example might be Freedman's density intensity hypothesis (Freedman 1975). The density intensity hypothesis stated that whatever emotion is extant when crowding occurs will be enhanced by the crowding. Crowding per se is not necessarily aversive. This was a nonintuitive but valuable finding. This phenomenon can be witnessed at most happy social gatherings. A group of people may have a whole house at their disposal, but one would seldom find them uniformly distributed about the premises. It is much more likely that they will be gathered together in two or three locations, happily interacting in close proximity. The reverse of this is also true, as can be seen in mob behavior, where the crowding amplifies the anger of the members. The important point for spacecraft design was that happy people in a cramped spacecraft would not have their good moods diminished by the crowding.

Just as the study of behavior in other extreme environments turned out to be valuable for designing spacecraft interiors and predicting behavioral patterns in eventual spaceflight, so too spaceflight simulators can be used to experiment with different aspects of spacecraft interiors such as color, lighting, noise, and temperature in order to determine optimal designs before committing to a final configuration. In fact, this method began with simple cardboard and wood mockups (primitive types of simulators); then, beginning in the 1950s with the advent of simulators, it became commonplace in the aerospace industry and at NASA. More importantly however, spaceflight simulators can also be used to explore social behavior in extreme environments. Experiments can be conducted with procedures designed to facilitate the prosocial behavior of the occupants. In the early days of the space program, when anecdotal studies of life in extreme environments such as submarines were all we had, these studies proved valuable and served us well. But spaceflight simulators can be used to create situations more specific to spaceflight and do so in a laboratory setting in which extraneous variables can be much better controlled.

Of course, spaceflight simulators on Earth cannot simulate weightlessness. That is unfortunate because the higher the fidelity of the simulator, the better the transfer to real-world situations. We have seen in aviation that extremely high-fidelity flight simulators can be so effective for training that airline pilots transitioning to a new airplane can take all of their training in a simulator and then go out and fly the airplane successfully the first time. However, the inability to simulate weightlessness by no means destroys the usefulness of spaceflight simulators as research tools for the study of human behavior. NASA currently uses them effectively to train astronauts for life on the International Space Station. Not every variable involved in an extreme environment has to be present in a simulation for an analogue to be able to generalize usefully from one setting to another. If that were the case, the early studies from real-world extreme environments would have been useless to NASA. But they were not.

Now, as we enter this new era of spaceflight, we need to use simulators to improve our understanding of the variables involved in successfully coping with the new challenges that will confront us. How such simulators can be used as research tools as opposed to training devices is not intuitively obvious. Since there have been few such studies involving civilian participants, the general public knows little of what goes on in such a study. Therefore, I will describe a study conducted in my laboratory that will demonstrate how simulator studies can address both applied and theoretical research questions. This study is particularly relevant to this chapter because it has given us important information about managing negative interpersonal interactions in a setting simulating those that would be found in future civilian spaceflight.

This study was conducted for McDonnell Douglas Aerospace in Huntington Beach, California, in the spring of 1996. The final report was submitted to the company and not published in the academic press. Here is an overview of the study conducted in my laboratory that shows how we might change the course of spaceflight design for the next phase in the history of life off planet Earth.

6.2 The Aerospace Psychology Laboratory Simulator Experiment

I will begin by setting the stage for what took place. McDonnell Douglas Aerospace (now Boeing Space Systems) in Huntington Beach, California, was in the process of developing a new, single-stage-to-orbit rocket to replace the Space Shuttles. This vehicle would take off vertically the way the Shuttles do, but instead of gliding in for a landing, it would land vertically using the thrust of its engines the way the Moon landers did in the Apollo program. The rocket, which was to be called the *Delta Clipper*, was first conceived of as a cargo vehicle. Soon, engineers began thinking about having both a cargo bay and, interchangeable with it, a passenger compartment The passenger compartment was to accommodate six passengers and a crew of two for a 2-day orbital flight. Former astronaut Charles "Pete" Conrad was then a vice president of McDonnell Douglas Aerospace and a key player in the development of the *Delta Clipper*. At the time, all of the McDonnell Douglas designers were fully occupied with work under a NASA contract on the design of what would eventually become the International Space Station. Dr. William Gaubatz, who headed the *Delta* Clipper program, had hired one of my graduate students as part of the team developing the vehicle. She gave Dr. Gaubatz and Pete Conrad a copy of my book, *Human Factors in the Design of Spacecraft* (Wichman 1992). After reading the book and engaging me in several interviews, they selected my laboratory to design the passenger compartment for the *Delta Clipper*.

Pete Conrad had determined that a spaceship in orbit about Earth at the inclination then being commonly flown by the Shuttles would be able to see most of the parts of Earth that the passengers would want to see in daylight if the spacecraft orbited for two consecutive days (remember that when orbiting Earth, one is in darkness half the time).

Once the passenger compartment design was satisfactorily completed, there was considerable excitement among the McDonnell Douglas engineers about the idea of taking civilian passengers to space (no one spoke words such as "space tourism" yet at that time). The designers were excited about such ideas as not putting full fuel on board the vehicle for orbital flight but keeping it lighter, adding more passengers, remaining suborbital, and flying from Los Angeles to Tokyo in 40 min or Los Angeles to Paris in 38 min. However, when the euphoria of the daydreaming was over, the *Delta Clipper* team was left with the question, can you really take a group of unselected, relatively untrained civilians; coop them up in a cramped spacecraft for 2 days of orbital spaceflight; and expect them to have a good time?

A new academic year was about to begin, so the designers called a meeting and asked my lab to address this question in a simulated spaceflight.

We accepted the challenge and built a spaceflight simulator in our laboratory that had the same volume per person as the one designed for the *Delta Clipper* (see Figs. 6.1 and 6.2).

The Delta Clipper team wanted to know whether people such as those we would select could tolerate being enclosed in a simulator for 45 h, whether this experience could be an enjoyable space vacation adventure, and whether anything could be

Fig. 6.1 A scale model of the spacecraft simulator used in the study conducted in the Aerospace Psychology Laboratory at Claremont McKenna College

done prior to a flight to ensure a high quality of interpersonal interactions among the participants during the flight. Furthermore, they wanted us to measure the quality of the interactions among the participants. Answering these questions required that we conduct an experiment; we decided to conduct two simulated flights with equivalent groups. The flights would have to be essentially identical except that one group (the experimental group) would get preflight training in effective group behavior techniques, and the other group (the control group) would spend the same preflight time in a placebo treatment without group training.

The two groups would not know whether they were the experimental or control group. The groups would have to be observed during the simulation. They did know that they would be observed by cameras and microphones placed so as to survey the entire interior except in the airlock and the toilet. (Participants changed clothes and "bathed" in the airlock with moist towels that were warmed in a microwave oven). The participants could also be observed through a one-way window that appeared to be a mirror on their side. Participants soon were oblivious to being observed, as was often demonstrated when an observer on the outside would be startled by a participant suddenly using the one-way window as the mirror it appeared to be on the inside.

In an effort to recruit participants who would approximate the kinds of people who might book a spaceflight, we contacted a travel company that booked adventure travel tours such as to Antarctica and got from them the demographics of the people who book such tours. We then advertised in a local paper for volunteers to act as

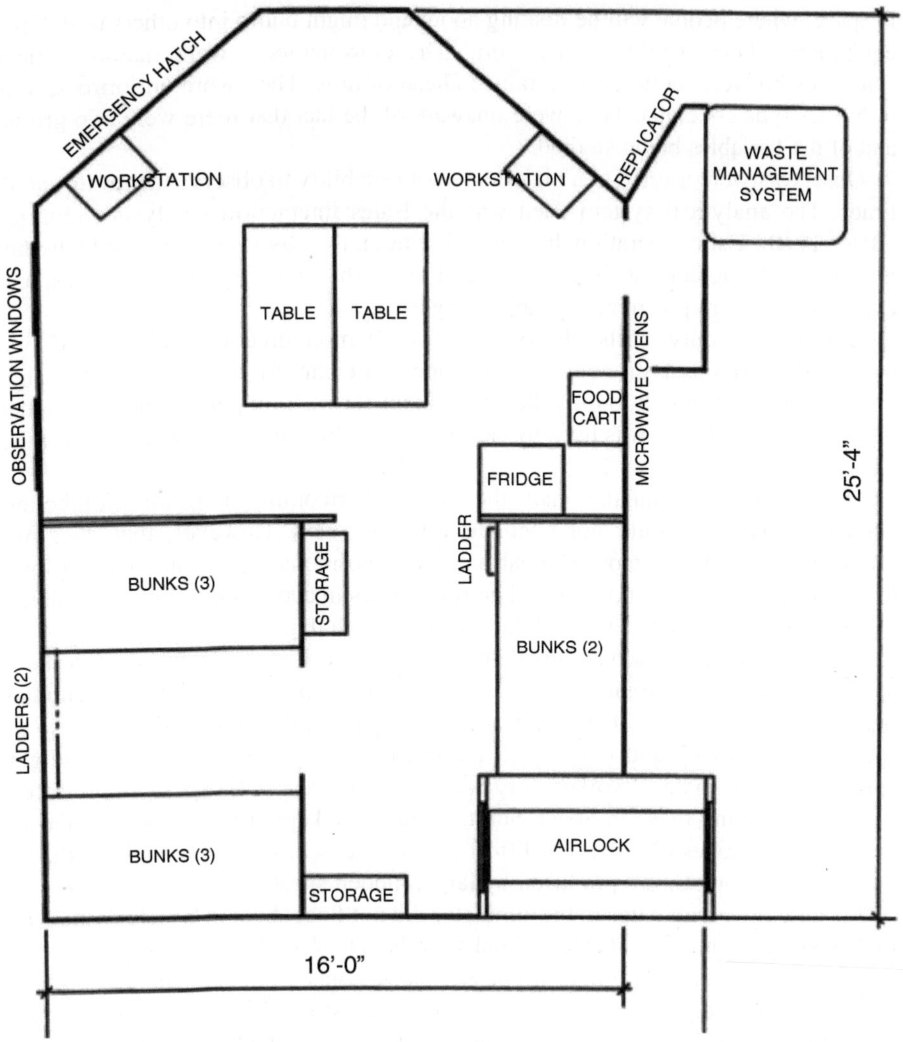

Fig. 6.2 A diagram of the interior of the spaceflight simulator pictured in Fig. 6.1

participants in a simulated space "vacation." Those applying would have to commit to participating for 48 h, from 5 p.m. on a Friday evening until 5 p.m. on the following Sunday evening. Six passengers were selected for each of the two groups: they ranged in age from 34 to 72, half of them were men and half were women, and each group had one married couple. In addition, each group had its own two-member crew, a white male and a black female. We knew of no spaceflight simulation study that involved such diversity of age, gender, and ethnicity involving civilians resembling those who might one-day be involved in space travel. Participants wore their own light sport clothing and soft slippers or warm socks because, as they were informed, in space, where people will be floating about and might bump into others or delicate equipment, shoes would not be worn. The crewmembers were mature college students who were recruited and trained ahead of time. They wore uniforms similar to NASA-type coveralls. They were unaware of the fact that there were two groups and of the variables being studied.

Observers were trained to a high degree of reliability to observe the groups at all times. The analytical system used was the Bales Interaction Analysis technique (Bales 1970). Using operationally defined criteria, the observers measured whether interpersonal interactions, both verbal and nonverbal (e.g., postures, gestures, and expressions), were positive, neutral, or negative.

During their duty shifts, the observers each monitored the behavior of two participants. An observer would monitor one participant for a 1-min period, assign a score, and then switch to the other participant for a 1-min period and assign that person a score. Then it was back to the first person for a minute and so on until the end of the shift.

Our spaceship simulator had the same shortcoming that all earthbound simulators have: it could not simulate weightlessness. However, that does not seem to be a critical factor. The astronaut and cosmonaut programs have a long history of using such simulators and getting results in actual spaceflight that match the behaviors observed in the simulators with a high degree of fidelity.

A simulator is, in a way, equivalent to a stage set. If it looks sufficiently like a spaceship and has the sounds and smells of a spaceship, and if the things that take place within it are those that take place in spaceflight, then the participants, so to speak, "buy into it" and experience the event as a spaceflight. Our spaceflight simulator seems to have worked very well in this respect. Loudspeakers produced sounds mimicking those in Space Shuttles and were kept at amplitudes similar to the Shuttle averages (72 dB). For liftoff and touchdown, very loud engine exhaust vibration and sound were produced by large, hidden speakers.

Because the participants in the simulator did not float about in weightlessness as they would in orbit, we had to have bunks for them to sleep in. During the simulated liftoff and insertion into orbital flight, the participants remained strapped in their bunks. The participants reported in postflight questionnaires that they felt they really had a sense of what a spaceflight would be like—that they often forgot that this was "make believe" and that they "really were living the real experience," to quote two of the participants. They reported being thrilled during the noisy liftoff and the powered landing.

The moment our experimental team knew for certain that we had been successful in creating a realistic spaceflight simulation occurred soon after the first group of participants was established in "orbit." The crewmembers were at their control stations and communicating with "Mission Control." The passengers had unstrapped from their bunks and were assembled in their seats facing forward toward the window area. Mission Control advised that they were preparing to remotely retract the radiation shield over the window and that everyone would soon have a view of Earth from space. By watching the changing postures of the participants, observers could easily see that tension was mounting during the

10-s countdown. Suddenly, a view of Earth taken from one of the Shuttle flights filled the window (actually a 27-in. television screen). One participant gasped and placed her hand to her mouth while staring at the scene. Another whispered aloud in awe, "Ohhhh myyy god." Another, holding her hands to her cheeks said tearfully, "Ohhh, isn't that beautiful." One of the men, gripping the sides of his chair, simply whispered, "Wow!" Another said, "Jeez, look at that!"

There is no doubt that the confinement, training, spacecraft routine, etc., were tolerated. All of the hourly questionnaires, the interaction measurements, and the postflight behavior of the participants clearly suggest to the most casual observer that the flight was not only tolerated but thoroughly enjoyed by the participants. It lived up to their preflight hopes, according to postflight questionnaire reports and their comments to us. Exit from the simulator was delayed in both flights because the participants took the unplanned-for time to trade telephone numbers and addresses before leaving. People who came to the experiment as strangers left as friends. Participants' moods during the simulated flights remained positive, and the number of negative interactions in both groups was small. In the year following the study, my lab received so many telephone calls from participants in both groups requesting a reunion that we felt compelled to go back to McDonnell Douglas and request that they sponsor such an event. They did, and it was a very well-attended, robust party.

One might then wonder if perhaps the social situation produced happiness but the space aspect of it was not important. What we learned is that the participants returned home and presented themselves in their respective social groups as having had a virtual trip to space of such realism that for all practical purposes, it might as well have been real. They were now, in their respective social settings, authorities on space travel. For months after the experiment, the lab kept receiving calls from participants requesting answers to all sorts of space questions. It seemed that now that they were perceived by others as authorities on space, people called to ask space-related questions of them. When they could not answer them, they turned to us for the answers. What is important here is that this postsimulation experience gave us the opportunity to see how this simulation had changed the participants' lives in a positive, space-related way.

During the first of the two simulations, we had programmed in an hour during the early part of the "orbiting" for the participants to talk with "Mission Control." This was a question-and-answer period. We hoped they might find it instructive and even entertaining. As it turned out, the early part of their flight had generated many questions that they were eager to have answered or have an opportunity to discuss. They so enjoyed that hour that they pleaded to have another such hour added for later in the flight. Fortunately, this request was made on the first flight, because the experimental design required that the program schedules be the same in both flights. We did add the extra hour to both flights. In the postflight questionnaires, the participants of both flights indicated that the discussions with Mission Control while flying were the favorite parts of the trip. Very clearly, the participants enjoyed the spaceflight aspects of the simulation very much. All of the subsystems of the simulator worked as planned. No extraneous variables intruded, such as outside noises. From the standpoint of the equipment, the experiment was uneventful.

Table 6.1 The total number of interpersonal interactions and their emotional nature (positive, neutral, or negative) for participants in the experimental and control groups

Type of communication	Experimental group	Control group	Percent difference
Positive communication	354	282	20
Negative communication	8	19	58
Neutral communication	2,120	2,370	11
Total communication	2,482	2,671	7.1

Just before entering the simulator, the experimental group received a 2-h-long program designed to enhance interpersonal prosocial behavior. It was designed much like the type of program corporations provide for their executives in order to develop team building and enhance effective workplace interactions. The program was divided into two main sections, one emphasizing effective interpersonal behavior and the other dealing with understanding and resolving conflicts. In essence, the first part taught participants how to be sensitive to one another and get along, and the second part taught them how to recover if a falling-out did occur. Each participant in the experimental group received a bound 14-page pamphlet of notes about the course material. Much of the formal presentation was lecture/discussion in style, but about one-fifth involved group activities as well.

Just before entering the spacecraft, the control group was given a presentation that lasted the same amount of time but had nothing to do with effective group behavior.

This pair of simulation studies provided much useful information. First of all, as has already been indicated, the participants not only tolerated their confinement very well, but really did enjoy it as a simulated space adventure. The observation data showed that the emotional tone of most of the interpersonal interactions was neutral and that there were relatively few negative interactions in either group, but there were significantly more negative interactions in the control group than in the experimental group that had received the preflight training in effective group behavior. Since the total number of interactions differed only slightly between the two groups, that result also meant that the experimental group had more positive interactions than the control group that received the placebo training. Table 6.1 summarizes the interaction data from the study.

There were 2,482 total communications in the experimental group and 2,671 in the control group, a difference of only 7.1%.

In order to convey the basic meaning of data such as those above in a simplified manner that would also allow an easy comparison of the difference between two groups, we developed a metric called the index of amicability. This index compared the number of positive and negative interactions in the form of a ratio.[1] An amicability index of one means there are equal numbers of positive and negative interactions.

[1] The index of amicability is the ratio of positive to negative interactions for a given group, or, $A_I = P/N$, where A_I is the amicability index, P is the number of emotionally positive interactions, and N is the number of emotionally negative interactions.

An index smaller than one (e.g., 0.75) indicates that there are more negative than positive interactions. An index greater than one (e.g., 25.0) shows that there are more positive than negative interactions. In our simulations, the experimental group index of amicability was 44.3. The control group had an index of amicability of 14.8. Thus, using the difference in index of amicability between the two groups as a measure of the efficacy of the preflight training, we find a very large improvement in social functioning of 299% from a small investment of 2 h in a training program.

It is important to remember that the total number of negative interactions was low in both groups and that both groups enjoyed the experience very much. There were no nasty incidents in either group. However, the group with the preflight training had an index of amicability nearly three times greater than that of the placebo group.

It is necessary to report publicly the above civilian spaceflight simulation study at this time in order to employ its results in discussing the management of negative interpersonal interactions. As the impetus for space tourism ramps up, more such studies will be forthcoming that will attempt to replicate these findings. They will also greatly broaden the range of variables explored. We now have useful information, especially about simulated civilian spaceflight, to use in discussing all interpersonal interaction—positive, neutral, and negative. One of the purposes of this chapter is to excite other scientists to conduct such research. Bales and others have given us the tools to be able to perform interpersonal interaction studies. This report of the study conducted in my laboratory demonstrates that such studies can be accomplished and produce valuable results.

Shortly before we conducted the simulation study in my laboratory, Sandal, Vaernes, and Ursin reported a simulation study of long-duration spaceflights (30 and 60 days) that they had conducted for the European Space Agency (ESA) (Sandal et al. 1995). This group of researchers used decompression chambers at a naval base as spaceflight simulators. This study, too, used the Bales Interaction Analysis technique. We patterned our design after theirs so that our data could be compared. Prior anecdotal evidence from extreme environments (e.g., wintering over in Antarctica) suggested that negative interpersonal interactions among persons in the environment tended to peak at the midpoint and shortly before the end of a stay (Harrison et al. 1991). If that were the case, it would be important information for flight managers and participants to know in understanding and managing spaceflight events.

Here was an illustration of using a simulator as a laboratory to subject anecdotal analogue information to experimental testing to establish more reliably the anecdotal information. The researchers in the ESA study in fact found that the anecdotal information was true and held, independent of the duration (30 or 60 days).

The *Delta Clipper* team was interested in much shorter timespans than those with which the long-duration studies had dealt. They wanted to know if this same phenomenon held for short periods, too, such as the 2-day simulation we were conducting for them. We designed our study to test whether the negative interactions in our groups peaked at the midpoints and just before the ends of the simulated flights. We found that our short-duration experimental study corroborated the findings of both the long-duration experimental studies and the anecdotal studies.

We then had two experimental studies that confirmed the anecdotal findings that negative interpersonal interactions peak at the middle and near the ends of group activities in which the durations of the events are known to the participants. What is more, the finding was independent of the durations involved. It held for short periods, as found in the study reported here, and longer periods, as reported in the ESA study. This information was useful beyond the realm of spaceflight and probably generalizes to all social epochs such as family vacations and school semesters, even if the participants are not confined.

Both of these studies reported that the great preponderance of interpersonal interactions was neutral and that positive interactions were much more frequent than negative ones. Both studies reported that people got along quite well in simulations of differing durations, with more formal and less formal social structure, and in private as well as government settings. The results of the study in my laboratory should be good news for the neophyte space tourism industry. It showed not only that relatively unselected, minimally trained civilians can tolerate the extreme environment of a spaceflight simulator, but also that they find the experience profoundly pleasing. In fact, these results hint that one could build a viable business out of just selling simulated spaceflights.

Another encouraging finding of the study reported here is the powerful effect of preflight group dynamics training on reducing negative interactions and increasing those that are positive. We are reminded by Freedman's density intensity hypothesis that crowding is not necessarily an aversive stimulus, but that it does tend to amplify whatever emotion is extant in a group.

Research that answers questions invariably raises new ones, and that is true of this research. Here is a sample of some questions raised by this research:

1. For how long does the effect of preflight training last? In this project, it only had to persist for 45 h.
2. Does the effect of preflight training end abruptly, or does it taper off?
3. Does a small increment of booster training return preflight training to its original effectiveness?
4. Which of the variables involved in the preflight training are responsible for the effect it produced?
5. Are there other variables that could be added to the preflight training syllabus that would increase its positive effect or duration or both?

6.3 A Call for a Different Type of Thinking

As the history of spaceflight unfolds, I contend that now we are at a transition point between the exploratory and settlement stages of spaceflight that is similar to the opening of the American West in the United States. The early exploration of the West was conducted by a relatively few brave and hardy explorer sorts with an emphasis on daring and pushing back frontiers. There was much ambiguity about the challenges and dangers that lay in uncharted territory. These beginning forays into the unknown were followed by the incursion of hardy trappers, hunters, miners, and various tradesmen. Settlers soon followed, and eventually tourists did as well. In parallel with the western movement of people, technology was improving to facilitate the western expansion—transportation evolved from stage coaches and Conestoga wagons to steamboats and trains.

So too, the early stages of spaceflight were conducted by heroic persons such as Yuri Gagarin and John Glenn. Space stations were eventually established in the frontier, and people learned to adapt to life in space. Now we are planning to return to the Moon and establish a permanent settlement there. The newest frontier dream, the planet Mars, is several orders of magnitude distant from the Moon. We have an International Space Station with a short but significant history of long-duration multinational crews. It has even been visited several times by tourists. In the beginning, the attitude about equipment design was simple: get there, survive, and get back in one piece. So too with astronaut selection: prove that you have "the right stuff"—which translated to "be a high-performance test pilot."

Our technology is much more sophisticated now than it was in the early days of the space program. Now we are designing vastly larger and more complex space stations, Moon colonies, long-duration spaceships for the journey to Mars, and space hotels. The people we will be sending to these sites will be scientists, technicians, service persons, and tourists. I believe that the shift in emphasis implied by these changes requires a shift in the way we think about space equipment and the personnel who will use that equipment.

6.4 Thinking About Equipment

> It is no longer enough to design to survive.
> The time has come to design to thrive.

In the beginning of the space program, engineers were not eager to have people on board space vehicles. The design spirit seemed to be something like, "We are confident this will work, so let's use it. If it uncomfortable or it is difficult to operate, then find astronauts who can tolerate it and who can be trained to make it work." We now need to shift design thinking to a human factors and ergonomics point of view. This kind of shift in emphasis means designing the apparatus to match the capabilities and aspirations of those who will use it. For example, space vehicles are currently very noisy. The noise is due to the fact that warm air does not rise in weightless environments. Without convection currents, any air that is to be moved must be moved mechanically. The large number of fans and bends in ductwork create much of the noise. Spacecraft typically have sound levels of about 72 dB (Wichman 1995). This is about like driving a car at 100 km per hour (kph) with the windows rolled down. By comparison, a living room on Earth would be about 45 dB.

This is much too noisy for comfort over long durations. Such noise levels degrade performance, communication, and satisfaction (Bell et al. 2003). Another human factors issue is spaciousness and privacy. Once the Space Shuttles became the primary heavy-lift spacecraft for the United States, the size of their cargo bays became the limiting factor for space hardware. Thus, the U.S. Destiny module on the International Space Station is 4.3 m in diameter and 8.5 m long. All of the other modules are similar. Fortunately, we are on the verge of having space modules considerably larger in volume than those that existed on the Soviet space station *Mir* or those currently on the International Space Station. These are inflatable modules, such as NASA's Transhab (Dismukes 2012) and other structures based on it, that are being privately developed by Bigelow Aerospace in Las Vegas, Nevada. Bigelow's Genesis I (launched 12 July 2006) and Genesis II (launched 28 June 2007) (Haakonstad 2012) are currently in orbit and functioning as planned. NASA canceled the Transhab program in 2000, but development work (based on NASA's efforts) continues at Bigelow Aerospace, and that is encouraging. The Genesis modules are both prototype, proof-of-concept structures in flight at the present time. Both models of Genesis were launched on Russian rockets and then inflated in space. Having a crowded cabin on a spaceship transporting people to space is no problem; after all, it is only a 100-mile trip. But living for extended periods of months with little privacy and cramped quarters, while obviously tolerable (as on the International Space Station), is not comfortable. The efforts of Bigelow Aerospace to move beyond the limits of past equipment design is illustrative of the shift in thinking that I am proposing as timely.

6.5 Thinking About Spacefarers

> Effective behavior stems not from "good" people
> It is called forth from "good" environments

In the early days of the space program, little was known about the effects of spaceflight on humans, physically or mentally, and the equipment was rudimentary. At that time, it made sense to experiment only with rigorously selected individuals who were exceptional physical and mental specimens. Those days are now over. We are about to enter an era of space tourism. The great message of social psychology is that the behavior we usually attribute to our character is much more determined by our environment than we ever imagined. This finding was amply demonstrated in the following three projects.

In his famous study of obedience, Stanley Milgram showed that everyday Americans could be made to behave cruelly by the making of subtle changes in an academic-like environment (Milgram 1963).

In an infamous 1964 murder, a young woman named Kitty Genovese was slowly killed through the night while she pleaded for help, but no one came to her aid or even called the police. Many of the people living in her apartment building admitted hearing her but were not motivated to help. Shortly thereafter, John Darley and Bibb Latané began their classic studies of bystander intervention and clarified the social and environmental variables that call forth or inhibit bystander intervention no matter who the bystander may be (Darley and Latané 1968).

Finally, Philip Zimbardo, in his classic 1971 Stanford Prison experiment, showed how social circumstances could cause a randomly assigned group of Stanford students to be so cruel to another randomly assigned group that the study had to be terminated (Meyers 2006).

The three projects described are dramatic because they dealt with negative behavior. But in the study reported in this chapter, with only a 2-h training program for essentially unselected people of a wide range of ages, we were able to produce an index of amicability in one group that was nearly three times greater than that in an equivalent group. The argument here is not against any selection. Obviously passengers in an airliner want their pilot to have good vision and a healthy heart. The argument here is for a shift in emphasis toward making it possible for a broad spectrum of people to become space tourists by briefly but effectively developing in them the social skills necessary for a safe and pleasant experience in space.

The spaceflight simulator is an excellent tool, both in which to conduct the necessary social psychology research to show what needs to be done and as the vehicle for conducting the training exercises to bring that about. Interestingly, in terms of selection, the spaceflight simulator provides people with an opportunity to see if they will like such an experience; if not, they will select themselves out without jeopardizing the safety or happiness of others on a real spaceflight.

There is a subset of social psychology theory referred to as attribution theory. Much of the research in this area indicates that humans have a tendency to attribute people's behavior to their character. This is known as the fundamental attribution error. It is the tendency to over-attribute the motivation for a person's behavior to that person's character and underestimate the effect of situational factors. When we emphasize selecting the right "type" of person for spaceflight instead of creating the right "type" of social and environmental factors, we are committing the fundamental attribution error. We have seen in the three social psychology research projects cited above how "good" people could be made to do "bad" things by simple manipulation of situational circumstances. We have also seen in my Aerospace Psychology Laboratory study presented here that people in one group similar to people in another group could have the negative behaviors they would be expected to produce dramatically reduced (58%; see Table 6.1) by a small amount of focused training. Emphasizing selection will diminish the number of prospective spacefarers and inhibit the growth of space tourism. However, emphasizing environmental design and training instead will have the effect of broadening the spectrum of potential spacefarers, facilitating the growth of space commercialization, and, finally, increasing the satisfaction everyone experiences from spaceflights. The primary tool available for fulfilling this shift in emphasis is the spaceflight simulator.

In late 2011 the European Space Agency completed a 520-day simulated Mars mission. Let us hope that this is only the beginning of a long series of studies that will reflect a fundamental change in the way the aerospace industry thinks about the behavior of people in space vehicles and habitats.

References

Bales, R.F.: Personality and Interpersonal Behavior. Holt, Rinehart and Winston, New York (1970)

Bell, P.A., Greene, T.C., Fisher, J.D., Baum, A.: Environmental Psychology, 5th edn. Harcourt Brace, Fort Worth (2003)

Darley, J., Latané, B.: J. Pers. Soc. Psychol. **8**, 377–383 (1968)

Dismukes K. (curator): Transhab Concept. International Space Station History, http://spaceflight.nasa.gov/history/station/transhab/. Accessed 25 Mar 2012

Freedman, J.L.: Crowding and Behavior. W. H. Freeman, San Francisco (1975)

Haakonstad E.: Genesis II Different from Genesis I. http://spacefellowship.com/news/art1851/bigelow-aerospace-genesis-ii-different-from-genesis-i.html. Accessed 25 Mar 2012

Harrison, A.A.: Spacefaring: The Human Dimension. University of California Press, Berkeley (2001)

Harrison, A., Clearwater, Y.A., McKay, C.P.: From Antarctica to Outer Space: Life in Isolation and Confinement. Springer, New York (1991)

Meyers, D.G.: Social Psychology, 9th edn. McGraw-Hill, New York (2006)

Milgram, S.: Behavioral study of obedience. J. Abnorm. Soc. Psychol. **67**, 371–378 (1963)

Sandal, G., Vaernes, R., Ursin, H.: Interpersonal relations during simulated space missions. Aviat. Space Environ. Med. **66**(7), 17–24 (1995)

Wichman, H.A.: Human Factors in the Design of Spacecraft. State University of New York [SUNY] Research Foundation, Stony Brook (1992)

Wichman H.A.; Designing user-friendly civilian spacecraft. paper 95–604. In: Proceedings of the 6th International Space Conference of Pacific Basin Societies 91 (1995). available online at http://www.spacefuture.com/archive/designing_user_friendly_civilian_spacecraft.shtml. Accessed 25 Mar 2012

Chapter 7
Gender Composition and Crew Cohesion During Long-Duration Space Missions

Jason P. Kring and Megan A. Kaminski

Abstract A major factor in the success of future long-duration space missions is the psychosocial functioning of the crew. An individual's psychological health and well-being has a major impact on how well he or she adapts to the demands of isolation, confinement, and workload associated with complex missions. Although each crewmember possesses a unique combination of knowledge, skills, and abilities that influence their capacity to adapt, in this chapter we argue that mission success also relies on how well an individual functions in the larger social context of the mission. More specifically, interactions between crewmembers, as well as between the crew and ground personnel, play a significant role in the crew's overall performance. Although many variables affect crew interactions, such as opportunities for personal space and privacy afforded by the spacecraft's architecture, we contend that the most prominent factor is the crew's composition. Beyond the size of the crew, the mixture of cultural and ethnic backgrounds, and the blend of professional expertise, the most salient crew composition variable is gender.

Since even before Valentina Tereshkova's flight in 1963, women have played an integral role in the history of human spaceflight. As of April 2010, for instance, 53 different women have flown in space, many as part of mixed-gendered crews aboard Russian space stations or the International Space Station (ISS). The April 2010 flight of Space Shuttle Discovery to the ISS set a record for the most women in space at one time as three female crewmembers aboard Discovery—Dorothy Metcalf-Lindenburger, Stephanie Wilson, and Naoko Yamazaki—joined Station resident Tracy Caldwell Dyson in orbit. As the number of mixed-gender crews will likely increase in the future, including those taking voyages back to the Moon and

J.P. Kring (✉)
Embry-Riddle Aeronautical University, Daytona Beach, FL, USA
e-mail: jason.kring@erau.edu

M.A. Kaminski
George Mason University, Fairfax, VA, USA

D.A. Vakoch (ed.), *On Orbit and Beyond*, Space Technology Library 29,
DOI 10.1007/978-3-642-30583-2_7, © Springer-Verlag Berlin Heidelberg 2013

then to Mars, it is prudent to ask if there are any potential limitations to men and women working together for extended periods of time.

This chapter reviews findings from mixed-gender crews in spaceflight as well as relevant analogues like aviation, Antarctic research bases, and other complex environments to highlight how gender composition moderates crew interactions and performance. To explore this relationship, we focus specifically on the variable of cohesion, or the degree to which crewmembers are committed to each other and to the crew's shared task, and offer recommendations for the optimal gender composition for future space missions in terms of this important crew variable.

7.1 Introduction

In 2004, the Bush administration's Vision for Space Exploration refocused the U.S. human spaceflight program on returning people to the Moon by 2020 and then sending a crew to Mars. This plan, ambitious on many fronts, will require the development of a host of new technologies, from improved launch and propulsion systems to a completely new crew vehicle. Scientists and engineers must also work diligently to design systems and mission activities to protect against physiological risks associated with long-duration spaceflight (LDSF), including radiation exposure, bone degradation, and muscle loss. However, beyond the technical and physiological challenges, a major obstacle to LDSF is the psychosocial environment during the mission. In conjunction with individual responses to isolation and confinement, researchers contend that problems associated with crew interactions may be a significant limiting factor for extended space missions (Kanas and Manzey 2003). Jack Stuster emphasizes this point in his review of literature from spaceflight and similar domains like polar bases and stations. Based on the writings of behavioral scientists and accounts from explorers and Antarctic personnel, he concludes that "... the smooth functioning of the group contributes greatly to mission success and can be essential to survival under emergency conditions" (Stuster 1996, p. 165).

The factors influencing something as dynamic and complex as human interaction are, of course, numerous. For even the simplest one-on-one conversation, the personalities of the individuals, their motivations, their organizational roles (e.g., leader versus follower), and the context of the conversation affect each person's perception and interpretation of the interaction. Nonverbal cues, such as body posture, and paralinguistic cues, like the tone of voice, also shape the interaction, altering the degree to which the conversation is deemed pleasant, threatening, or productive. However, in the context of a long-duration space mission with three or more crewmembers, the complexity of human interaction increases significantly. For one thing, interactions occur in the context of a high-workload and high-stress environment. The crew is under tremendous pressure to perform tasks correctly and according to strict timelines with little room for error, creating a setting ripe for tension. Furthermore, beyond their own internal interactions, the crew must also routinely communicate with numerous groups on the ground. These can range from flight-related personnel (e.g., flight controllers, engineers, medical staff), to family members, to even representatives of the media and governments around the world.

Nevertheless, the most prominent factor affecting crew interactions is the composition of the crew itself. Findings from the behavioral and social sciences, spaceflight, and similar settings indicate that the number of people on a team or crew and their individual characteristics are influential to the team's interactions and success. Evidence from spaceflight and analogous settings like Antarctica and submarines indicate, for instance, that the size of the crew has a major impact on crew interactions. Harrison concludes that larger crews possess several advantages over smaller crews, such as a greater range of skills and abilities, as well as providing more opportunities to form friendships and create a more interesting social experience (Harrison 2001). Larger crews also appear to get along better, exhibit less hostility, be more stable, and make better and more efficient decisions, particularly if they are odd-numbered, because in the event of a tie, one crewmember can cast the deciding vote (Kanas and Manzey 2003).

In addition to the size of the crew, crew composition also refers to the characteristics of the individual members. Each crewmember brings his or her own unique qualities to the crew based on his or her experiences; attitudes; personality; motivation; and combination of knowledge, skills, and abilities. For example, differences related to national culture and ethnic background are important crew composition considerations. Although not a significant problem, there have been cases in which cultural diversity led to difficulties in crew interactions. For instance, U.S. astronauts cited cultural factors related to personal hygiene and housekeeping practices as partially responsible for incidents of miscommunication and interpersonal conflict before, during, and after international Space Shuttle missions based on responses to a survey conducted by Patricia Santy and colleagues (1993). Nine respondents—astronauts from flights between 1981 and 1990—reported over 40 incidents of misunderstanding and interpersonal friction related to culture, with at least five rated as having a high impact on the mission.

This chapter focuses on the most salient crew composition characteristic that influences crew interactions: the gender of the individual crewmembers. For this discussion, it is worthwhile to reiterate the distinction between the terms "sex" and "gender." As Stephen Davis and Joseph Palladino note, "sex" is a biological classification of male or female, whereas "gender" refers to the cultural and social expectations about what is masculine or feminine (Davis and Palladino 2007). In the context of interpersonal relations, these gender-based expectations significantly influence how a man or woman interacts with others, from styles of verbal and nonverbal communication to the expression and interpretation of emotions. Although gender differences have the potential to affect a number of variables related to how crews interact and perform, we will address how gender, specifically in the mixture of men and women, influences the cohesion of the crew. Cohesion, defined here as the degree to which individuals in a crew or on a team are committed to each other (interpersonal cohesion) and to the goals of the team's task (task cohesion), has received considerable empirical attention. Although there is some disagreement over the specific relationship between cohesion and performance, the general consensus is that teams possessing higher levels of cohesion function more effectively and exhibit better performance than low-cohesion teams.

This chapter reviews findings from mixed-gender crews in spaceflight as well as relevant analogues like aviation, Antarctic research bases, and other complex environments to highlight how gender composition moderates crew cohesion and, ultimately, performance. The discussion also addresses challenges associated with interpersonal relationships during LDSF and whether guidelines are needed to limit or prohibit romantic relationships. We first describe what is known about mixed-gender teams in space and similar settings and then later turn to the issue of team cohesion.

7.2 Women and Men in Space and Analogous Settings

Even before Valentina Tereshkova's 1963 flight aboard Vostok 6, women have played an integral role in the history of human spaceflight. According to NASA's History Division, 54 different women have flown in space as of April 2011, including Soviet/Russian cosmonauts, American astronauts, and citizens of other countries, with 48 of these women flying with (NASA History Division 2010). Even before the U.S. program got off the ground, the Woman in Space program proved that women could endure the rigors of astronaut selection. Thirteen female pilots passed the same physical examinations at the Lovelace Clinic in New Mexico used to screen and select male pilots for the Mercury program in the late 1950s and early 1960s (NASA History Division 2012). In space, it was not until the early 1980s that the number of female astronauts and cosmonauts began to rise. In fact, nearly 20 years passed between Tereshkova's record-making flight and Svetlana Savitskaya's 1982 mission aboard a Russian Soyuz, which was followed closely by the flight of the first U.S. woman in space, Sally Ride, in 1983. Today, it is not uncommon for women to conduct extravehicular activities; pilot or command the Space Shuttle, as Eileen Collins has done twice; or command the ISS, as in the case of Peggy Whitson.

With the increase in female astronauts and cosmonauts, the number of mixed-gender crews has also risen. As of 2009, seven crews of men and women had worked together during long-duration space station missions aboard the USSR's *Salyut* station, the Soviet/Russian station *Mir*, or the ISS. As the number of mixed-gender crews will likely increase in the future, including for voyages back to the Moon and then to Mars, it is prudent to ask if there are any potential limitations to men and women working together for extended periods of time.

On the surface, this mixture would seem ideal as each gender offers unique and complimentary skills and abilities to a mission. The behavioral literature is replete with studies of gender differences in cognition, sensation and perception, and team performance in business settings; however, few studies have focused specifically on gender differences in the context of extended space missions. One question is how the mixture of men and women in space crews affects overall crew performance. In the general team literature, for example, findings suggest that men and women do work in slightly different ways that may influence team performance such as leadership style and reactions to stress (Eagly et al. 2003). Research also suggests that the unique contributions from each gender often improve team performance in settings such as health care, manufacturing, and extreme environments, thereby supporting the use of mixed-gender teams (Bishop 2004).

For instance, in some contexts, all-male teams make less accurate and more overly aggressive decisions than mixed-gender teams (LePine et al. 2002). However, these and other results are qualified by the fact that gender differences in team performance are often moderated by other factors such as the type of task and the personality composition of the individual team members. With regard to personality in team sports, for example, male and female athletes exhibit different personality profiles and attitudes toward recreational activities (Kirkcaldy 1982). Furthermore, gender heterogeneity may influence the development of team factors that contribute to successful team performance like cohesion and trust. As noted above, cohesion is a team's commitment to a shared task and attraction between team members, whereas trust refers to attitudes held by team members regarding their emotional closeness with, and the reliability of, other members of the team. How men and women respond to stress, for instance, can influence both cohesion and trust, particularly at the interpersonal or emotional level. N. S. Endler (2004, p. C25) notes that men tend to cope with stress using "fight or flight" strategies, whereas women employ a "tend or befriend" approach. This latter strategy may therefore evoke more emotional closeness among crewmembers.

In spaceflight, mixed-gender crews have flown successfully since the 1980s; however, a majority of these missions were short-duration flights of 1–2 weeks. For example, although no performance issues were attributed to Svetlana Savitskaya's gender when she visited the Soviet *Salyut 7* station for 8 days in 1982, cosmonaut Valentin Lebedev's (1988, p. 171) account of the visit suggests that gender stereotyping did occur. After presenting her with a floral print apron upon her arrival, he declared, "Look, Sveta, even though you are a pilot and a cosmonaut, you are still a woman first. Would you please do us the honor of being our hostess tonight?" For longer-duration missions lasting between 5 and 7 months, anecdotal reports from two mixed-gender missions aboard the ISS indicate that the crew also got along and functioned effectively (Kanas and Manzey 2003a).

More scientific observations of mixed-gender crews are available from research in space analogues—extreme settings that possess similar environmental and social features. In her review of team literature in extreme environments, S. L. Bishop (2004, p. C17) found support for using mixed-gender crews to improve long-duration performance, concluding, "The presence of both men and women appears to normalize group behavior in ways that promote individual and group functioning." This result may stem from differences each gender possesses that benefit team performance in specific situations. For example, groups composed of both men and women perform well on tasks requiring diverse perspectives and interpersonal skills, and cohesion appears to increase due to women's interpersonal style and ability to involve all group members in the task (Bishop 2004).

Additional support for employing mixed-gender teams comes from Stuster's findings that although the inclusion of women at U.S. Antarctic stations in the early 1980s resulted in some minor conflicts, in general, gender diversity had a positive influence on morale and productivity (Stuster 1996). E. Rosnet et al. found a comparable beneficial effect of mixed-gender groups at a French polar station; however, some women reported problems related to rude behavior from their male colleagues and instances of sexual harassment (Rosnet et al. 2004). Interpersonal relationships between men and women and sexual issues may also impact mixed-gender crews during extended isolation. Crewmembers of an experiment termed the Simulation of Flight of International Crew on Space Station (SFINCSS) reported increased crew tension after an incident in which a male Russian commander from one group attempted to kiss a Canadian female from another group during a New Year's Eve celebration in the enclosed space station mockup (Sandal 2004).

In summary, despite the potential for conflict and tension, evidence from spaceflight and related environments suggests that the inclusion of men and women on long-duration space missions will benefit individual and crew functioning. However, more detailed analyses are needed to identify team performance issues specifically influenced by the gender of the crewmembers. One important question is how gender heterogeneity affects the development of crew cohesion.

7.3 Cohesion

In general, cohesion refers to the closeness and solidarity of a group or team of individuals. However, researchers have long debated the specifics of the construct, particularly the number of associated factors or dimensions. Early researchers used a multidimensional approach. L. Festinger (1950, p. 274), for instance, defined cohesiveness as "...the resultant of all the forces acting on the members to remain in the group. These forces may depend on the attractiveness or unattractiveness of either the prestige of the group, members in the group, or the activities in which the group engages." In other words, cohesion was seen to result from one or more of three sources: group prestige, interpersonal attraction, or attraction to the group's tasks. Similarly, C. W. Langfred conceptualized cohesion as the degree to which group members feel a part of the group and their desire or motivation to remain in the group (Langfred 1998). In a military context, G. L. Siebold and D. R. Kelly (1988, p. 1) posited that cohesion "...is a unit or group state varying in the extent to which the mechanisms of social control maintain a structured pattern of positive social relationships (bonds) between unit members, individually and collectively, necessary to achieve the unit or group's purpose." In contrast, some have argued that cohesion only encompasses one dimension. Cartwright and others, for example, defined cohesion simply as the degree to which group members desire to remain in the group (Cartwright 1968). Similarly, Kenneth Dion (2004, p. C39) defined group cohesion as "...the social glue that binds members of a group and keeps them together in the face of internal and external threats."

More recently, several authors have begun differentiating between the social or interpersonal aspects of cohesion and those related to the group task. Interpersonal cohesion includes dimensions such as interpersonal attraction and the intensity and positive nature of relationships (Carless and Paola 2000). Task cohesion, in contrast, refers to the attraction or commitment to the group and task. Task-cohesive groups, according to S. J. Zaccaro, J. Gualtieri, and D. Minionis (1995, p. 79), also "...care about the success of other group members because their own goal attainment is often inextricably bound to the collective achievement. They will exert strong effort on behalf of the group and their fellow members to facilitate group processes." Combining these two dimensions, we can define cohesion as the combination of task cohesion, referring to the degree to which group or team members are committed to the task, and interpersonal cohesion, the degree to which individuals are attracted to each other and have positive relationships. Before addressing this construct in terms of gender differences, it is worth briefly noting what is known regarding cohesion and performance.

After decades of research, dating back to the 1950s, the relationship between cohesion and team performance continues to generate debate. Early efforts concluded that group productivity and cohesiveness were not clearly related (Steiner 1972). For example, R. M. Stogdill found, in his review of 34 studies, that roughly a third of the studies showed cohesive groups to be more productive, with a third reporting that cohesive groups were less productive and the remaining third showing no difference (Stogdill 1972). However, none of the studies referenced by Stogdill used the same definition for group cohesion, and many made no attempt even to measure cohesiveness. Later research showed support for a positive correlation between cohesion and performance, but some have argued that these effects were often moderated by additional variables (Gully 1995). The current view is that cohesion does have some influence over team processes and how well a team performs, but that the effect often depends on the type of cohesion, the type of task, and the interaction with other team variables. In the case of group performance, a majority of authors cite task cohesion as the critical component in the cohesion-performance effect. For example, task cohesion has been related to better performance for teams making decisions under temporal stress (Zaccaro et al. 1995). Particularly for additive tasks, for which individual efforts are combined to complete an overall group task, S. J. Zaccaro and C. A. Lowe (1988) found that high task cohesion increased performance, but that interpersonal cohesion had no effect.

Such is not the case with disjunctive tasks, for which group members must work together to produce a collective product. Zaccaro and M. C. McCoy had groups rank 15 items in order of importance to group survival in a simulated survival situation task. Results indicated that high task and high interpersonal cohesion groups outperformed groups either high on one type but low on another, or low on both types. For disjunctive tasks, Zaccaro and McCoy (1988, p. 846) noted, "High task-based cohesion increases the likelihood that high ability members will contribute to the group problem-solving, whereas high interpersonal cohesion facilitates the procurement, recognition, and acceptance of high quality contributions." In addition, better-performing teams competing in a complex business simulation game, a disjunctive-type task, were more cohesive, as represented by higher scores on measures of interpersonal and task cohesion. The authors maintained that cohesive teams "...are better performers because they are able to satisfy the social needs of the team members while simultaneously demonstrating a shared commitment to the team task" (Miesing and Preble 1985, p. 336).

Despite the apparent support that task cohesion and, to a lesser extent, interpersonal cohesion positively influence team and group performance, part of the difficulty in defining the cohesion-performance effect is attributed to the influence of moderator variables. C. Evans and K. Dion (1991), in their meta-analyses of over 372 groups, showed that group cohesion led to increased performance; however, the effect was relatively small and appeared to depend on other factors. One example is A. Tziner and Y. Vardi's (1983) finding that performance for three-person tank crews, performance, effectiveness, and cohesiveness were correlated only when studied in combination with the command style of tank commanders. Highly cohesive teams exhibited better performance only if the command style emphasized an orientation toward the task and the team members. For command styles only emphasizing team member orientation, low cohesiveness was related to better performance. Another perspective is that cohesion is associated with performance, but that high levels of team cohesion may negatively affect a team, as in the case of groupthink or in teams whose norms do not support productivity (Porter and Lilly 1996). If the predominate group norm is a slow work pace, cohesiveness might actually reduce performance.

Similar to the effect of moderator variables, another complicating factor in studying the cohesion-performance effect is determining which comes first. There remains significant controversy over the causal nature of the relationship. In the sports domain, Daniel Landers, Michael Wilkinson, Brad Hatfield, and Heather Barber (1982, p. 171) commented, "Even when the same measuring instruments are employed for interacting team sports, some studies demonstrate a reciprocal causality between the two variables (i.e., cohesion affects performance outcome and vice versa), whereas other studies find that performance outcome affects cohesion, but cohesion does not influence performance."

Taken together, the above studies suggest that there is a relationship between the cohesion of a team or crew and its performance, although the specifics remain unclear. In the context of LDSF, it is even more difficult to research the construct given that crew cohesion is not stable over the course of the mission. Dion (2004), for example, cites several studies showing that cohesion declines in the middle and later stages of a mission as reflected by increases in crew tension and conflict. Likewise, Nick Kanas (2004) found that cohesion levels were significantly higher during the first few weeks of missions than in later stages.

In summary, despite methodological differences between cohesion-performance studies, the influence of moderator variables, and disagreement over the direction of the relationship, several conclusions are possible. First, both task and interpersonal cohesion may improve performance and group processes, but task cohesion more consistently predicts performance. Second, on additive-type tasks, high interpersonal cohesion can have a negative effect on performance due to more non-task-relevant conversations between team members, but high levels of both task and interpersonal cohesion benefit performance on disjunctive tasks. What these studies do not demonstrate, however, is how the composition of the team with regard to gender affects the development and maintenance of cohesion.

7.4 Gender Composition and Crew Cohesion

With regard to gender specifically, the team literature suggests that men and women do work in slightly different ways, such as in their leadership styles and reactions to stress, that can influence cohesion. In addition, even though the unique contributions from each gender often improve team performance in some extreme environments analogous to spaceflight, there are concerns over how gender stereotypes and disagreements negatively affect crew interactions. If we assume that teams possessing higher levels of cohesion, in general, perform more effectively than teams with low cohesion levels, then a primary consideration when selecting crews for LDSF is defining the optimal combination of individuals to maximize cohesion. In addition to size, experience, and culture, an important question is what mixture of men and women is ideal for a long-duration space mission? Preferably, we would rely on results from empirical studies on gender and cohesion; however, few researchers have addressed this relationship specifically, fewer still in the context of LDSF. Nevertheless, if we focus first on team performance in general, findings from the business and military domains offer some insight into the effects of gender. For example, manufacturing teams with a larger number of women taking on informal leadership roles within the team received higher supervisor ratings than teams with fewer female leaders (Neubert 1995). Likewise, as women were added to technical teams at a Fortune 500 aerospace company in a stepwise fashion, the addition of one or two women did not adversely affect team performance in comparison to that of all-male teams, and adding three or four women produced a slightly positive effect on team performance (Burris 2002). In contrast, some studies suggest that gender heterogeneity is problematic for teams, at least in terms of ratings of team effectiveness, particularly when there is diversity in composition variables beyond gender, like ethnic and cultural background. Gayle Baugh and George Graen (1997) found that members of project teams in a state regulatory agency rated their teams as less effective when the members varied in terms of gender and race than when the team members were all male or all white. However, ratings of the teams by outside evaluators showed no differences with regard to gender and race.

As already described, there are few studies focused exclusively on gender and cohesion; however, available research does suggest that the inclusion of women on a team, at the very least, does not negatively affect cohesion and in some cases actually improves cohesion. In a military context, arguably a better analogue to spaceflight than business, a Women in Combat Task Force Study Group concluded that women had either a positive or a neutral effect on the type of cohesion present in military units. This model of cohesion takes into account team factors such as interdependence, unit identity, personnel stability, communications, and leadership

(Saimons 1992). Similarly, Robert Vecchio and Donna Brazil's (2007) survey of nearly 2,000 U.S. armed services cadets indicated that increases in the number of women in a squad were not associated with any decreases in squad-level measures of cohesion. In the aviation domain, an even better analogue to spaceflight, the relationship between gender and cohesion is less clear. For example, four-person teams of students from a Florida university, flying an F-22 flight simulation, exhibited higher levels of interpersonal cohesion when the teams were of the same gender. Furthermore, this increased interpersonal cohesion helped to enhance coordination between team members, leading to improved performance (Panzer 2003).

Clearly, additional empirical investigations are warranted to more completely define how gender heterogeneity in teams affects cohesion. In lieu of controlled research, a tenable approach is to survey personnel from space and analogue settings, although very few studies using this approach have specifically focused on cohesion. Personnel at polar stations, for example, reported that the presence of women improved the general group climate but was also a source of stress for members of both sexes (Rosnet et al. 2004). Similarly, a former commander of the U.S. Naval Support Force Antarctica, who had experienced both all-male winter-over missions and gender-integrated stays, reported that women had a stabilizing effect on personnel and believed these heterogeneous groups were more productive than all-male groups (Stuster 1996). In space, an American woman who lived aboard the ISS reported that the crew interacted well and achieved all mission goals (Kanas and Manzey 2003). On the other hand, reports from space analogues such as offshore oil rigs, naval vessels, and Antarctic bases indicate that interpersonal problems related to mixed-gender crews had a negative effect on crew performance (Stuster 1996). Stuster (1996, p. 177) notes that "...on closer inspection, however, the problems appear to have been not directly attributable to mixed crews, but rather to the behavioral consequences of immaturity, faulty personnel selection, and inadequate pre-mission training for both male and female members of the crews."

Another approach is to consider how gender heterogeneity affects factors closely related to the development of cohesion. Recall that Bishop's (2004) conclusion after reviewing literature from extreme environments was that the presence of women in mixed-gender crews appeared to promote crew behaviors that improved its functioning. Likewise, Endler's (2004) finding that women employ a more interpersonal and caring approach when dealing with stress may significantly improve the interpersonal atmosphere within a crew, improving interpersonal cohesion by bringing members closer together.

7.5 Conclusion

The success of future human space missions rests squarely on the shoulders of the men and women who will venture into space for months, possibly years, at a time. In this chapter, we argued that despite inherent differences in the behaviors and abilities of men and women, mixed-gender crews have performed effectively, both in space and in similar settings like Antarctica. In most cases, teams composed of both men and women function as well as or better than all-male teams. We also supported the conclusion that crew interactions, specifically the level of cohesion within the crew, are extremely important to the crew's overall performance. Available evidence, albeit limited in scope and size, indicates that gender-homogenous crews exhibit similar, and at times higher, levels of cohesion than gender-heterogeneous crews in settings similar to spaceflight.

In this chapter, we concentrated on just one piece of this puzzle related to the gender composition of the crew and how the mixture of men and women affects the crew's task and interpersonal cohesion. Our conclusion, based on available research from space, space analogues, and other team-related activities, is that a crew composed of both women and men is the right choice for extended missions to the International Space Station, a return to the Moon, and, one day, the first human mission to Mars.

References

Baugh, G.S., Graen, G.B.: Effects of team gender and racial composition on perceptions of team performance in cross-functional teams. Group Organ. Manag. **22**(3), 366–383 (1997)

Bishop, S.L.: Evaluating teams in extreme environments: from issues to answers. Aviat. Space Environ. Med. **75**(7, sect. II), C14–C21 (2004)

Burris, J.W.: The impact of gender diversity on technical team effectiveness. Dissertation Abstracts International: Section B: The Sciences and Engineering 62, Ann Arbor, Michigan (MI), USA no. 10-B: 4715 (2002)

Carless, S.A., De Paola, C.: The measurement of cohesion in work teams. Small Group Res **31**(1), 71–88 (2000)

Cartwright, D.: The nature of group cohesiveness. In: Cartwright, D., Zander, A. (eds.) Group Dynamics: Research and Theory, 3rd edn. Harper and Row, New York (1968)

Davis, S.F., Palladino, J.J.: Psychology. Pearson, Upper Saddle River (2007)

Dion, K.L.: Interpersonal and group processes in long-term spaceflight crews: perspectives from social and organizational psychology. Aviat. Space Environ. Med. **75**(7, sect. II), C36–C43 (2004)

Eagly, A.H., Johannesen-Schmidt, M.C., van Engen, M.L.: Transformational, transactional, and laissez-faire leadership styles: a meta-analysis comparing men and women. Psychol. Bull. **129**(4), 569–591 (2003)

Endler, N.S.: The joint effects of person and situation factors on stress in spaceflight. Aviat. Space Environ. Med. **75**(7, sect. II), C22–C27 (2004)

Evans, C., Dion, K.: Group cohesion and performance: a meta-analysis. Small Group Res. **22**(2), 175–186 (1991)

Festinger, L.: Informal social communication. Psychol. Rev. **57**(5), 271–282 (1950)

Gully, S.M., Devine, D.J., Whitney, D.J.: A meta-analysis of cohesion and performance: effects of levels of analysis and task interdependence. Small Group Res **26**(4), 497–520 (1995)

Harrison, A.A.: Spacefaring: The Human Dimension, p. 138. University of California Press, Berkeley (2001)

Kanas, N.: Group interactions in space, Aviat. Space Environ. Med. **75**(7), C3–C5 (2004)

Kanas, N., Manzey, D.: Space Psychology and Psychiatry. Microcosm Press/Kluwer Academic, El Segundo (2003)

Kirkcaldy, B.D.: Personality profiles at various levels of athletic participation. Pers. Individ. Differ. **3**(3), 321–326 (1982)

Landers, D.M., Wilkinson, M.O., Hatfield, B.D., Barber, H.: Causality and the cohesion-performance relationship. J. Sport Psychol. **4**(2), 170–183 (1982)

Langfred, C.W.: Is group cohesiveness a double-edged sword? An investigation of the effects of cohesiveness on performance. Small Group Res **29**, 124–143 (1998)

Lebedev, V.: Diary of a Cosmonaut: 211 Days in Space. Bantam Books, New York (1988)

LePine, J.A., Hollenbeck, J.R., Ilgen, D.R., Colquitt, J.A., Ellis, A.: Gender composition, situational strength and team decision-making accuracy: a criterion decomposition approach. Organ. Behav. Hum. Decis. Process. **88**(1), 445–475 (2002)

Miesing, P., Preble, J.: Group processes and performance in a complex business simulation. Small Group Behav. **16**, 325–338 (1985)

NASA History Division. Lovelace's Woman in Space Program. NASA. http://history.nasa.gov/flats.html. Accessed 25 Mar 2012

NASA History Division. Women in Space, NASA. http://history.nasa.gov/women.html. Accessed 25 Mar 2012

Neubert, M.J.: Too much of a good thing or the more the merrier? Exploring the dispersion and gender composition of informal leadership in manufacturing teams. Small Group Res. **30**(5), 635–646 (1995)

Panzer, F.-J.: The influence of gender and ethnic diversity on team effectiveness. ProQuest ETD Collection for Florida International University, Paper AAI3085015. Doctoral Dissertation, available: http://search.proquest.com/docview/305233519/previewPDF?accountid=27203. pp. 1–151 (2003)

Porter, T.W., Lilly, B.S.: The effects of conflict, trust, and task commitment on project team performance. Int. J. Confl. Manag. **7**(4), 361–376 (1996)

Rosnet, E., Jurion, S., Cazes, G., Bachelard, C.: Mixed-gender groups: coping strategies and factors of psychological adaptation in a polar environment. Aviat. Space Environ. Med. **75** (7, sect. II), C10–C13 (2004)

Saimons, V.J.: Women in combat: are the risks to combat effectiveness too great? Monograph Report No. AD A258 247,(U.S. Army Command and General Staff College, School of Advanced Military Studies, Fort Leavenworth (1992)

Sandal, G.M.: Culture and tension during an international space station simulation: results from SFINCSS'99. Aviat. Space Environ. Med. **75**(7, sect. II), C44–C51 (2004)

Santy, P., Holland, A., Looper, L., Marcondes-North, R.: Multicultural factors in the space environment: results of an international shuttle crew debrief. Aviat. Space Environ. Med. **64**, 196–200 (1993)

Siebold, G.L., Kelly, D.R.: Development of the Combat Platoon Cohesion Questionnaire. ARI Technical Report 817, ADA 204917. U.S. Army Research Institute for the Behavioral and Social Sciences, Alexandria (1988)

Steiner, I.D.: Group Processes and Productivity. Academic, New York (1972)

Stogdill, R.M.: Group productivity, drive, and cohesiveness. Organ. Behav. Hum. Perform. **8**, 26–43 (1972)

Stuster, J.: Bold Endeavors: Lessons from Polar and Space Exploration. Naval Institute Press, Annapolis (1996)

Tziner, A., Vardi, Y.: Ability as a moderator between cohesiveness and tank crew's performance. J. Occup. Behav. **4**, 137–143 (1983)

Vecchio, R.P., Brazil, D.M.: Leadership and sex-similarity: a comparison in a military setting. Pers. Psychol. **60**(2), 303–335 (2007)

Zaccaro, S.J., Lowe, C.A.: Cohesiveness and performance on an additive task: evidence for multidimensionality. J. Soc. Psychol. **128**(4), 547–558 (1988)

Zaccaro, S.J., McCoy, M.C.: The effects of task and interpersonal cohesiveness on performance of a disjunctive group task. J. Appl. Soc. Psychol. **18**(10), 837–851 (1988)

Zaccaro, S.J., Gualtieri, J., Minionis, D.: Task cohesion as a facilitator of team decision making under temporal emergency. Mil. Psychol. **7**(2), 77–93 (1995)

Chapter 8
The Risk for Groupthink During Long-Duration Space Missions: Results from a 105-Day Confinement Study

Gro Mjeldheim Sandal, Hege H. Bye, and Fons J. R. van de Vijver

Abstract On a mission to Mars the crew will experience high autonomy and interdependence. "Groupthink," known as a tendency to strive for consensus at the cost of considering alternative courses of action, represents a potential safety hazard. This chapter addresses two aspects of "groupthink": the extent to which confined crew members perceive increasing convergence in personal values, and whether they attribute less tension to individual differences over time. It further examines the impact of personal values for interpersonal compatibility. These questions were investigated in a 105-day confinement study in which a multinational crew ($N = 6$) simulated a Mars mission. The Portrait of Crew Values Questionnaire was administered regularly to assess personal values, perceived value homogeneity, and tension attributed to value disparities. Interviews were conducted before and after the confinement. Multiple regression analysis revealed no significant changes in value homogeneity over time; rather the opposite tendency was indicated. More tension was attributed to differences in hedonism, benevolence and tradition in the last 35 days when the crew was allowed greater autonomy. Three subgroups, distinct in terms of personal values, were identified. No evidence for "groupthink" was found. The results suggest that personal values should be considered in composition of crews for long-duration missions.

Adaptation of Sandal et al. (2011), reprinted with permission from Elsevier.

G.M. Sandal (✉) • H.H. Bye
University of Bergen, Bergen, Norway
e-mail: gro.sandal@psych.uib.no

F.J.R. van de Vijver
Tilburg University, Tilburg, The Netherlands

North-West University, South Africa

D.A. Vakoch (ed.), *On Orbit and Beyond*, Space Technology Library 29,
DOI 10.1007/978-3-642-30583-2_8, © Springer-Verlag Berlin Heidelberg 2013

8.1 Introduction

Future human space missions to Mars raise significant challenges with regard to maintenance of the health, performance and safety of crew members (Manzey 2003). Such missions will not be comparable to any other undertaking humans have ever attempted because of the long distance of travel, the duration of permanent living under dependence of automated life-support systems, and the lack of short-term rescue possibilities in case of emergencies. Due to these factors and the restricted possibilities of psychological support from Earth during the stay in Martian orbit or on Martian surface, crew autonomy will inevitably be higher than ever before. The lack of a visual link to Earth is likely to add to the feelings of isolation and autonomy (Kanas and Manzey 2008). A major challenge for mission planning is to ensure the ability of the crew to function efficiently and safely under these conditions. Long-duration space missions are a challenge for interpersonal relationships and feelings of coherence of crewmembers. Problems in this area can take on two forms. On the one hand, interpersonal tensions can become so large that the group disintegrates and the mission's aim is jeopardized. On the other hand, the coherence can become so strong that crewmembers lose their independence and critical evaluation skills.

An unequivocal recommendation from experts in space psychology is that emphasizing interpersonal compatibility in crew composition is important for a successful completion of any mission. While historically, research attention has been directed towards the potential impact of personality differences and complementarities on how crew members relate to each other, the potential impact of values have remained unexplored. On the basis of past theoretical and empirical research on psychological need fulfilment and value congruence in organisations in urban settings (Cable and Edwards 2004), it has been suggested that constellations of individuals should be avoided whose values, needs and beliefs are competitive or incongruent. Yet the empirical basis for this assumption in isolated and confined settings is weak as longitudinal studies that empirically relate similarities in individual values to crew cohesion and tension are missing. Studies on whether confined individuals over time develop more homogeneity in values and thus increasingly tend to share their views and perspectives also seem to be lacking. We addressed these issues during a 105-day confinement study in which a multinational crew simulated a number of scenarios related to launch, the outbound and return journey to Mars, and transfer to and from the Martian surface. This chapter will take a closer look at findings from this study.

8.2 Personal Values

We adopted the theory developed by Schwartz and Bilsky (1990) which defines values as desirable, trans-situational motivational goals of varying importance, and which serves as guidelines for action. The model derives ten types of values:

Self-direction, stimulation, hedonism, achievement, power, security, conformity, tradition, benevolence, and universalism. The theory has been tested in more than 200 samples from over 60 countries. In the vast majority of samples, both the distinctiveness of the ten values and the structure of their relations have been verified. The set of ten values has been used to explain a wide variety of attitudes, behaviours, and subjective states across many nations. Being closely linked to motivation, values guide attention and action to intrinsically rewarding social, intellectual, and emotional opportunities (Schwartz 2006a). As values serve as standards for judging the behaviour of self and others, they are likely to play an important role in tension between crewmembers and for subgroup formation during human space missions. This idea is supported by a large body of evidence within the field of social psychology showing that attitudinal similarity is a powerful determinant of interpersonal attraction.

The relative stability of values across context and time demonstrated in large samples makes them useful psychological constructs. Nonetheless, to some degree individual value systems have been found to be affected by experiences from spaceflights (Suedfeld 2006) and polar expeditions (Leon et al. 2011). Suedfeld (2006) conducted a pilot study, involving 12 astronauts, to provide a quantitative content analysis of how the completion of space missions affected their personal values. He found that in the period after the space experience, Achievement decreased in importance, as did Enjoyment and Benevolence; in contrast, Transcendence became by far the dominant value. Transcendence is an aspect of Universalism in the value system of Schwartz. Similar changes were found in content analyses of the autobiographies of four early astronauts (Suedfeld 2004). Leon and her colleagues (2011) demonstrated change in value hierarchies among participants in a two-man North-Pole expedition team. These studies point to the relevance of values for understanding the psychological reactions to life in extreme environments.

8.3 Value Congruence, Cohesion, and Groupthink

Extensive studies of humans in a variety of organizational settings and national cultures indicate that values relate to preferences in areas including leadership, appropriate gender relationships, structuring of tasks, sources for work motivation, and self-reliance (Hofstede 2001). Anecdotal (Burrough 1998) and empirical (Santy and Holland 1993; Lozano et al. 1996) evidence underscore that differences in these areas have been a source of disagreements that have strained the ability of international space crews to mount an optimal, unified performance. *Person-job fit*, the match between the technical skills, knowledge, and abilities of a person and work tasks inherent in the job, has received extensive attention in astronaut selection since the early days of human spacefight. With longer duration missions and increasingly heterogeneous crews, the importance of fit between crewmembers and their teammates become more salient. *Person-group fit* refers to the compatibility between individuals and their work groups (Werbel and Johnson 2001; Kristof-Brown et al. 2005). Sharing beliefs and values (i.e., supplementary person-group fit) is theorized to facilitate effective interpersonal interactions (Kristof-Brown et al. 2005). Sarris (2008) found that subjective fit with station culture predicted satisfaction with station membership among Antarctic expeditioners. Results from one space station simulation study suggested that cultural differences in values played a role in many misunderstandings and conflicts that took place within and between two confined crews; both national and organizational factors were implicated. This resulted in a breakdown of cohesion, and group rancour that occurred during the mission affected not only the isolated crews but also the participating agencies (Sandal 2004; Tomi 2001). However, one inherent limiting factor of studies involving small sample sizes is that we cannot isolate the effects of culture from individual idiosyncrasies. Mindful of this consideration, we solely focus on values at the individual level in our study.

While interpersonal tension is an expected correlate of prolonged confinement, cohesion sometimes increases over time as people adjust to one another. In a study of seven men and women participating in a 3-week Arctic scientific expedition, Palinkas and his colleagues (1995) reported significantly higher tension levels prior to the start of the mission than during the mission itself, where the crewmembers seemed to adapt to their situation. Similarly, in a 135-day Mir space station simulation study, Kanas and his colleagues (1997) found significantly less tension during the last half of the isolation than during the first half although other researchers (Sandal 2001) failed to demonstrate this effect. Sharing a unique experience and mutual excitement over the mission are factors that seem to enhance crew member communication in space (Kelly and Kanas 1992). Interestingly, researchers have noted that in the process of cohesive group formation, crewmembers begin to regard each other as very "similar" or "close" in terms of sharing common values and beliefs (Gushin et al. 1997a). Simultaneously, those crewmembers who do not perceive themselves as close and who do not attempt to understand or share common group values run the risk of becoming a "stranger" or an "alien", and to become socially excluded from group activities (Sandal 2001; Palinkas 1989; Sandal et al. 1995).

Social pressure is essential to establish a set of agreed rules capable of influencing the team members' behaviour and facilitating crew adaptation and coordination. However, too much cohesion can also be a threat to the mission. A concern is that highly cohesive crews during long-duration missions are vulnerable to a phenomenon known as "groupthink" (Janis 1972) which is characterized by illusions of invulnerability, reluctance to express concerns or disagreement about decisions, pressure towards conformity, and stereotyped views of people outside the group. "Groupthink" has been associated with low quality performance, in particular under decisional stress, since members are too concerned with getting along and are reluctant to express disagreement. It therefore represents a risk to the safety of space missions. While we are not aware of studies on "groupthink" among space crews, evidence exists from ground based operations. For example, Ferraris and Carveth (2003) performed a textual analysis on reported information made public during the investigation of the explosion of the NASA Columbia shuttle, and concluded that the faulty decision making by the management team was an expression of "groupthink". It has also been advocated that "groupthink" was involved in the Challenger space shuttle disaster more than 15 years earlier.

A comparatively large body of research is available showing that workgroup performance might profit from a heterogeneous team composition (Shore et al. 2009). The creative tensions associated with diversity may encourage mutual inspiration and ensure richness of input that in turn may facilitate problem solving when dealing with complex tasks. This capacity is likely to gain increasing importance as crews become more autonomous such as during a future mission to Mars.

How does diversity in values relate to crew cohesion and psychological compatibility during long-duration missions? This was the main question we addressed in our study as part of the Mars 105 simulation. We were also interested in investigating the extent to which crew members either adopt more similar personal values over time leaving them more vulnerable to "groupthink" or that the group disintegrates and tension between cliques evolves. Perceived homogeneity in values does not necessarily correspond to actual similarities. Whereas the former aspect seems most relevant to crew cohesion, the latter aspect may be more important for the crew's problem solving capacity. In our study, we considered both aspects. Finally, we were interested in the question to what extent crew members view interpersonal tension as being associated with value heterogeneity. Based on the literature review three hypotheses were tested: (a) Members of subgroups within the crew show similarities in personal values; (b) Crew members show increasing convergence of values over time; and (c) Individual differences in values become less salient as sources of intra-crew tension over time. We used a combination of quantitative and qualitative methods. This mixed-methods approach enabled us to combine the rigor of statistical methods with the richness in detail of qualitative methods.

8.4 The Mars 105 Study

The Mars 105-day confinement study was conducted at the Institute of Biomedical Problems (IBMP) in Moscow in 2009. This was one of six simulation studies aimed to investigate psychological aspects of long-duration space missions sponsored by the European Space Agency since 1990. The Mars 105 study preceded a full simulation of a mission to Mars, lasting for 520 days, conducted the following year. The multinational crew consisted of crew of six men: two Russian cosmonauts (one was the Commander), a Russian medical doctor, a Russian sports physiologist, a German mechanical engineer, and a French airline pilot. Their age ranged from 25 to 41 years. Three of the Russian crew members worked at the IBMP. Three of the Russian crew members had limited English skills and the two West European subjects had limited Russian skills. Four crew members were either married or in a long term relationship.

The environmental parameters (humidity, pressure, gas composition and temperature) in the chambers corresponded to International Space Station standards. The total volume of the chambers was 500 cubic meters. Crew members had separate sleeping compartments. The crew had to follow a daily schedule that specified work, physical training, recreation and the different experiments (including questionnaires) to be completed each day. They followed a set rotation with a 24 h night shift each 6th day. Sunday was a day off except for the crew member on duty. They were allowed to sleep on the following day until lunch time. The daily diet was fixed, and the crew was not to deviate from this. However, two crew members did not consume all the planned food items, and consequently their weight decreased significantly. Therefore, these crew members received alternative food after day 45. The French and the German crew members received news on a daily basis by video-clips from television news, whereas the Russian crew members received news in a written format three times each week. The mission was separated into three 35-day cycles, and after each cycle the research protocol was repeated. A high autonomy phase was implemented in the last 35 days. During the first two-thirds of the mission, the crew interacted in real time and as frequently as they wanted with outside monitoring staff ("mission control"), but during the last third (from day 70) they had more responsibility for monitoring and planning their own activities and experienced a 20 min communication delay with the outside (one direction), such as they might expect on the surface of Mars (a "high autonomy" condition). From day 70 no supplies from the outside was provided to the crew in the chambers. Communication with the mission control was in Russian.

8.5 Value Change over the Course of the Confinement

To monitor change in values over time, we asked the crew to answer the Portraits of Crew Values Questionnaire (PCVQ) once every month throughout the confinement period. The questionnaire was answered once 14–16 days before the start of the confinement, and once 8 days following the termination. All materials were provided in Russian and English. The PCVQ consists of three parts: *Part 1. Portrait Values Questionnaire (PVQ)* (Schwartz 2006b) was used to measure personal values. A short version of the PVQ was used that includes short verbal portraits of 21 different people, gender-matched with the respondent. Each portrait describes a person's goals, aspirations, or wishes that point implicitly to the importance of a value. For example: "Thinking up new ideas and being creative is important to him. He likes to do things in his own original way" describes a person for whom self-direction values are important. "It is important to him to be rich. He wants to have a lot of money and expensive things" describes a person who cherishes power values. For each portrait, respondents answer: "How much like you is this person?" on a six-point response scale from "very much like me (6)" to "not like me at all (1)." Respondents' values are inferred from their self-reported similarity to people described implicitly in terms of particular values. Studies in numerous countries

Table 8.1 Definitions of personal values

Value type	Defining motivational goal
Self-direction	Independent thought and action-choosing, creating, exploring
Stimulation	Excitement, novelty, and challenge in life
Hedonism	Pleasure or sensuous gratification for oneself
Achievement	Personal success through demonstrating competence according to social standards
Power	Social status and prestige, control or dominance over people and resources
Security	Safety, harmony, and stability of society, of relationships, and of self
Conformity	Restraint of actions, inclinations, and impulses likely to upset or harm others and violate social expectations or norms
Tradition	Respect, commitment, and acceptance of the customs and ideas that one's culture or religion provides
Benevolence	Preserving and enhancing the welfare of those with whom one is in frequent personal contact (the 'in-group')
Universalism	Understanding, appreciation, tolerance, and protection for the welfare of *all* people and for nature

demonstrate the reliability and validity of the values measured with the PVQ (Schwartz et al. 2001). Table 8.1 provides definitions of each value construct in terms of its central goal and lists exemplary items. *Part 2 and 3* of the PCVQ were designed for this study. Each part consists of a description of the values as defined in Table 8.1, with the exception of universalism, which was considered to be less relevant for group dynamics of confined crews. For each value, the crew member was asked to indicate how much he felt that the members of the crew differed from each other (part 2), and how much he felt that differences between crew members contributed to interpersonal tension within the crew (part 3). Answers were given on five point scales (not at all, a little, some, quite a bit, and very much).

Our first hypothesis dealt with one symptom of "groupthink", namely the tendency of crew members to regard each other as becoming more similar over time in terms of sharing the same values. More specifically, in operational terms the hypothesis predicts a decrease in mean scores and a reduction in standard deviation on the value diversity scales (PCVQ part 2). The second hypothesis dealt with another aspect of "groupthink", a concern about maintaining harmony within the crew at all costs. This would imply that individual differences in values are becoming less salient as perceived sources of tension among the crew members over the course of the confinement (PCVQ part 3). We tested these hypotheses by using multiple regression procedures. The independent variables were 6 measurement times (coded as 1 to 6). Dependent variables were scores on perceived differences (homogeneity) (PCVQ part 2) and tension attributed to value differences (PCVQ part 3). We also used standard deviations as dependent variables; more specifically, we computed standard deviations of the perceived homogeneity and tension per time point. The distribution of the perceived differences (PCVQ part 2) shows a bias toward positive values, which means that individual differences in values were seen by crew members as increasing and homogeneity as decreasing over the course of the confinement. The standard

Table 8.2 Mean value scores across time points and correlation between values and time point

Value	Average	Correlation
Achievement	−0.79	−0.89*
Benevolence	0.76	−0.81*
Conformity	−0.31	0.50
Hedonism	−0.11	0.52
Power	−1.26	−0.46
Security	−0.53	0.69
Self-direction	0.65	−0.61
Stimulation	0.98	0.76
Tradition	0.25	0.60
Universalism	0.42	−0.85*

$*p < 0.05$.

deviations do not show the expected decrease over time either. Taken together, these results do not support the first hypothesis; if anything, the data suggest the opposite pattern in which individual differences are seen as increasing over time.

The picture for tension attributed to value diversity (PCVQ part 3) also suggests salient differences between expected and obtained regression coefficients. More specifically, there are three strong, positive (and indeed significant) regression coefficients, indicating that tension attributed to three values (benevolence, hedonism, and tradition) increased over time (we come back to these changes in the qualitative part of the results). Finally, the standard deviations of the perceived tension scores did not show any decrease either. Thus, our second hypothesis was not confirmed.

We also analyzed individual values before and during the mission based on data from PVQ. We found slightly more strong increments and decrements in values than would be expected if values were unrelated to time. The changes we found were meaningfully related to events during the mission. For example, emphasis on hedonism among crewmembers having problems with the food went up over time whereas the benevolence of some crewmembers decreased over time. Table 8.2 presents the average scores of personal values across time and the correlation between the value and time. It was found that aggregated across all crewmembers, achievement, benevolence, and universalism decreased over time.

8.6 Qualitative Analysis (Post-Mission Interview)

Semi-structured interviews, each lasting for approximately 1.5 h, were conducted 2 or 3 days following the termination of the confinement. The interview guide was constructed based on previous simulations (Sandal 2004). A translator was used during interviews with three Russian crew members who were not fluent in English. Interviews were recorded and transcribed. The Nvivo 8 (QSR International 2009) software package was used for organizing and analyzing the qualitative data from the interviews. The transcription was done by two research assistants blind to the research hypotheses. The analytic approach was based on template analysis (Nadin and Cassell 2004) that involved a classification of responses into predefined categories or themes central to the research questions. New categories were created to capture information that did not fit any of the established categories. Finally, consistencies and discrepancies in the reports of crew members were registered. Below we describe the themes that came up as part of the interviews.

8.6.1 Subgroups

The crew members were asked about whom they felt closest to within the crew. Across the five interviews, there was strong agreement that crew members could be divided into three distinct subgroups that we refer to as A, B, and C. These subgroups cut across nationalities. An inspection of individual value scores on PVQ revealed similarities between members of the same subgroups. The two members of *subgroup A* shared a stronger emphasis on hedonism. One member of subgroup A obtained the highest scores on power among all six subjects suggesting that he possessed the strongest motivation for dominance. The two members of *subgroup B* shared a strong emphasis on tradition, conformity, and benevolence. *Subgroup C* was less distinct in terms of values although there was tendency that both subjects scored low on conformity and tradition relative to the four other crew members.

On the whole, low levels of intra-crew tension were described. Several crew members expressed that they had been conscious about avoiding confrontations. Nonetheless, sources of interpersonal tension described by the majority of the subjects involved differences in acceptance of the diet, work coordination, and the leadership of the Commander. We shall review some of the comments made by crew members in each of these areas.

Dissatisfaction with the diet was presented as a major stressor for crew members, in particular members of subgroup A. One commented: "For me it was a big stressor. And all the time, especially the first month, I was hungry all the time. And angry." This situation was described as having a negative impact on interactions between members in the crew as a whole. "For me it was just difficult to work with some of the others because I thought they were moody. (…) I really felt that food had a strong impact." Some negative impacts were also described after members of subgroup A was offered a different diet, around day 45 of the confinement. One of them noted that: "It was bad food…. And when we lost more than 10 kg, our chiefs changed our meal plans and added some food for us. It was trouble because other crew members looked at us with hungry eyes and we tried to eat separately from our crew members." However, another subjected commented that: "In the last part of the study they had more food and it was much better for all of us I think."

8.6.2 *Obedience to Instructions*

Several incidents of tension were described that involved differences between crew members in compliance to written instructions. One of them noted that this could be due to their cultural backgrounds: "It was in some situations when Earth, gave us some instructions, and Russians made their own way to how it was correct. But Europeans followed the instructions of the Earth like blind." Differences in professional experience were emphasized by one crew member: "For cosmonauts it's easy, because we are used to work with onboard documentation. It is a rule on board that if you are working according documentation it's not your fault if something is going wrong."

Being meticulous, organized, and committed to procedures appeared to be distinct characteristics of members of subgroup B. One of them commented that: "He thinks very slowly because he is very organized. And like this he cannot always be in time. So I was nearby and tried to help him. I think that we were in a symbiotic relation." Several incidents of tension between this subgroup and other crew members were described. One of the subjects in subgroup B described his frustration with what he described as a lack of responsibility on the part of the Commander for work coordination (for example, cleaning), and for giving priority to personal entertainment in front of work.

8.6.3 *Relationships with the Outside*

The relationship between crewmembers and Mission Control personnel was generally described as good with no major conflicts. Nonetheless, several situations were described in which the Mission Control personnel were viewed as not being sufficiently sensitive to the specific demands and needs of a crew, or not providing justifications for their decisions. One crew member commented:

> In the crew we had…one doctor, one doctor in physiology, one captain in the German army, one airline captain. We are a population that are responsible and can do a lot of things as long as we know why. We do not just do things.

Two crew members expressed frustration about the response of the Mission Control when the crew asked for permission to change the diet for subjects who were loosing weight. The introduction of increased autonomy and less frequent contact with the Mission Control was described by the majority of crew members as a positive shift. Sleep deprivation experienced by some crew members became less of an issue in the autonomous phase because those crew members who coped by sleeping in the afternoon (3 pm–5 pm) were woken up less frequently by the phone.

8.7 What Have We Learned?

Researchers have voiced concerns that the high autonomy and isolation that crews will experience during a mission to Mars may lead to the development of "groupthink" which can seriously impair crew performance and interactions with the ground. This chapter addressed two aspects of "groupthink." Firstly, we examined the extent to which crew members adopt more homogeneity in values over time. Secondly, we investigated whether individual differences in values become less salient as perceived sources of tension. Our analysis of data from Mars 105 revealed no such effects. On the contrary, crew members tended to regard each other as more different in personal values with time. Also, interpersonal tension attributed to value differences increased significantly over time. This was in particular the case with regard to hedonism, tradition, and benevolence. Interestingly, similarities and differences in the two former values also seemed to play a role in the formation of the subgroups within the crew.

8.7.1 Convergence in Values and Intra-Crew Tension

Interpersonal tension attributed to value differences showed a marked increase around the time when the crew was allowed higher autonomy. During these last 5 weeks of the confinement they had more responsibility for monitoring and planning their own activities and also experienced a 40-min two-way communication delay with the outside. It is important to note that our results do not necessarily mean that intra-crew relationships in general became less harmonious over time. Rather the results may suggest that individual differences became more salient when strains from outside factors were reduced. However, we shall also discuss other possible contributing factors.

Several crew members reported having sleeping problems in the initial period of the confinement. These difficulties decreased as crew members adapted to the living and working conditions, such as continuous noise and the work rest-cycle. Dislike of the food, and subsequent reduction in food intake of two crew members was presented as a major stressor in the first third of the confinement by all crew members. This situation was reported to have a strong negative impact not only on the mood and well-being of the two crew members but it also created tension within the crew as a whole and between the crew and the Mission Control. As an attempt to resolve this situation, a different diet was offered only to these crew members. Interestingly, we observed a significant rise in tension attributed to individual differences in hedonism and benevolence in our data collected in the measures following this event. One possible explanation indicated by crew members in the post-mission interview, is that differential treatment of crew members caused irritations and resentment.

Data from the post-confinement interviews suggested that the crewmembers' perceptions of stress decreased when the crew was allowed greater autonomy. Several subjects indicated rising tension and frustrations in relation to Mission

Control before the high autonomous conditions were implemented, and described the reduction in the contact with the outside as "a relief." Most crewmembers described the atmosphere within the chambers as being "calmer" during the high autonomy condition, which allowed them to focus more on their work. Likewise, observations by other investigators suggest that increased autonomy was well-received by the crew and that mission goals were accomplished (Kanas et al. 2010). This is in line with previous research showing that during long-term conditions of isolation and confinement, crewmembers want to separate from the influences of outside monitoring personnel (Gushin et al. 1997b). Although evidence from space is largely anecdotal, strict adherence to timelines has been reported to be fatiguing and stressful for some astronauts during spaceflights (Kanas and Manzey 2008).

A frequent observation has been that crew members express apparently irrational anger at people monitoring the mission which has been interpreted as a displacement of intra-crew tension and negative dysphoric emotions to safer, more remote individuals on the outside (Gushin et al. 1997a; Nadin and Cassell 2004). A matter of speculation is whether reduced contact with the outside world would increase interpersonal tension between crewmembers. It is noticeable that the average individual scores on benevolence decreased significantly during the autonomy period suggesting that crew members put less emphasis on preserving and enhancing the welfare of others. Thus growing crew autonomy may be an important factor for intra-crew relationships during long-duration missions.

While the introduction of the higher autonomy represents one possible explanation for the salience of individual differences as source of tension, another contributing factor might be that the crew foresaw the termination of the confinement. Analysis of communication during previous simulation studies suggested a rise in negative emotional expressions between crew members towards the end of the missions (Sandal et al. 1995). Whereas intra-crew communication at the beginning of the "mission" appeared to be quite normal with many bilateral interactions among the different crew members, it changed dramatically throughout the course of confinement. At the end of these confinement studies, most of the bilateral communications between crewmembers had broken down, almost all communication went through the commander of the crew, and one of the subjects was totally isolated. Even though other results reported in the literature do not always show the same strong effects, at least episodes of latent or open hostility between crew members, and tendencies to form cliques have usually been observed after some time in analogue environments and during spaceflight (Kanas and Manzey 2008; Sandal 2001; Sandal et al. 1995).

8.7.2 *Values and Interpersonal Compatibility*

Individual differences in tradition, hedonism, and benevolence not only appeared to cause more interpersonal tension over time, but the two former values also seemed to play important roles in the formation of subgroups. Information from the post-mission interviews showed three distinct subgroups. Members comprising one subgroup (A) shared a strong motivation for hedonism (pleasure and sensuous gratification). The refusal of these crew members to eat certain food products, and their subsequent weight loss, is consistent with behavioural expressions associated with hedonism. Their forceful reaction to their dislike of the food, that ultimately necessitated a change in the predetermined diet, may to some extent reflect the strong power motive of one of these crew members (indicated by scores on the PVQ). A second subgroup (B) shared conformity values that may have been expressed through a strong emphasis on structure and commitment to task accomplishment according to specified standards. Several incidents of tension between members of this subgroup and other crew members were reported, and may be reflected in the rise in interpersonal tension from individual differences in tradition/conformity observed in the last third of the confinement. For example, one of them accused the Commander for not taking sufficient responsibility for the hygienic standard within the chambers and the coordination of tasks between crew members. The third subgroup (C) was less distinct in terms of value congruence on the PVQ, but they were noticeable lower on conformity and tradition than subgroup B. Their psychological compatibility may primarily be related to similarities in personality attributes.

While a concern has been that clique formation in multinational crews will occur along national/cultural/language lines, it is noticeable that members of two subgroups (B and C) were of different nationalities. One of them consisted of members with limited abilities to communicate in a common language. Within the crew as a whole, only one of the Russians spoke fluent English and the Russian language skills of the German and French members were limited. Although fluency in a common language has been claimed to represent a basic requirement for interpersonal compatibility (Kanas and Manzey 2008), such differences were not presented as a challenge for crew interaction in this study. Astronauts and cosmonauts have endorsed the need for a common language during space missions, and language differences have been found to affect crewmember compatibility in space analogue environments on Earth (Kelly and Kanas 1992). For example, language differences were implicated in some of the group disintegration that was observed during the SFINCSS space simulation study. The sensitivity to language problems may be stronger when intra-crew tension is high. That such effects were not observed in this study might be due to that personal sympathies were sufficiently strong to cross-cultural and linguistic barriers.

8.8 Conclusions

Caution is required in generalizing these results due to the small sample size and the limitations connected with ground based simulations. Clearly, simulation studies cannot completely reproduce the living conditions experienced by astronauts and cosmonauts in space. For example the experiment took place in full gravity with no worries about radiation exposure from deep space. One must also question whether the dynamics among people who are cognizant that they are psychological subjects in a study and who have the ability to leave the chambers on short notice are comparable to the psychological reactions of crews during long-duration space missions. Despite these possible limitations, however, simulations have allowed the identification of behavioural issues observed during actual spaceflights and among personnel operating in other isolated and confined settings. A further strength of this study was that the crew consisted of two trained cosmonauts and two others had been candidates for the recent astronaut selection campaign in Europe. Thus, they may be relatively representative for the population from which space crews are drawn.

Our study as part of the Mars 105 confinement makes several contributions to the space psychology literature. Specifically, we suggest that values become more important for interpersonal compatibility with increasing mission duration and autonomy. Future space simulation studies should disentangle the relative impact of these factors, for example by introducing high autonomy conditions earlier in the confinement. Our findings suggest that convergence in values should be considered in crew composition in order to enhance team performance and smooth interpersonal interactions during long-term space missions. Nonetheless, removal of all potential sources of interpersonal tension may not be an optimal goal due to the risk for "groupthink." Although we did not find evidence of "groupthink" in this confinement study lasting for 105 days, we cannot exclude the possibility that this phenomenon will occur on a future mission to Mars that is likely to last for a much longer time. As the next step, further validation of our findings will emerge from the Mars500 study that was recently terminated.

Acknowledgements The authors would like to thank the European Space Agency and the Institute for Biomedical Problems in Moscow for the support of our participation in the Mars 105 study. They also express their gratefulness for the contributions of the crew members.

References

Burrough, B.: Dragonfly: NASA and the Crisis Aboard Mir. HarperCollins, New York (1998)

Cable, D., Edwards, J.R.: Complementary and supplementary fit: a theoretical and empirical integration. J. Appl. Psychol. **89**, 822–834 (2004)

Ferraris, C.L., Carveth, R.: NASA and the Columbia disaster: decision-making by Groupthink? In: Proceedings of the 2003 Association for Business Communication Annual Convention. Albuquerque, New Mexico (2003)

Gushin, V.I., Efimov, V.A., Smirnova, T.M., Vinokhodova, A.G.: Dynamics of subjective perception of in-crew interactions during long-term isolation and confinement. Aerospace Ecol. Med. **31**(4), 23–29 (1997a)

Gushin, V.I., Zaprisa, N.S., Kolinitchenko, T.B., Efimov, V.A., Smirnova, T.M., Vinokhodova, A.G., et al.: Content analysis of the crew communication with external communicants under prolonged isolation. Aviat. Space Environ. Med. **68**, 1093–1098 (1997b)

Hofstede, G.: Culture's consequences. Comparing values, behaviors, institutions, and organizations across nations, 2nd edn. Sage, Thousand Oaks (2001)

Janis, I.L.: Victims of Groupthink. Houghton Mifflin, Boston (1972)

Kanas, N., Manzey, D.: Space Psychology and Psychiatry, 2nd edn. Springer, Dordrecht (2008)

Kanas, N., Weiss, D.S., Marmar, C.R.: Crewmember interaction during a Mir space station simulation. Aviat. Space Environ. Med. **10**, 969–975 (1997)

Kanas, N., Saylor, S., Harris, M.A., Neylan, T., Boyd, J., Weiss, D.S., et al.: High versus low crewmember autonomy in space simulation environments. Acta Astronaut. **67**, 731–738 (2010)

Kelly, A.D., Kanas, N.: Crewmember communication in space: a survey of astronauts and cosmonauts. Aviat. Space Environ. Med. **63**, 721–726 (1992)

Kristof-Brown, A.L., Zimmerman, R.D., Johnson, E.C.: Consequences of individuals' fit at work: a meta-analysis of person-job, person-organization, person-group, and person-supervisor fit. Pers. Psychol. **58**(2), 281–342 (2005)

Leon, G.R., Sandal, G.M., Fink, B., Ciofani, P.: Positive experiences and personal growth in a two-man north pole expedition team. Environ. Behav. **43**, 710–731 (2011)

Lozano, M.L., Wong, C.: Concerns for a multicultural crew aboard the International Space Station, CSERIAC Gateway (1996)

Manzey, D.: Study of the survivability and adaptation of humans to long-duration interplanetary and planetary environments. ESA/ESTEC, Nordwijk (2003)

Nadin, S., Cassell, C.: Using data matrices. In: Catherine Cassell and Gillian Symon (eds) Essential guide to qualitative methods in organizational research. Sage, London (2004)

Palinkas, L.A.: Sociocultural influences on psychosocial adjustment in Antarctica. Med. Anthropol. **10**, 235–246 (1989)

Palinkas, L.A., Cravalho, M., Browner, D.: Seasonal variation of depressive symptoms in Antarctica. Acta Psychiatr. Scand. **91**, 423–429 (1995)

QSR International: QSR Nvivo Version 8.0.340.0 SP4 (2009)

Sandal, G.M.: Crew tension during a space station simulation. Environ. Behav. **33**(1), 134–150 (2001)

Sandal, G.M.: Culture and crew tension during an International Space Station simulation; results from SFINCSS'99. Aviat. Space Environ. Med. **75**(1), 44–51 (2004)

Sandal, G.M., Bye, H.H., van de Vijver, F.J.R.: Personal values and crew compatibility: Results from a 105 days simulated space mission. Acta Astronautica **69**, 141–149 (2011)

Sandal, G.M., Værnes, R., Ursin, H.: Interpersonal relations during simulated space missions. Aviat. Space Environ. Med. **66**, 617–624 (1995)

Santy, P.A., Holland, A.W.: Looper BS multicultural factors in the space environment: results of an international shuttle crew debrief. Aviat. Space Environ. Med. **64**, 196–200 (1993)

Sarris, A.: Applying organisational theory to isolated, confined and extreme settings. Aust. N. Z. J. Organ. Psychol. **1**, 1–6 (2008)

Schwartz, S.: Basic human values: their content and structure across countries. In: A. Tamayo and J.B. Porto (eds) *Valores e comportamento nas organizacões* [Values and behavior in organizations], pp. 21–55. Vozes, Petrópolis (2006a)

Schwartz, S.H.: A theory of cultural value orientations. Explications and applications. Comp. Sociol. **5**, 137–182 (2006b)

Schwartz, S., Bilsky, W.: Toward a theory of the universal content and structure of values: extensions and cross cultural replications. J. Personal. Soc. Psychol. **58**, 878–891 (1990)

Schwartz, S.H., Melech, G., Lehmann, A., Burgess, S., Harris, M.: Extending the cross-cultural validity of the theory of basic human values with a different method of measurement. J. Crosscult. Psychol. **32**, 519–542 (2001)

Shore, L.M., Chung-Herrera, B.G., Dean, M.A., Ehrhart, K.H., Jung, D.I., Randel, A.E., et al.: Diversity in organizations: where are we now and where are we going? Hum. Resour. Manag. Rev. **19**, 117–133 (2009)

Suedfeld, P.: The impact of outer space on inner space. Aviat. Space Environ. Med. **75**(7), C6–C9 (2004). Suppl

Suedfeld, P.: Space memoirs: value hierarchies before and after missions—a pilot study. Acta Astronaut. **58**(11), 583–586 (2006)

Tomi, L.: The role of cross-cultural factors in long-duration international space missions: lessons from the SFINCSS study. In: Baranov, V.M. (ed.) Simulation of extended isolation: advances and problems. Firm Slovo, Moscow (2001)

Werbel, J.D., Johnson, D.J.: The use of person-group fit for employment selection: a missing link in person-environment fit. Hum. Resour. Manag. **40**(3), 227–240 (2001)

Part III
Cross-Cultural Dimensions of Space Exploration

Chapter 9
Psychology and Culture During Long-Duration Space Missions

Nick Kanas, Gro Mjeldheim Sandal, Jennifer E. Boyd, Vadim I. Gushin, Dietrich Manzey, Regina North, Gloria R. Leon, Peter Suedfeld, Sheryl L. Bishop, Edna R. Fiedler, Natsuhiko Inoue, Bernd Johannes, Daniel J. Kealey, Norbert O. Kraft, Ichiyo Matsuzaki, David Musson, Lawrence A. Palinkas, V.P. Salnitskiy, Walter Sipes, Jack Stuster, and Jun Wang

---

Adaptation of Kanas et al. (2009), reprinted with permission from Elsevier. This chapter is a shortened version of a study group report sponsored by the International Academy of Astronautics. The complete study group report was printed by the DLR Institute of Aerospace Medicine, Cologne, Germany as: Kanas (2009).

N. Kanas (✉) • J.E. Boyd
University of California/San Francisco, and VA Medical Centre, San Francisco, USA
e-mail: nick.kanas@ucsf.edu

G.M. Sandal
University of Bergen, Norway

V.I. Gushin • V.P. Salnitskiy
Institute for Biomedical Problems, Russia

D. Manzey
Berlin University of Technology, Germany

R. North • W. Sipes
NASA Johnson Space Center, USA

G.R. Leon
University of Minnesota, USA

P. Suedfeld
University of British Columbia, Canada

S.L. Bishop
University of Texas Medical Branch, Galveston, USA

E.R. Fiedler
Aerospace Psychology Consultants, Tucson, AZ, USA

N. Inoue
JAXA Tsukuba Space Center, Japan

B. Johannes
German Aerospace Center (DLR), Germany

D.J. Kealey
Centre for Intercultural Learning, Canada

N.O. Kraft
NASA Ames Research Center, USA

D.A. Vakoch (ed.), *On Orbit and Beyond*, Space Technology Library 29,
DOI 10.1007/978-3-642-30583-2_9, © Springer-Verlag Berlin Heidelberg 2013

Abstract The objective of this chapter is twofold: (a) to review the current knowledge of cultural, psychological, psychiatric, cognitive, interpersonal, and organizational issues that are relevant to the behavior and performance of astronaut crews and ground support personnel and (b) to make recommendations for future human space missions, including both transit and planetary surface operations involving the Moon or Mars. The focus will be on long-duration missions lasting at least 6 weeks, when important psychological and interpersonal factors begin to take their toll on crewmembers. This information is designed to provide guidelines for astronaut selection and training, in-flight monitoring and support, and post-flight recovery and re-adaptation.

9.1 Introduction

During the early years of human space exploration, the impact of psychological and interpersonal factors on astronaut behavior and performance was minimal. Missions were relatively short-term and were operated by one space organization, and crews were homogeneously comprised of males from a similar national and professional background. However, with the advent of on-orbit space stations, missions became longer, and crews became more multinational and heterogeneous in terms of gender, cultural background, and professional training. Therefore it was not surprising that anecdotal reports began to hint at the existence of psychosocial problems in space crews in both the American and Russian programs (North et al. 1991; Ball 2001; Kanas and Manzey 2008). For example, during one Salyut mission a visiting cosmonaut from Czechoslovakia felt socially isolated and complained of being restricted from doing productive work by his Russian crewmates, who were concerned that this foreign "guest" might inadvertently make an operational error. Similarly, a Russian cosmonaut reported discomfort in the crew environment when a visitor from France came to the Salyut, even though the crewmembers all got along during pre-flight training (Kanas and Manzey 2008; Fig. 9.1).

I. Matsuzaki
University of Tsukuba, Japan

D. Musson
McMaster University, Canada

L.A. Palinkas
University of Southern California, USA

J. Stuster
Anacapa Sciences, USA

J. Wang
Astronaut Center, China

Fig. 9.1 Long-duration space travel can produce a sense of isolation and separation from family and friends. One can feel insignificant in the cosmos, with resultant loneliness and homesickness. This plate is from the first great star atlas, *Uranometria*, by Johannes Bayer, which was first published in 1603. It depicts the constellation Bootes, the herdsman, who was placed in the heavens by the ancient Greeks. In a sense, he became one of the first "astronauts" in space (Courtesy of *Star Maps: History, Artistry, and Cartography*, N. Kanas, Springer-Praxis, 2009)

But long-duration stays aboard an orbital space station represent but one step in the evolution of human space exploration, which also aims at much more ambitious endeavors (e.g., an outpost on the Moon or an expedition to Mars), and these can produce serious psychological and interpersonal consequences. For example, even though some aspects of Mars missions are shared by other settings (e.g., long-duration stays on orbital space stations, historical expeditions to unknown parts of the Earth, wintering-over in Antarctica), there are major differences, mostly associated with the enormous distances involved and the long mission duration (Kanas and Manzey 2008). According to currently discussed mission scenarios, a mission to Mars may last up to 3 years. Depending on the distance between the Earth and Mars and their relative orbital positions, a one-way audio or video transmission between these two planets may take some 22 min to accomplish. As a consequence, no real time two-way communication will be possible. Furthermore, there will be no possibility for emergency re-supply or rescue flights. As a consequence, crews on a Mars mission will be exposed to a much higher level of autonomy and long-term confinement and isolation than any previous space crew. At the same time, most strategies of ground-based support that currently are used to foster crew morale and psychological well-being during orbital space missions will be ineffective during expeditions to Mars. This may not only increase the psychological risks related to individual crewmember performance, but it also will produce new psychological challenges never before experienced. One of these challenges has been referred to as the "Earth-out-of-view phenomenon" and is related to the fact that astronauts traveling to Mars will be the first human beings put in a situation where their home planet will be reduced to an insignificant-looking dot in space (Kanas and Manzey 2008). The human response to such a situation is unknown. However, given the positive value of looking at the Earth for astronaut well-being (Ihle et al. 2006), it seems almost certain that seeing the Earth as yet another dot in space will impact negatively on the psyche of Mars travelers.

This chapter will review the current knowledge of cultural, psychological, psychiatric, cognitive, interpersonal, and organizational issues that are relevant to the behavior and performance of astronaut crews and ground support personnel. Recommendations will be offered to space agencies for future long-duration human space missions.

9.2 The Impact of Cultural Issues

It is important to keep in mind that cultural issues may impact on each of the areas that are discussed below, in several ways. First, differences in personality styles of coping with stress can occur within any space crew, but the effects are more complicated if the crew is also multicultural. This is because some characteristics, such as emotional expressivity, may be common in some cultures but relatively unusual in others. Second, mental health issues may also manifest differentially across cultural groups. For example, depressed mood may be more likely to co-occur with anxiety among Americans but with fatigue among Russians (Ritsher et al. 2007). Third, cognitive and decision-making styles, along with individual behavior norms such as privacy expectations and personal grooming habits, may also vary by culture. Fourth, cultural differences in social behavior norms, such as how hosts are expected to treat guests or whether everyone is expected to socialize together at mealtime, can also impact on tension and cohesion during missions (Kozerenko et al. 1999).

In addition, there are additional psychological issues that are unique to culturally heterogeneous crews. For example, international missions may suffer when there are individual differences in language skills or where crewmembers are in the cultural minority or from a different national culture than mission control (Inoue et al. 2004; Sandal 2004). In addition to national culture, other relevant types of cultural heterogeneity include professional culture, organizational culture, and the unique culture that a particular group forms over time. Even though diversity may represent an asset to the extent that it adds to the crew's behavior and performance repertoire, having to cope with such differences also adds to the overall level of effort required from crewmembers.

A thorough review of known differences between specific sociocultural groups on these dimensions is beyond the scope of this report. Such differences have been well documented by the psychology, sociology, anthropology, and business literatures regarding the general population, and to some extent regarding airline crews, but they are only just beginning to be documented for space station crews (Ritsher et al. 2007; Kraft et al. 2003).

A number of specific cultural issues related to space missions involving humans are imbedded in the four sections that follow (see for example Sects. 9.5.5 and 9.6.3). Our cultural upbringing affects all of us and impacts upon our personality, behavior, cognition, and ways we affiliate in groups. Consequently, it seemed more feasible in this report to integrate culture as an overarching concept into these sections than to expand it into a larger independent section that was separate from the others.

9.3 Personality, Coping, and Adaptation

The success of most human space missions and the numerous examples of complicated tasks that have been accomplished are generally taken as evidence of the ability of most astronauts to perform and cope in space, both as individuals and as teams. Yet there is considerable anecdotal and behavioral evidence that many crewmembers have experienced psychological and interpersonal difficulties arising from the myriad stressors inherent in space missions, especially those involving longer-durations (Shayler 2000). Psychological and behavioral reactions have included lapses of attention, sleeping problems, emotional lability, psychosomatic symptoms, irritability toward crewmates and/or mission control staff, and a decline in vigor and motivation (Suedfeld 2003). Most of these reactions are not specific to spacefarers but also occur in other isolated and confined environments (Rivolier et al. 1988; Sandal et al. 1996; Suedfeld 2005). During space missions, psychological difficulties not only affect individual crewmembers, but they may also affect the entire crew, with potentially disastrous effects on the mission (e.g., a clinically depressed astronaut might be unable to perform required tasks in an emergency situation).

By definition, adaptation presumes an accommodation over time. Observations during Russian long-duration space missions (Myasnikov and Zamaletdinov 1998) and over-wintering in Antarctic research stations (Rohrer 1961; Stuster et al. 2000) have suggested that adaptation may occur in stages over time and that the most serious problems are likely to occur around the third quarter. However, empirical evidence for the existence of specific critical phases has been equivocal (Sandal et al. 1996; Kanas et al. 2006; Wood et al. 2000; Palinkas et al. 2000). Also, recent studies in space have been limited to on-orbit missions with durations of 4–6 months. The issue of critical periods in adaptation is important, given that such knowledge may enable both crewmembers and mission control personnel to prepare for problems and intervene before maladjustments result in operational impact.

9.3.1 Coping Strategies During Short-Duration Versus Long-Duration Space Missions

In the context of long-duration space missions, it may be that the most severe stressors involve monotony and boredom resulting from low workload, hypo-stimulation, and restricted social contacts due to isolation from family and friends. A number of individual factors may impact on the coping ability of crewmembers in dealing with these living conditions. They include: individual experiences, personality, leisure activities, coping strategies, and the kind of social and emotional support available. Also, coping and adaptation are strongly related to interpersonal factors. According to evidence from space and other isolated and confined settings, intra-crew tension, leadership styles, and group dynamics are key factors responsible for exacerbating or ameliorating stress, or facilitating coping and adaptation (Kanas and Manzey 2008; Sandal et al. 1995).

Efficient strategies to ameliorate stress levels during short-term missions may become problematic or impossible during interplanetary voyages. For example, support provided to ISS crews that includes private family conferences via audio or Internet channels in real time will not be possible during a Mars mission due to the communication delays resulting from the great distances involved. Coping based on self-reliance and autonomy will represent important psychological challenges during such missions. There are stable cultural and individual differences in how people habitually deal with stress. Such strategies may be more or less efficient during long-duration isolation and confinement. Although social support may act as a "buffer" against stress, there is anecdotal evidence that the confinement of a small group may lead to openness about a variety of personal matters, and that in some cases such self-disclosure may produce discomfort and regret (Suedfeld 2005). Social support factors operate differently in short-duration missions as compared with long-duration missions. In the latter, the burden of support will have to shift away from the "absent" network of family and friends to members of the present crew. This is an area that needs further attention and research.

9.3.2 What Is "the Right Stuff" Personality?

Is it possible to predict from pre-mission assessments how an individual will adapt and perform during long- duration missions? A number of studies addressing the "right stuff" for long-duration space missions have involved analyses of personality factors as possible predictors of adaptation and performance among personnel operating in space analogue environments, including Antarctic stations (Sandal 2000) and polar crossings (Atlis et al. 2004). Several studies have linked superior performance to a personality profile characterized by a combination of high levels of instrumentality and expressivity along with lower levels of interpersonal aggressiveness (Sandal et al. 1996, 1999; Chidester et al. 1991). Research in which personality traits have been validated against criteria of astronaut effectiveness during short-term space missions and training sessions have indicated high "agreeableness" and low aggressiveness as general characteristics of high performers (Rose et al. 1994). Yet, the "optimal" personality for short- and long-term missions may differ as a function of mission duration. Research conducted among over-winterers in Antarctic stations suggest that the ideal candidates for long-duration missions have low levels of neuroticism (emotional lability), extraversion and conscientiousness, and show a low desire for affection from others (Palinkas et al. 2000). Although the implications of such characteristics for coping and adaptation during long-term space missions have not yet been investigated, studies like these represent a first step in providing empirical data for defining the "right stuff" for such missions. There is also a need to determine how to translate behavioral data collected in other isolated and confined settings to the space program.

9.3.3 Implications for Mission Operations

9.3.3.1 Selection

At the individual level, the objectives of selection strategies are twofold: to eliminate unfit or potentially unfit applicants, and to select from otherwise qualified candidates those who will perform and cope optimally based on their basic aptitudes, personality characteristics, attitudes, and prior experiences. A distinction is therefore drawn between "select-out" and "select-in" criteria. In contrast to select out, select-in criteria need to be developed in relation to specific aspects of the mission, and the weighting of criteria needs to be based on systematic analyses of issues involving mission objectives, duration, and crew composition. In the context of assigning crewmembers for multinational space missions, communication and interpersonal skills, interpersonal compatibility, and cross-cultural competence must be regarded as mandatory. While political and public relations issues will always be contributing factors in assignments for multinational missions, this must not detract from the importance of psychologically-guided crew composition.

Assessment tools used for selection purposes usually include performance tests, personality questionnaires, analyses of biographical data, interviews, and behavioral observations. The combination of several tools is likely to result in the most valid judgment. Particular emphasis should be put on the development and validation of behavioral testing tools (i.e., group exercises, isolation chamber tests) that are suitable to assess interpersonal skills and to predict individual reactions to stress, confinement, and isolation. To date, the absence of formal and valid criteria for astronaut performance, coupled with limited research opportunities, have made it difficult to evaluate the efficiency of crew selection strategies. Such evaluation also requires that select-in criteria not be used in the initial selection process until they have been found to reliably predict astronaut performance. One potential bias in validating selection criteria on astronauts who have already gone through a formal selection process is related to uniformity in the range of personality scores. In multinational pools of applicants, the possible impact of cultural differences in scale usage and response styles also must be considered when personality measures are used in selection. Even though the different space agencies may agree on criteria, a common selection procedure may be difficult to achieve due to differences in preferred methodology.

9.3.3.2 Training

Psychological training should be provided to both space crewmembers and selected mission control personnel in order to prepare members of both groups for coping with psychological issues during long-term missions. General organizational research and studies from aviation suggest that personality may determine the effects of training programs (Chidester et al. 1991). The greatest benefit can be expected if selection is combined with pre-mission training that focuses on further development of the coping abilities of individual crewmembers and crews. Training programs should be implemented by each space agency based on international agreements. These programs should include self-care and self-management, team-work and group living, leadership and followership, and cross-cultural aspects in the preparation for long-term missions. It will be essential for pre-flight training to include a mission simulation that will allow for hands-on training, evaluation, and coaching of the crew under conditions of confinement. An important consideration is that training needs to be tailored to meet the needs of individual crewmembers and specific crews based on differences in personality, cultures, and previous training backgrounds.

9.3.3.3 Monitoring and Support

Remote monitoring of crewmember behavior and performance represents an important basis for the early detection of impairments and for providing ground-based counseling and advice to the crew. During the mission, crewmembers may be reluctant to give information regarding emotional stress and adaptation. Willingness to discuss personal matters may be reflective of personality differences as well as expectations of the privacy of information. Experience from multinational airlines suggests that there are cultural differences in attitudes regarding discussion of personal problems and fallibility. These issues need to be addressed prior to the mission.

For monitoring purposes, subjective reports used in combination with more objective and non-invasive methods may be most useful. Such methods include computerized performance tests, video recordings of crew meetings, and analyses of speech (including voice analysis). So far, such methods have been utilized mostly in research, and more experience from operational applications is needed.

Approaches should take into account the operational constraints of the different mission scenarios. For example, monitoring from the ground, and implementing interventions when problems are detected, might be impossible during future interplanetary missions. During such missions, it will be critical that crews are provided with sufficient training and tools with regards to self-monitoring and resolving their own problems.

During future long-duration on-orbit missions, provision of in-flight support to crewmembers is likely to be an important countermeasure to stabilize emotional state, ensure optimal well-being, and maintain a close contact between space crew and ground. The methods utilized for this purpose will rely on the availability of effective space–ground communication systems (audio–video transmissions) and of re-supply flights that are used for sending support items to the crew. The limited availability of on-site medical or counseling help emphasizes the necessity of accurate and easily available telemedicine/telepsychology consultation. Attention should be given to enhancing individually tailored leisure time activities that take into account changing interests and needs over the course of the mission (Kanas and Manzey 2008; Kelly and Kanas 1994). Other important support activities include private psychological conferences, informal space–ground contact and news from Earth (preferably in the crewmember's native language and from homeland news sources), and opportunities to maintain close contact with family and friends on Earth on a regular basis.

An important issue that needs careful consideration concerns issues related to death, either of a fellow crewmember or of a friend or family member on Earth. An on-board death will affect all of the crewmembers and may best be dealt with at the crew level, perhaps with group counseling mediated by experts on Earth. Bad news from home will primarily affect a single crewmember, and this should be dealt with individually with support from fellow crewmembers. The news should probably be transmitted only after any near-term mission-critical operation has been completed. Providing support for families during the mission can contribute to maintaining the crewmembers' concentration on the objectives of the mission by relieving them of considerations about possible problems at home and feelings of responsibility. In addition, families should be coached in interacting with their in-space family member and be prepared for possible psychological changes during the mission as a result of a psychological crisis.

9.3.3.4 Re-adaptation to Earth

Family support must not be limited to the in-flight phase. Evidence from various sources (e.g., submarines, Antarctica) has indicated a persistent incidence of reintegration problems for family members absent for long periods of time (Kanas and Manzey 2008). Participation in long-term space missions may have similar psychological effects on the participant, which may make it difficult to re-adjust to daily life on Earth and to reintegrate into "normal" family and work.

9.3.4 Recommendations

Based on the above, a number of recommendations are suggested for space agencies to deal with issues related to personality, coping, and adaptation:

- More attention needs to be paid to human factors in the planning of future long-duration space missions.
- Psychological countermeasures must be implemented before, during, and after space missions that involve crewmembers and their families, as well as relevant ground support personnel.
- To determine the generalization of reported experimental findings for crews in space, the results from analogue studies must be compared with those from actual space missions.
- More empirical work is needed on defining individual characteristics (e.g., personality, attitude, motivation, skills, and coping strategies) and group characteristics that promote optimum coping and adaptation during different kinds of multinational space missions, both short- and long-duration.
- A major challenge in evaluating the efficiency of psychological countermeasures is the use of valid and reliable performance criteria against which they can be tested. The ISS partners should identify and agree on a set of common standards and procedures for the selection, training, support, and evaluation of multicultural crews working in missions involving the ISS, the Moon, or Mars and beyond.

9.4 Behavioral Health and Psychiatry

In the context of space travel, behavioral health issues are normal psychological and interpersonal reactions to the conditions of off-Earth environments. In contrast, psychiatric issues are abnormal responses to these conditions. In a given person, genetic, constitutional, and developmental vulnerabilities may contribute to the presence of psychiatric difficulties. However, operational mission stressors (e.g.,confinement, danger, microgravity, and radiation) and psychosocial factors (e.g., crew tension, cohesion, leadership issues, cultural and language differences) also may play a role and need to be taken into account (Kanas and Manzey 2008).

Before considering these psychiatric issues, it is important to state that isolated and confined environments can also be growth enhancing and salutogenic (Ihle et al. 2006; Suedfeld 2005; Suedfeld and Weiszbeck 2004). For example, people in polar environments or space may experience increased fortitude, perseverance, independence, self-reliance, ingenuity, comradeship, and even decreased tension and depression. Some astronauts and cosmonauts in space have reported transcendental experiences, religious insights, or a better sense of the unity of mankind as a result of viewing the Earth below and the cosmos beyond (Connors et al. 1985; Kanas 1990). In his diary, cosmonaut Lebedev (Lebedev 1988) stated that his Earth photography experiences from the Salyut 7 space station were restful and positive, and he hoped that they would help him gain an advanced degree after he returned from his 211-day mission. Thus, involvement in space missions and related environments can be positive for some people.

9.4.1 *Psychiatric Issues in Space*

Mental health problems affect both genders and occur across cultures (although the way they are manifested may vary from one culture to another). Some of these problems are more frequent than others during space missions (Kanas and Manzey 2008; Kanas 1998). Most commonly reported are adjustment reactions, which are abnormal responses to internal or external stressors. For example, one astronaut beginning a long-duration space mission had symptoms of clinical depression due to the isolation he felt on-orbit and his separation from his wife and family (Kanas and Manzey 2008). These symptoms resolved as he adjusted to his new environment. Psychosomatic reactions also have been reported from space. For example, a cosmonaut wrote in his diary that he experienced tooth pain following anxious dreams he had of a tooth infection (Kanas 1990). Problems related to major psychotic disorders (e.g., bipolar or manic-depressive disorder and schizophrenia) have not been reported during space missions, probably because potential space travelers are well screened psychiatrically for predispositions to these psychotic conditions based on genetic and family history background. However, such severe psychiatric disorders have been reported in astronaut applicants, and they occur in up to 5% of people working in space analogue environments, such as submarines and the Antarctic (Kanas and Manzey 2008; Kanas 1998).

Post-mission personality changes and psychiatric problems also have affected returning space travelers. These have ranged from positive changes, such as new insights into the meaning of life and the unity of mankind, to substance abuse, anxiety, and major depression that have necessitated psychotherapy and psychoactive medications (Kanas 1990). Readjustment difficulties also have been reported to affect the families of people returning from long absences (Isay 1968).

9.4.2 *Asthenia*

According to Russian space psychologists and flight surgeons, asthenia is another important psychiatric condition that may affect people in space. This syndrome is defined as a weakness of the nervous system that may result in fatigue, irritability and emotional lability, attention and concentration difficulties, restlessness, heightened perceptual sensitivities, palpitations and blood pressure instability, physical weakness, and sleep and appetite problems (Kanas 1991; Kanas et al. 2001a). In the Russian space program, asthenia is viewed as a problem that affects most cosmonauts participating in long-duration space missions. Although "asthenization" is carefully monitored, and a number of countermeasures are employed to prevent it from progressing, empirical evidence for its existence as a discrete pathological entity has been equivocal (Kanas et al. 2001a). In addition, although the full syndrome was first described as "neurasthenia" in the late 1800s by the American George Beard, there is controversy as to its existence in the US, and it is not recognized in the current American psychiatric diagnostic system. However, neurasthenic spectrum disorders appear in the international diagnostic system used in Europe, Russia, and China. Thus, if flight surgeons only use the US system, they may under-detect asthenic distress in crewmembers, or label problems from a different perspective than that used in other countries. Indeed, there is some evidence that patterns of mood states are systematically different among astronauts versus cosmonauts (Ritsher et al. 2007).

9.4.3 Implications for Mission Operations

9.4.3.1 Selection

In terms of selection, formal psychiatric examinations, and psychological testing typically occur only at the time when candidates are screened in their application to become astronauts (Santy 1994). The emphasis is on selecting-out people who have a history of mental problems, who have documented psychopathology, or who are likely to decompensate under the stressful conditions characteristic of the space environment. However, for a specialized expedition-type mission (e.g., to Mars or beyond), potential crewmembers may be under special scrutiny due to the long-duration of the mission and to the unusual stressors to be expected. It is likely that some sort of select-out procedure will be conducted specific to this mission. In addition, relevant select-in procedures may need to be utilized that take into account not only personality traits but also interpersonal skills, with the aim of establishing a compatible, cohesive group.

Formal tests for interpersonal compatibility exist and have been used to enhance cohesion in space analogue environments. Examples include the fundamental interpersonal relations orientation—behavior (FIRO-B) test, sociometric questionnaires, and the personality characteristics inventory (PCI) (Kanas and Manzey 2008; Chidester et al. 1991). Careful attention should be given to cultural differences and minority status in crew composition so that an individual does not feel isolated on the basis of cultural background, gender, or work role. Potential crews should be observed in simulation and other relevant group activities prior to launch to test for compatibility and performance. For missions conducted during programs where crewmembers in space are joined by others in a staggered rotation, a record could be kept of each individual's score on interpersonally-oriented tests, and the total score for a prospective crew could be examined for compatibility before additional members are launched into space. Finally, it is essential that all of the crewmembers are fluent in the language that is considered to be common for the mission.

9.4.3.2 Training

Crewmembers and mission control personnel need to be involved in pre-mission training, sometimes together, since these two groups are not only mutually dependent in conducting the activities of the mission, but they also may be involved in maladaptive communication patterns. Potential training topic areas for briefings include: the effects of mission duration on crew tension and cohesion, the relationship between crewmembers and mission control personnel, the impact of cultural differences on interpersonal interactions, and the appropriate use of different leadership roles. Team building experiences in aircraft cockpit simulators using cockpit resource management (CRM) and line-oriented flight training (LOFT) have been useful in exposing inter-personal conflicts and performance problems in flight teams (Helmreich et al. 1990), and these kinds of activities need to be adapted for space missions and evaluated empirically in the future. Crewmembers could be taught techniques such as relaxation training, meditation, biofeedback, or autogenic training to calm themselves and to lower anxious arousal by controlling autonomic functions.

9.4.3.3 Monitoring and Support

In-Flight Monitoring

During future space missions, the interactions of crewmembers in space need to be monitored, and conflicts resulting from psychosocial issues need to be dealt with as they arise. Russian space experts on the ground traditionally have tracked crew-ground audio communications, observed video behavior, and held private conferences to assess crewmember well-being. Formal voice analysis of speech patterns and frequencies has not yet proven to be specific and sensitive enough to be practically useful (Older and Jenney 1975), although recent findings in this area have shown some promise (Johannes et al. 2000). It is important for crewmembers to monitor themselves, especially during an expedition-type mission to Mars, where two-way communications with Earth could be delayed by times of up to 45 min. Although there likely will be behavioral and medical experts on-board who are trained to be sensitive to psychosocial issues, what happens if these people become impaired or are part of the deviant behavior? It is important for all crewmembers to be aware of psychosocial issues and to be trained to respond to them when they begin to impact negatively on crew behavior. Also, NASA doctors are evaluating the use of computerized tests that may help crewmembers assess their cognitive state and their ability to perform certain behaviors at various times during the mission.

In-Flight Support

In past space missions, supportive activities have focused on providing increased novelty and stimulation for space crews during times of monotony or asthenia (Kanas and Manzey 2008; Kanas 1991). Such activities have included delivering surprise presents and food via resupply vehicles, increasing on-board music and lighting, and providing contact with people on Earth in real time. For expedition-type missions, the long distances involved will negate some of these supportive activities, and crews will need to use on-board supplies and facilities for support. Crewmembers should have time set aside to discuss interpersonal differences ("bull sessions") and to debrief problems and other critical incidents that have occurred. Attention should be given to leisure time activities and work/rest schedules that take into account changing interests and needs (Kelly and Kanas 1994). Finally, family members on Earth should be supported while their loved ones are on long-duration space missions, either via formal discussions sponsored by the space agencies or via informal groups led by trained counselors or the family members themselves.

Counseling

Crewmembers should be monitored for symptoms and signs of developing psychiatric disturbances. In near-Earth orbital missions, counseling sessions or crisis intervention can occur between individuals in the crew and therapists on the ground using private two-way audiovisual links. During deeper space missions (such as a trip to Mars), the distance involved will result in communication delays. As a consequence, supportive encounters will depend on the skills of on-board culturally-sensitive crewmembers who are trained in counseling and the use of psychoactive medications. Facilities also need to be available on-board to seclude and restrain a potentially suicidal, violent, or impulsive crewmember.

It is unlikely that a psychiatrist or clinical psychologist will be a member of the crew in early missions involving a lunar base or a trip to Mars. However, it is likely that a physician or some other medically trained person would be a crewmember. This individual should have a knowledge of: (1) individual psychopathology and small group behavior; (2) the individual and interpersonal effects of stressors to be expected during the mission; (3) crisis intervention techniques and the facilitation of group awareness, cohesion, and team-building; and (4) the appropriate use of tranquilizers and other psychoactive medications, including their usefulness and side effects under conditions of microgravity (Kanas and Manzey 2008; Kanas 1998).

Computer-Interactive Intervention Programs

Computer-interactive intervention programs show promise for adaptation to long-duration space flights. A computerized format for presenting prevention and intervention information may be more comfortable for some crewmembers than disclosing highly personal information to others. Outcome research suggests that computer-based intervention programs applying cognitive-behavioral and self-help instruction may be as effective as face-to-face intervention for dealing with mild to moderate depression, anxiety, and other types of psychopathology (Cavanagh and Shapiro 2004; Greist et al. 2002; Leon et al. 1999; Proudfoot et al. 2003).

Psychoactive Medications

Medical kits are available on-board during manned space missions that contain supplies to help the crewmembers cope with space motion sickness, illnesses, and injuries. Psychoactive medications have been part of the formulary as well. For example, Space Shuttle flights have included medications to counter anxiety, pain, insomnia, depression, psychosis, and space motion sickness (Pavy-LeTraon et al. 1997). Santy and her colleagues (Santy et al. 1988) have reported that 78 % of Space Shuttle crewmembers have taken medications in space, primarily for space motion sickness (30 %), headache (20 %), insomnia (15 %), and back pain (10 %). Newer psychoactive medications are being added, such as the so-called "atypical" antipsychotics (e.g., olanzapine and risperidone) and the selective serotonin reuptake inhibitor (SSRI) antidepressants. Prescribing practices and usage rates may vary across cultures, and this issue needs to be studied further in order to better serve the needs of multicultural crews.

Physiological changes due to microgravity and other effects of space may change the pharmacokinetic characteristics of psychoactive medications, influencing both their dosage and route of administration (Saivin et al. 1997). In microgravity, blood flow increases in the upper part of the body and decreases in the lower part. Relative disuse of muscle groups can cause atrophic changes as well. As a result of these two effects, the blood available and the amount of atrophy that has taken place at a possible injection site will influence the bioavailability of medication from an intra-muscular injection. For example, intra-muscular promethazine for motion sickness usually is given in the arm rather than in the hip in space, with good results (Pavy-LeTraon et al. 1997).

Other physiological changes also may affect medication absorption and metabolism. The movement of oral medications out of the stomach may be decreased by the weightlessness of the gastric contents in space, and intestinal absorption rates may be reduced by blood and other fluid shifts to other areas of the body. Fluid shifts may also affect the bioavailability of medications sensitive to the first pass effect in the liver, where metabolism occurs (Saivin et al. 1997). Finally, renal excretion rates also may be influenced by microgravity.

9.4.3.4 Re-adaptation to Earth

Readjustment to life on Earth after a mission needs to be addressed through debriefings at both the individual and the crew level. Some crewmembers may have had unpleasant psychosocial experiences in space that need to be debriefed. For example, a crewmember who was scapegoated during a mission may have unresolved feelings post-flight that may affect future interactions with his or her former crewmates. Some returning space travelers have experienced psychological problems or personality changes as a result of being in space, in some cases becoming more humanistic, religious, or spiritual after observing the oneness of people on Earth or the infinity of the Cosmos (Kanas 1990). Other returning individuals may have difficulty dealing with the resulting fame and glory of their mission, especially during a first-of- its-kind mission such as a trip to Mars. Family reentry also may be difficult. For example, studies have shown that many wives of male submariners learned to adjust to the absence of their sailor husband when he was on sea patrol. However, over half experienced depression or marital strife after he returned and tried to reinsert himself back into the family dynamics, which led to the expression: "submariners' wives syndrome" (Isay 1968). Thus, support activities continue into the post-mission period and need to take into account not only crewmembers and mission control personnel but also family members and loved ones who may have their own adjustment issues.

9.4.4 Recommendations

Based on the above, a number of recommendations are suggested for space agencies to deal with issues related to behavioral health and psychiatry:

- More attention should be paid to identifying and treating possible psychiatric problems that could occur in space, especially during long-duration missions away from the Earth's immediate neighborhood, such as the Moon or Mars.
- More psychosocial research should be conducted in space, especially in the psychiatric area.
- One area of research would be to help define asthenia as a discrete entity in space and determine if its occurrence is dependent upon one's cultural background.
- Research needs to be done on the effectiveness of voice analysis and telemedicine to diagnose psychiatric conditions and to treat them through counseling or computer-interactive programs.
- Cultural factors related to psychotherapy (e.g., need for counselor–patient cultural match) should be explored, both through research and operational interventions.
- Further studies should be performed on the influence of microgravity on the effects and side effects of psychoactive medications, both in space (e.g., the ISS) and in microgravity simulations on Earth, such as bed rest and water immersion.

9.5 Cognition and Complex Performance Skills

The work of astronauts in space usually includes a variety of tasks, like operating complex technical systems (e.g., docking or undocking a spacecraft and controlling robot arms), conducting scientific experiments, or performing specific tasks during extravehicular activities. Such tasks place demands on different cognitive and psychomotor functions and skills. As a consequence, maintaining a high level of performance efficiency in astronauts during their stay in space represents an important pre-condition of success and safety of long-duration missions.

9.5.1 Origins of Performance Disturbances in Space

The lack of the usual gravitational force in space has been shown to affect different brain mechanisms. One particularly important effect is the disruption of congruence between vestibular signals and other (e.g., visual and tactile) receptors, as well as between the vestibular otolith and semicircular receptors caused by the altered signals from the gravity-sensitive otoliths. Another direct effect of microgravity is related to mechanical and proprioceptive changes during the execution of movements, leading to a disruption of the usual relationships among efferent and afferent signals that has been referred to as a state of "sensorimotor discordance" (Bock 1998). Both of these effects can be expected to require complex adaptive processes (e.g., a reweighting of afferent information) and to thus affect the efficiency of cognitive and perceptual-motor skills that have been established on the ground.

In addition to these specific effects of microgravity, the working and living conditions of space can induce stress states in astronauts that in turn can lead to degradations of cognitive and psychomotor performance. Examples of such stress states include decreased alertness and fatigue, high workload, and emotional stress due to interpersonal tension or the long-term effects of confinement and isolation. Performance impairments in these states may result from a shift in the pattern of physiological activation (arousal) into a region that is non-optimal for efficient performance, or from a compensatory performance adjustment aimed at actively coping with stress and high workload by applying less effortful performance strategies (Hockey 1997).

9.5.2 Effects of Microgravity on Cognitive and Psychomotor Functions

The impact of microgravity on cognitive and psychomotor functions has been primarily addressed in experiments focusing on the effects of microgravity-induced changes in the vestibular and sensorimotor system. The main effect of changes in the vestibular system involves a sensory conflict that seems to be the basic mechanism underlying visual illusions of self and surround motions that are induced by head movement disturbances of spatial orientation and space motion sickness (Reschke et al. 1998). Although information from the vestibular system and other graviceptors seems to affect different parts of the visual cortex, higher processes of visual perception and cognition have been found to remain more or less unimpaired in space, or at least do not exhibit overt performance decrements (Leone 1998).

The effects of microgravity on the sensorimotor system involve mechanical and proprioceptive changes, as well as a reduced efficiency of central motor programs which have been established under 1 g conditions. Performance impairments resulting from these changes include a loss of precision of voluntary movements, or a slowing of movement times and changed kinematics during execution of these movements, as compared to 1 g conditions (Bock et al. 2001). Since adaptation to most of the microgravity-induced changes in the sensorimotor system can be achieved very rapidly in space, the emergence of these performance decrements can be expected to be limited to an early flight phase (Manzey 2000).

9.5.3 Maintenance of Cognitive and Psychomotor Performance in Space

Empirical studies addressing the possible effects of spaceflight-related stressors on human cognitive and psychomotor performance have focused primarily on elementary cognitive and psychomotor tasks. Most of them have been conducted during short-term (<30 days) spaceflights (Manzey 2000). In spite of the comparatively small number of studies using different methodological approaches, they have revealed a fairly consistent pattern of effects. Whereas performance in elementary cognitive tasks such as memory search, logical reasoning, or mental arithmetic seem to remain largely unimpaired in space, significant disturbances have been found in perceptual-motor and attentional tasks (Manzey and Lorenz 1998). The origin of these effects (i.e., whether they reflect effects of microgravity on the central nervous system or unspecific stress effects) has remained unclear. Whereas impairments of perceptual-motor performance during an early phase of the mission most likely are related to direct effects of microgravity on the sensorimotor system, most of the other effects seem to be related to non-specific stress effects arising from physiological changes, inadequate work–rest schedules, or sleep disturbances during adaptation to microgravity and the extreme living conditions in the space habitat.

Considering the impact of long-duration missions on perceptual, cognitive, and psychomotor performance, little systematic knowledge is available. A comprehensive performance monitoring study was conducted during a long-term mission involving one Russian cosmonaut during a 438-day stay on-board Mir (Manzey and Lorenz 1998). The results of this study suggest a remarkable resiliency of elementary cognitive and perceptual-motor functions to the impact of spaceflight-related stressors, even during an extraordinary long mission. Impairments of performance, alertness, and subjective well-being were found only during the first

2 weeks in space and back on Earth after the mission (i.e., during critical phases of adaptation to changed gravitational forces, which also were associated with comparatively high workload). After successful adaptation to the space environment, subjective mood and performance functions returned to pre-launch baseline levels and remained stable throughout the remaining 400+ days in space.

9.5.4 Maintenance of Complex Cognitive and Perceptual-Motor Skills

Nevertheless, maintenance of operational efficiency and complex skills may become a serious problem, since decrements in operational task performance have been observed during Russian isolation studies and space missions. For example, Salnitski et al. (1999) investigated the level of performance of cosmonauts in a simulated manual docking maneuver during their stay in space. They found a considerable loss of skill after a period of 3 months in space, which was mainly attributed to a lack of on-orbit refreshment training under changed gravity conditions. Other degradations of operational performance in cosmonauts (i.e., errors made by crew members in conducting mission tasks) have been related to non-specific stress effects arising from disturbances of the usual sleep-wake cycle, high workload, or psychosomatic discomfort (Nechaev 2001). But systematic research addressing acquisition and retention of complex cognitive and perceptual-motor skills during space missions is lacking. Apart from a few early studies of operational efficiency during Skylab and the Russian studies mentioned above, research efforts have been limited to just a few studies that were performed during ground-based simulations of spaceflight (Sauer et al. 1999). The results of these studies did not provide any consistent pattern of effects, and most of them suffered from methodological constraints, such as persistent learning effects throughout the experiment. Other areas of possible concern, like the effects of the space environment on higher order cognitive processes relevant for space operations (e.g., decision making), have not been studied to date.

9.5.5 Performance and Culture

The aforementioned research suggests that the specifics of the space environment and the different stressors astronauts are exposed to can cause detrimental effects on basic cognitive, attentional, and psychomotor processes. Much less is known about if and how these effects are moderated by culture. Concerning the effects of microgravity on performance, it does not seem likely that these effects are different for members of different cultures. However, performance may depend on the cultural background of astronauts in other ways. One example regards the effect of lack of privacy and crowding in a confined environment, like a space habitat. Whether lack of privacy is perceived as a stressor, and thus entails detrimental effects on mood and performance, largely depends on the cultural background of individuals (Raybeck 1991). Similar effects may be assumed for other psychosocial stressors as well, like monotony and boredom, time pressure, or workload. In addition, culture might be a factor influencing higher-order cognitive processes, like decision-making or the use of schemes in information processing.

9.5.6 Implications for Mission Operations

9.5.6.1 Selection

Current practices of psychological selection of astronauts already ensure that only individuals are accepted as astronaut candidates who show comparatively high levels of cognitive capabilities. Given the fact that attentional and psychomotor skills in particular appear to be prone to performance degradations in space, these kinds of capabilities should be given special weight in future selection activities.

9.5.6.2 Training

During pre-flight training, astronauts should be briefed about likely performance decrements in space. In particular, astronauts need to be made aware of the possibility of impaired psychomotor performance during their adaptation to the space environment (e.g., the precision and speed of their arm and hand movements may be impaired). These effects also must be taken into account in the planning and scheduling of motor-tasks during the mission, such as the setting up of experimental hardware. In addition, astronauts should be briefed on the importance of sleep and adherence to work–rest schedules in order to protect themselves against any resultant performance decrements. Rotational training on the ground may also help astronauts learn to adjust to spatial disorientation in space.

9.5.6.3 Monitoring and Support

It has been shown that performance impairments might be expected during adaptation to the space environment and during specific stress states induced by sleep problems, high workload, and/or emotional burden. In order to detect such adverse effects and to provide appropriate support and countermeasures (e.g., re-scheduling of mission tasks and accommodation of work–rest schedules), tools are needed which can discern subtle performance decrements before they escalate. One approach is to monitor the cognitive and psychomotor capabilities of astronauts during their stay in space by repeatedly assessing their performance in specific screening tests, and to compare the results with a self-referenced baseline established during pre-flight training. One example of such a tool that is currently being used on the ISS is the WinScat (Windows-Spaceflight Cognitive Assessment Tool (Kane et al. 2005)). The current focus on cognitive probing tasks should be expanded by including tasks that measure attention and psychomotor functions that have been found to be particularly sensitive to microgravity and stress-related effects during space missions.

With respect to in-flight support and maintenance of performance functions, the currently available results point to the importance of sleep and appropriate work–rest schedules as a significant pre-condition for performance efficiency. In addition, in-flight training of critical perceptual-motor tasks will become more and more important if mission lengths increase. The importance of these factors has been emphasized by the investigations of a severe accident that occurred on the former Mir station when one of the modules was hit and damaged during the manual re-docking of a supply spacecraft. Among the factors that were identified as contributing to this accident were issues of maintaining the operator's docking skills throughout the stay in space and decrements in his performance due to fatigue.

9.5.6.4 Re-adaptation to Earth

Based on the currently available database, no permanent problems are expected with respect to basic cognitive processes and performance during re-adaptation to Earth. The only exception relates to psychomotor skills that might be impaired for a limited period of time due to re-adaptation to Earth's gravity.

9.5.7 Recommendations

Based on the above, a number of recommendations are suggested for space agencies to deal with issues related to cognition and complex performance skills:

- The range of cognitive performance assessment techniques employed in space research has been restricted. To obtain a full description and monitoring of the nature of cognitive and psychomotor deficits encountered in microgravity, an integrated test battery should be developed and applied which not only assesses the behavioral aspects of information processing but also includes subjective and psychophysiological measures (Johannes et al. 2003).
- The relative influence of microgravity stress versus the stress resulting from working in isolated and confined conditions on basic cognitive and psychomotor functions, as well as complex skills, remains unclear. More research is necessary to resolve the comparative influence of microgravity and work-related stress.

- The effects of radiation on cognitive (and psychological) functioning in space have not been adequately determined, and further research is necessary to evaluate these effects during longer-duration missions beyond the Earth's immediate environment.
- Current research in space and space analogue environments has only addressed a narrow range of basic cognitive performance tasks. Examples of issues which have not been studied systematically, but which are of concern for long-duration space flight, include the impact of the space environment on higher-order cognitive processes like decision- making, the impact of culture on performance, and the impact of transient exposure to artificial gravity on mental functions, which will be important if artificial gravity is considered as a countermeasure for future interplanetary space missions (Young et al. in press).

9.6 Interpersonal and Organizational Issues

The ways in which crewmembers interact and function as a group can affect productivity and the accomplishment of mission goals. In order to optimize these interactions, it is important to understand several issues that may interfere with crew cohesion. These include: (1) crew tension resulting from environmental stress and factors related to crew heterogeneity; (2) time-related drops in cohesion, especially during the second half of the mission; (3) poor leadership skills (e.g., failure to support the crew during times of low morale); and (4) stress due to cultural and language differences (Kanas and Manzey 2008).

In addition, during long-duration space missions there is evidence that time may influence the impact of these psychosocial issues. There have been reports that crewmembers in space or in space analogue environments may experience psychological and interpersonal difficulties after the halfway point of a mission (Kanas and Manzey 2008; Sandal et al. 1995; Gushin et al. 1997a). According to this view, a sense of relief that half of the mission is over is outweighed by the realization that another half is yet to come. Another time model incorporates Rohrer's three sequential phases, which include: (1) initial anxiety, (2) mid-mission depression, and (3) terminal euphoria (Rohrer 1961; Kanas and Feddersen 1971). Other time models have emphasized decrements in particular quarters of the mission, especially the third quarter (Sandal et al. 1995; Gushin et al. 1997a; Bechtel and Berning 1991). However, others have failed to find such time effects in space and space analogue environments (Kanas et al. 2006, 2001b; Wood et al. 2000).

The above issues can result in psychological and interpersonal problems if they are not understood and addressed. These include: (1) decreased crew morale and compatibility, (2) withdrawal or territorial behavior as crewmembers cease to interact with each other, (3) the scapegoating of an individual as a "solution" to group conflict, and (4) the formation of subgroups that compete with each other and destroy crew unity.

9.6.1 Alienation

Group tension can lead to feelings of alienation and personal isolation. Although mission goals may be achieved, tension could still be stressful for the crewmembers, influencing their mood, sleep quality, and general satisfaction with spaceflight conditions.

Post-flight debriefings have revealed differences in values among crewmembers arising from their differences in professional culture, age or nationality. This corresponds to findings from studies of small isolated groups, where it was shown that in the process of cohesive group formation, crewmembers began to regard each other as very "similar" or "close", sharing common values and belief (Andreeva 1998; Gushin et al. 1996, 1997b). On the contrary, crewmembers who did not perceive themselves as close and did not make attempts to understand and share common group values ran the risk of becoming the "stranger" or the "alien" in the group (Gushin et al. 1997c, 1998; Helmreich et al. 1980; Leon et al. 1997; Penwell 1990).

The "stranger's" incapability or lack of motivation to transform his or her own system of values and attitudes can be rationalized in two ways: others might be regarded as (1) conforming insincerely or (2) searching for and exaggerating objective factors (which might actually influence understanding). Among these factors, astronauts and cosmonauts who participated in Mir/NASA studies mentioned language barriers, cultural differences, gender differences, professional variation, and differences in perspective (Morphew et al. 1997; Kozerenko et al. 1997). Interestingly, in several isolation studies (Gushin et al. 1997b; Kelly and Kanas 1993), the existence of an individual seen as the "alien", either inside or outside (e.g., mission control) of the group, could be a uniting factor that supported group cohesiveness. This is consistent with the sociological literature (Dentler and Erickson 1959).

Cultural, religious, professional and other differences have played the role of triggers and formed a rationale for tension appearing in space crews. These have, however, been overcome during the course of international space missions (Kozerenko et al. 1997; Bluth 1984; Grigoriev et al. 1986, 1987; Santy et al. 1993). It could be that the key factor that supports successful adaptation of the group is the individual's capability to share the general values and aims of the group and to establish empathetic relations with partners. In studies simulating the influence of space flight, crewmembers more favorably perceived those they referred to as "supportive", "loving", "warm", "socially competent", and who had an orientation toward empathetic interrelations and the social side of life (Gushin et al. 1997b, c, 1998).

9.6.2 The "Host–Guest" Problem

Problems involving "guest" cosmonauts have been observed during Russian on-orbit space missions, when a visiting crew joins the more permanent crew, whose members have already adapted to space flight conditions and are executing their complicated flight program. In one example on the Mir space station where visiting crewmembers came for a short period of 2 weeks, their working capacity was low due to adaptation to microgravity. However, playing the "hospitable host," the primary crewmembers had to postpone the completion of their own tasks to help the newcomers, which substantially delayed the execution of their flight program. The "host-guest" situation has been a source of psychic tension and decreased effectiveness of interaction on- orbit, and not necessarily only in cases where foreign astronauts participated (Kanas et al. 2006, 2000a, b).

Another aspect of the problem could be discussed from the point of view of Altman's ecological approach (Altman and Haythorn 1967; Altman et al. 1971; Harrison and Connors 1984). A cohesive crew, already adapted to life on a space station, could regard the visiting crew's appearance as a psychological invasion. In the course of space station habitation, the cosmonaut or astronaut forms his or her private psychological space, which feels psychologically comfortable. The arrival of a visiting crew, however, redistributes the station's territory. Even the presence of a newcomer in one's "area" might be regarded as an encroachment and cause a certain discomfort.

9.6.3 Minority Status and Organizational Culture

Cultural differences were examined during two large international studies that were funded by NASA that involved missions to the Mir space station and to the International Space Station (Kanas and Manzey 2008; Kanas et al. 2006, 2001b, 2000a, b). The Mir study sample consisted of five American astronauts, eight Russian cosmonauts, and 42 American and 16 Russian mission control personnel. The ISS study sample consisted of eight American astronauts, nine Russian cosmonauts, and 108 American and 20 Russian mission control personnel. The findings suggested that compared with the Americans, the Russians experienced more direction, support, and self-discovery during the Mir missions; more tension/anxiety during the ISS missions; and less pressure on the job in both studies. These findings may have reflected cultural and attitudinal differences between these two groups of people. But there also were differences in the mission profiles that might have influenced these results.

For example, in the Mir program the missions took place in a Russian space station, were largely Russian-operated using the Russian language, and always had a Russian commander. The American astronaut was always in the minority in the three-person crew and had less of an operational relevance. Thus, it is understandable that American participants would have felt less comfortable and supported that the Russian participants.

In contrast, the ISS program had more variable crew compositions, with both sides having clearer work roles but with a decidedly American style of operations. This might have created tension or anxiety on the part of the Russian crew and ground participants, possibly from perceived differences in organizational culture between the two space agencies. For example, it has been suggested that compared with

American organizational practices, the Russian space program typically utilizes fewer written procedures and relies more on expert opinion to deal with problems that arise (Committee on Space Biology and Medicine/Space Studies Board 1998; Ritsher et al. 2005; Clement et al. 2005). Thus, it is possible that the Russian participants in the ISS study may have been reacting to what they perceived as a less comfortable work environment in terms of familiar organizational practices.

In terms of work pressure being higher in US subjects during the Mir program, working on a foreign space station (Mir) or mission control center (Russian mission control) with relatively unclear goals might have created more job pressure for the Americans, who were also less experienced in conducting long-duration space missions. In the ISS program, being primarily responsible for the complex operations involved with the construction of a new space station would be expected to impact more on the workload of US participants.

9.6.4 Psychological Closing, Autonomization, and Displacement

Psychological closing in space is a phenomenon manifested by decreased communication intensity as well as increased filtration and limitation of the scope and content of crew communication. It has been observed that people in space tend to hide medical and psychological problems that occur in flight from mission control. They also have demonstrated preferences in contacts with mission control personnel, openly rejecting contacts with some people and giving preferences to others (Gushin 1995; Gushin et al. 2004). This phenomenon is attributed to the combined influence of the "forced" character of communication sessions (e.g., limited time and choice of people to talk with), sensory deprivation and monotony in space, and the intention of space crews to maintain their socially desirable image as heroes.

Autonomization in space is expressed by crew egocentrism and an attempt to deal with an out-group (i.e., mission control). This kind of development in isolated and confined individuals is a natural and even necessary stage of the formation of a cohesive group. Unfortunately, the manifestation of this process sometimes results in mission control personnel being perceived as opponents, not partners, and it sometimes is the source of mutual tension and misunderstanding. The crew becomes more critical in their discussions of operations with mission control, frequently pointing out the discrepancies in their positions (Gushin et al. 1997a, b, 2004, 1993; Penwell 1990; Kelly and Kanas 1993; Grigoriev et al. 1986). The most vivid examples of autonomization of the crew in space are the famous Skylab strike and the discrepancy in opinions between mission control and one of the Mir crews concerning the need for an EVA to repair some solar batteries.

Another phenomenon of group behavior in space crews that is closely connected to autonomization is *displacement*. This occurs when crewmembers in space experience high levels of anxiety, negative emotions, and interpersonal conflict that cannot be resolved directly. These unpleasant affects may be transferred externally to people in mission control, resulting in a belief that they are unempathic and non-supportive (Kanas and Manzey 2008; Kanas et al. 2006, 2001b; Lebedev 1988; Kanas and Feddersen 1971). Evidence for the occurrence of displacement has been found during both Shuttle/Mir (Kanas et al. 2001b) and ISS missions (Kanas et al. 2007). Some space psychologists regard this phenomenon as a coping strategy, allowing people in an isolated group to avoid open conflict by draining out negative emotions, thus not perturbing their psychological and interpersonal climate. Whereas this strategy might be effective in the short run, in the long run it encourages negative feelings to fester and build, resulting in blocked emotions, territorial behavior, and poor cohesion.

9.6.5 *Crew Autonomy*

The relationship between crewmembers and people in mission control also relates to crew autonomy. In future space missions involving a lunar base or an expedition to Mars, crewmembers will have to function more autonomously than before due to the long distances involved and the types of activities required in setting up camp and exploring an alien surface (Kanas and Manzey 2008). In all likelihood, crews will need to plan most of their day-to-day activities without mission control input, and should a problem occur they would have to deal with it themselves. This planning extends to leaving time to sleep, eat and exercise, since nutritional and physiological factors are important in maintaining crewmember health and well-being (Heer et al. in press), and they should not be compromised during the course of a busy work day. In addition, crewmembers will have to trust their on- board resources for monitoring life support and providing them with the information they will need to carry out their mission, so the human-machine interface will become even more important. It is time to start thinking about the issue of increased crew autonomy and the role of mission control during such non-orbital activities beyond the Earth's immediate environment.

9.6.6 *Implications for Mission Operations*

9.6.6.1 Selection

From the pool of highly motivated persons with well-developed coping strategies in the astronaut corps, psychologically compatible personalities are selected for a particular crew. It would be useful to have an internationally-approved test battery available that predicted not only the interpersonal compatibility of the crew-members but also their ability to perform in a group. One such measure being used for crew selection in Star City in Russia is the Homeostat (Gushin et al. 1997b, c, 1998; Savilov et al. 1997).

9.6.6.2 Training

People working in space missions need to recognize and deal with psychosocial problems before they deteriorate crew performance and behavior. Training countermeasures could include formal lectures and briefings that address important psychological and interpersonal issues in space, accompanied by training in conflict resolution, cultural sensitivity, and team building (Kanas and Manzey 2008). Because the crew-ground relationship is important, especially during on-orbit missions, it would be useful to include members from both groups during some of the training sessions.

9.6.6.3 Monitoring and Support

Internationally approved procedures for the in-flight monitoring of crewmember interactions and group performance should be created, especially for on-orbit missions. Recent experiences in Russia suggest that the monitoring of crew communication sessions by mission control, including voice frequency analysis, content analysis of speech, and face expression dynamics, gives valuable information about the psychosocial status of the crew. However, such intense monitoring will be less productive during non-orbital missions, such as a lunar base or an expedition to Mars, due to the long distances involved and the need for crews to be more autonomous in carrying out their missions. In addition, the long distances will delay audio or video transmission times, thus hampering traditional supportive strategies that have depended on real-time crew-ground communication, and the possibility of sending emergency supplies or evacuating a sick crewmember will be low.

9.6.6.4 Re-adaptation to Earth

After space missions, both crewmembers and mission control personnel follow a strategy of keeping positive attitudes towards their teammates and reuniting crewmembers with their families. The role of space psychologists is to facilitate this process.

9.6.7 Recommendations

Based on the above, a number of recommendations are suggested for space agencies to deal with issues related to interpersonal and organizational issues:

- To encourage crew cohesion during future manned space missions, crewmembers should undergo survival training together in a variety of extreme conditions, such as mountain climbing, polar wintering-over, desert expeditions, parachute jumping, etc.

- It is important that future space flight crewmembers train together during pre-launch operational training. We believe that the longer they train together, the smoother will be their future interactions.
- Some pre-launch training should involve people from mission control in order to enhance crew-ground cohesion and improve the communication between crewmembers and people on the ground.
- Group sensitivity training for astronauts and cosmonauts could reduce the influence of personal, cultural, national, and other peculiarities of behavior during the mission. This training needs to be executed by experienced social and behavioral experts who can supply astronauts and cosmonauts with complete and precise information about their individual and group behaviors.
- Communication and conflict resolution training already used in modern industry (e.g., airline CRM and LOFT programs) could be tailored for use among space crews, with special attention paid to cultural differences. This includes the need to study the language of one's foreign crewmates, with special focus on words and terms describing everyday life related to leisure time and social activities.
- The opinion of crewmembers should be solicited before and during space missions regarding operational issues such as access to logistics, fair workload distribution, and leisure time, so that potential sources of quarrels can be identified and offset. Crewmembers should have many lines of communication, both among themselves and with people on Earth.
- Methods used in the social and behavioral sciences should be applied in the monitoring of intra- and inter-group relations, with the goal of providing a continual supply of objective data about the psychosocial climate of crewmembers in space.
- Crewmembers working on the lunar surface or participating in an expedition to Mars will be more autonomous and less dependent on mission control direction and support than crewmembers engaged in an on-orbit mission. Research needs to done involving the effects of increased crew autonomy during space missions.

References

Altman, I., Haythorn, W.W.: The ecology of the isolated group. Behav. Sci. **12**, 169–182 (1967)

Altman, I., Taylor, D.A., Wheeler, I.: Ecological aspects of group behavior in social isolation. J. Appl. Soc. Psychol. **1**, 76–100 (1971)

Andreeva, G.M.: Social Psychology, 3rd edn. Astok Press, Moscow (1998)

Atlis, M.M., Leon, G.R., Sandal, G.M., Michelle, G.: Decision processes and interactions during a two-woman traverse of Antarctica. Environ. Behav. **36**, 402–423 (2004)

Ball, J.R.: Safe Passage: Astronaut Care for Exploration Missions. Institute of Medicine, National Academy of Sciences, Washington, DC (2001)

Bechtel, R.B., Berning, A.: The third quarter phenomenon: do people experience discomfort after stress has passed? In: Harrison, A.A., Clearwater, Y.A., McKay, C.P. (eds.) From Antarctica to Outer Space. Springer, New York (1991)

Bluth, B.J.: The benefits and dilemmas of an international space station. Acta Astronaut. **2**, 149–153 (1984)

Bock, O.: Problems of sensorimotor coordination in weightlessness. Brain Res. Rev. **28**, 155–160 (1998)

Bock, O., Fowler, B., Comfort, D.: Human sensorimotor coordination during spaceflight: an analysis of pointing and tracking responses during the "Neurolab" space shuttle mission. Aviat. Space Environ. Med. **72**, 877–883 (2001)

Cavanagh, K., Shapiro, D.A.: Computer treatment for common mental health problems. J. Clin. Psychol. **60**, 239–251 (2004)

Chidester, T.R., Helmreich, R.L., Gregorich, E., Geis, C.E.: Pilot personality and crew coordination: implications for training and selection. Int. J. Aviat. Psychol. **1**, 25–44 (1991)

Clement, J., Ritsher, J.B.: Operating the ISS: cultural and leadership challenges. In: Proceedings of the 56th International Astronautical Congress, IAC-05-A1.5.05, Fukuoka, 17–21 October 2005

Committee on Space Biology and Medicine/Space Studies Board: National Research Council Commission on Physical Sciences, Mathematics, and Application, A Strategy for Research in Space Biology and Medicine in the New Century. National Academy Press, Washington, DC (1998)

Connors, M.M., Harrison, A.A., Akins, F.R.: Living Aloft: Human Requirements for Extended Spaceflight, NASA SP-483. NASA, Washington, DC (1985)

Dentler, R.A., Erickson, K.: The functions of deviance in groups. Soc. Probl. **7**, 98–107 (1959)

Greist, J.H., Marks, I.M., Baer, L., Kobak, K.A., Wenzel, K.W., Hirsch, M.J., Mantle, J.M., Clary, C.M.: Behavior therapy for obsessive-compulsive disorder guided by a computer or by a clinician compared with relaxation as a control. J. Clin. Psychiatr. **63**, 138–145 (2002)

Grigoriev, A.I., Kozerenko, O.P., Myasnikov, V.I., Egorov, A.D.: Ethical problems of interaction between ground-based personnel and orbital station crewmembers, IAF 86–398. In: Proceedings of the 37th Congress of the International Astronautical Federation, New York (1986)

Grigoriev, A.I., Kozerenko, O.P., Myasnikov, V.I.: Selected problems of psychological support of prolonged space flights. In: Proceedings of the 38th Congress of the International Astronautical Federation, Washington, D.C. (1987)

Gushin, V.I.: Problems of psychological control in prolonged space flights. Earth Space Rev. **4**(1), 28–31 (1995)

Gushin, V.I., Kholin, S.F., Ivanovky, Yu.R.: Soviet psycho-physiological investigations in experiments with simulated isolation—some results and prospects. In: Bonting, S. (ed.) Advances in Space Biology and Medicine: European Isolation and Confinement Study, vol. 3, pp. 5–14. Elsevier Science (1993)

Gushin, V.I., Efimov, V.A., Kolinitchenko, T.B., Davies C.: Psychological evaluation and support during EXEMSI. In: Bonting, S. (ed.) Advances in Space Biology and Medicine: European Isolation and Confinement Study, vol 5, pp. 283–295. Elsevier Science (1996)

Gushin, V.I., Zaprisa, N.S., Kolinitchenko, T.B., Efimov, V.A., Smirnova, T.M., Vinokhodova, A. G., Kanas, N.: Content analysis of the crew communication with external communicants under prolonged isolation. Aviat. Space Environ. Med. **68**, 1093–1098 (1997a)

Gushin, V.I., Efimov, V.A., Smirnova, T.M., Vinokhodova, A.G.: Dynamics of subjective perception of in-crew interactions during long-term isolation and confinement. Aerospace Ecological Med. **31**(4), 23–29 (1997b)

Gushin, V.I., Efimov, V.A., Smirnova, T.M., Kanas, N., et al.: Subject's perception of the crew interaction under prolonged isolation. In: Proceedings of the 68th Annual Meeting of the American Aerospace Medical Association, Chicago (1997)

Gushin, V.I., Efimov, V.A., Smirnova, T.M., Vinokhodova, A.G., Kanas, N.: Subject's perception of the crew interaction dynamics under prolonged isolation. Aviat Space Environ. Med. **69**, 556–561 (1998)

Gushin, V.I., Yusupova, A., Popova, I.: Crew-ground control communication styles: preliminary results. In: Proceedings of the 55th International Astronautical Congress, IAC-04-G.5.a.06, Vancouver, 4–8 October 2004

Harrison, A., Connors, M.: Groups in exotic environments. Adv. Exp. Soc. Psychol. **18**, 49–87 (1984)

Heer, M., Smith, S.: Integrated physiology on long-duration space flight: nutritional aspects. International Academy of Astronautics Study Group Report (in press)

Helmreich, R.L., Wilheim, J.A., Runge, T.E.: Psychological considerations in future space missions. In: Cheston, S., Winter, D. (eds.) Human Factors of Outer Space Production. American Association for the Advancement of Science (1980)

Helmreich, R.L., Wilhelm, J.A., Gregorich, S.E., Chidester, T.R.: Preliminary results from the evaluation of cockpit resource management training: performance ratings of flight crews. Aviat. Space Environ. Med. **61**, 576–579 (1990)

Hockey, G.R.J.: Compensatory control in the regulation of human performance under stress and high workload: a cognitive-energetical framework. Biol. Psychol. **45**, 73–93 (1997)

Ihle, E.C., Ritsher, J.B., Kanas, N.: Positive psychological outcomes of spaceflight: an empirical study. Aviat. Space Environ. Med. **77**, 93–101 (2006)

Inoue, N., Matsuzaki, I., Ohshima, H.: Group interactions in SFINCSS-99: lessons for improving behavioral support programs. Aviat. Space Environ. Med. **75**, C28–C35 (2004)

Isay, R.A.: The submariners' wives syndrome. Psychiatr. Q. **42**, 647–652 (1968)

Johannes, B., Salnitski, V.P., Gunga, H.-C., Kirsch, K.: Voice stress monitoring in space—possibilities and limits. Aviat. Space Environ. Med. **71**(Suppl. 9), A58–A65 (2000)

Johannes, B., Salnitski, V.P., Polyakov, V.V., Kirsch, K.: Changes in the autonomic reactivity pattern to psychological load under long-term microgravity-twelve men during 6-month spaceflights. Aviakosmicheskaia i Ekologicheskaia Medits **37**, 6–16 (2003)

Kanas, N.: Psychological, psychiatric, and interpersonal aspects of long-duration space missions. J. Spacecraft Rockets (AIAA) **27**, 457–463 (1990)

Kanas, N.: Psychosocial support for cosmonauts. Aviat. Space Environ. Med. **62**, 353–355 (1991)

Kanas, N.: Psychiatric issues affecting long-duration space missions. Aviat. Space Environ. Med. **69**, 1211–1216 (1998)

Kanas, N. (ed.): Psychology and culture during long-duration space missions. IAA Position Paper International Academy of Astronautics, Paris (2009)

Kanas, N., Feddersen, W.E.: Behavioral, Psychiatric, and Sociological Problems of Long-duration Space Missions, NASA TM X-58067. Johnson Spacecraft Center/National Aeronautics and Space Administration, Houston (1971)

Kanas, N., Manzey, D.: Space Psychology and Psychiatry, 2nd edn. Microcosm Press/Springer, El Segundo/Dordrecht (2008)

Kanas, N., Salnitskiy, V., Grund, E.M., Gushin, V., Weiss, D.S., Kozerenko, O., Sled, A., Marmar, C.R.: Social and cultural issues during Shuttle/Mir space missions. Acta Astronaut. **47**, 647–655 (2000a)

Kanas, N., Salnitskiy, V., Grund, E.M., Gushin, V., Weiss, D.S., Kozerenko, O., Sled, A., Marmar, C.R.: Interpersonal and cultural issues involving crews and ground personnel during Shuttle/Mir space missions. Aviat. Space Environ. Med. **71**(Suppl. 9), A11–A16 (2000b)

Kanas, N., Salnitskiy, V., Gushin, V., Weiss, D.S., Grund, E.M., Flynn, C., Kozerenko, O., Sled, A., Marmar, C.R.: Asthenia—does it exist in space? Psychosom. Med. **63**, 874–880 (2001a)

Kanas, N., Salnitskiy, V., Weiss, D.S., Grund, E.M., Gushin, V., Kozerenko, O., Sled, A., Bostrom, A., Marmar, C.R.: Crewmember and ground personnel interactions over time during Shuttle/Mir space missions. Aviat. Space Environ. Med. **72**, 453–461 (2001b)

Kanas, N.A., Salnitskiy, V.P., Ritsher, J.B., Gushin, V.I., Weiss, D.S., Saylor, S.A., Kozerenko, O. P., Marmar, C.R.: Human interactions in space: ISS versus Shuttle/Mir. Acta Astronaut. **59**, 413–419 (2006)

Kanas, N.A., Salnitskiy, V.P., Boyd, J.E., Gushin, V.I., Weiss, D.S., Saylor, S.A., Kozerenko, O. P., Marmar, C.R.: Crewmember and mission control interactions during International Space Station missions. Aviat. Space Environ. Med. **78**, 601–607 (2007)

Kanas, N., et al.: Psychology and culture during long-duration space missions. Acta Astronautica. **64**, 659–677 (2009)

Kane, R.L., Short, P., Sipes, W., Flynn, C.: Development and validation of the spaceflight cognitive assessment tool for Windows (WinSCAT). Aviat. Space Environ. Med. **76**(Suppl. 1), B183–B191 (2005)

Kelly, A.D., Kanas, N.: Communication between space crews and ground personnel: a survey of astronauts and cosmonauts. Aviat. Space Environ. Med. **64**, 795–800 (1993)

Kelly, A.D., Kanas, N.: Leisure time activities in space: a survey of astronauts and cosmonauts. Acta Astronaut. **32**, 451–457 (1994)

Kozerenko, O.P., Sled, A.D., Salnitsky, V.P.: Psychological support: Russian experience. In: Proceedings of the 68th Annual Meeting of the American Aerospace Medical Association, Chicago (1997)

Kozerenko, O.P., Gushin, V.I., Sled, A.D., Efimov, V.A., Pystinnikova, J.M.: Some problems of group interaction in prolonged space flights. Human Perform. Extrem. Environ. **4**, 123–127 (1999)

Kraft, N.O., Lyons, T.J., Binder, H.: Intercultural crew issues in long-duration spaceflight. Aviat. Space Environ. Med. **74**, 575–578 (2003)

Lebedev, V.: Diary of a Cosmonaut: 211 Days in Space. Phytoresource Research Information Service, College Station (1988)

Leon, G.R., Koscheyev, V.S.: Cross-cultural polar expedition teams as an analog to long-duration space missions. In: Proceedings of the 68th Annual Meeting of the American Aerospace Medical Association, Chicago (1997)

Leon, G.R.: Select-in and countermeasure considerations for long-duration crews. In: Proceedings of the 29th International Conference on Environmental Systems, SAE Technical Paper Series 1999-01-2095, SAE International, Warrendale (1999)

Leone, G.: The effect of gravity on human recognition of disoriented objects. Brain Res. Rev. **28**, 203–214 (1998)

Manzey, D.: Monitoring of mental performance during spaceflight. Aviat. Space Environ. Med. **7**, A69–A75 (2000)

Manzey, D., Lorenz, B.: Mental performance during short-term and long-term spaceflight. Brain Res. Rev. **28**, 215–221 (1998)

Morphew, M.E., MacLaren, S., Herring, L., Azar, B., Thagard, N.: Voyage of discovery: American astronauts aboard Russia's Mir space station. Human Perform. Extrem. Environ. **2**(1), 40–61 (1997)

Myasnikov, V.I., Zamaletdinov, I.S.: Psychological states and group interactions of crew members in flight (reprint). Human Perform. Extrem. Environ. **3**, 44–56 (1998)

Nechaev, A.P.: Work and rest planning as a way of crew member error management. Acta Astronaut. **49**, 271–278 (2001)

North, R.: Human requirements for long-duration missions: Antarctic and Arctic stations, planetary surface operations, and space transportation vehicles. Revista Portuguesa de Medicina Militar. **40**, 1–4 (1991)

Older, H.J., Jenney, L.L.: Psychological Stress Management Through Voice Output Analysis, NASA CR 147723, pp. 1–42. Planar Corp, Alexandria (1975)

Palinkas, L., Gunderson, E.K.E., Holland, A.W.: Predictors of behavior and performance in extreme environments: the Antarctic space analogue program. Aviat. Space Environ. Med. **71**, 619–625 (2000)

Pavy-LeTraon, A., Saivin, S., Soulez-LaRiviere, C., Pujos, M., Guell, A., Houin, G.: Pharmacology in space: pharmacotherapy. In: Bonting, S.L. (ed.) Advances in Space Biology and Medicine, vol. 6. JAI Press, Greenwich (1997)

Penwell, L.: Problems of intergroup behavior in human spaceflight operations. J. Spacecraft and Rockets **27**, 464–469 (1990)

Proudfoot, J., Goldberg, D., Mann, A., Everitt, B., Marks, I., Gray, J.A.: Computerized, interactive, multimedia cognitive- behavioral program for anxiety and depression in general practice. Psychol. Med. **33**, 217–227 (2003)

Raybeck, D.: Proxemics and privacy: managing the problems of life in confined environments. In: Harrison, A.A., Clearwater, Y.A., McKay, C.P. (eds.) From Antarctica to Outer Space. Springer, New York (1991)

Reschke, M.F., Bloomberg, J.J., Harm, D.L., Paloski, W.H., Layne, C., McDonald, V.: Posture, locomotion, spatial orientation, and motion sickness as a function of space flight. Brain Res. Rev. **28**, 102–117 (1998)

Ritsher, J.B.: Cultural factors and the International Space Station. Aviat. Space Environ. Med. **76** (6, Section II), B135–B144 (2005)
Ritsher, J.B., Kanas, N., Gushin, V.I., Saylor, S.: Cultural differences in patterns of mood states on board the International Space Station. Acta Astronaut. **61**, 668–671 (2007)
Rivolier, J., Goldsmith, R., Lugg, D.J., Taylor, A.J.W.: Man in Antarctica: The Scientific Work of the International Biomedical Expedition to the Antarctica. Taylor & Francis, London (1988)
Rohrer, J.: Interpersonal relationships in isolated small groups. In: Flaherty, B. (ed.) Psychological Aspects of Manned Spaceflight. Columbia University Press, New York (1961)
Rose, R.M., Fogg, L.F., Helmreich, R.L., McFadden, T.J.: Psychological predictors of astronaut effectiveness. Aviat. Space Environ. Med. **65**, 910–925 (1994)
Saivin, S., Pavy-LeTraon, A., Soulez-LaRiviere, C., Guell, A., Houin, G.: Pharmacology in space: pharmacokinetics. In: Bonting, S.L. (ed.) Advances in Space Biology and Medicine, vol. 6. JAI Press, Greenwich (1997)
Salnitski, V.P., Myasnikov, V.I., Bobrov, A.F., Shevchenko, L.G.: Integrated evaluation and prognosis of cosmonaut's professional reliability during space flight. Aviakosmicheskaia i Ekologicheskaia Medits **33**, 16–22 (1999)
Sandal, G.M.: Coping in Antarctica: is it possible to generalize results across settings? Aviat. Space Environ. Med. **71**, A37–A43 (2000)
Sandal, G.M.: Culture and tension during an International Space Station simulation: results from SFINCSS'99. Aviat. Space Environ. Med. **75**, C44–C51 (2004)
Sandal, G.M., Værnes, R., Ursin, H.: Interpersonal relations during simulated space missions. Aviat. Space Environ. Med. **66**, 617–624 (1995)
Sandal, G.M., Bergan, T., Warnche, M., Værnes, R., Ursin, H.: Psychological reactions during polar expeditions and isolation in hyperbaric chambers. Aviat. Space Environ. Med. **67**, 227–234 (1996)
Sandal, G.M., Endresen, I.M., Værnes, R., Ursin, H.: Personality and coping strategies during submarine missions. Mil. Psychol. **11**, 381–404 (1999)
Santy, P.A.: Choosing the Right Stuff: The Psychological Selection of Astronauts and Cosmonauts, pp. 81–96. Praeger Scientific, Westport (1994)
Santy, P.A., Kapanka, H., Davis, J.R., Stewart, D.F.: Analysis of sleep on Shuttle missions. Aviat. Space Environ. Med. **59**, 1094–1097 (1988)
Santy, P.A., Holland, A.W., Looper, L., Marcondes-North, R.: Multicultural factors in the space environment: results of an international shuttle crew debrief. Aviat. Space Environ. Med. **64**, 196–200 (1993)
Sauer, J., Hockey, G.R.J., Wastell, D.G.: Maintenance of complex performance during a 135-day spaceflight simulation. Aviat. Space Environ. Med. **70**, 236–244 (1999)
Savilov, A.A., Bayevsky, R.M., Bystrytskaya, A.F., Gushin, V.I., Manovtsev, G.A., Nichiporuk, I.A., Novikov, M.A., Ponomareva, I.P., Salnitsky, V.P.: On-board equipment-based study of the dynamic of psycho-physiological and biomedical responses of the operators during 135-day isolation in the Mir orbital station mock-up. Aerospace Ecol. Med. **5**, 28–38 (1997)
Shayler, D.A.: Disasters and Accidents in Manned Spaceflight. Springer/Praxis, Chichester (2000)
Stuster, J., Bachelard, J., Suedfeld, P.: The relative importance of behavioral issues during long-duration ICE missions. Aviat. Space Environ. Med. **71**(Suppl. 9), A17–A25 (2000)
Suedfeld, P.: Canadian space psychology: the future may be almost here. Can. Psychol. **44**(2), 85–92 (2003)
Suedfeld, P.: Invulnerability, coping, salutogenesis, integration: four phases of space psychology. Aviat. Space Environ. Med. **76**, B61–B66 (2005)
Suedfeld, P., Weiszbeck, T.: The impact of outer space on inner space. Aviat. Space Environ. Med. **75**(Suppl. 7), C6–C9 (2004)
Wood, J.-A., Hysong, S.J., Lugg, D.J., Harm, D.L.: Is it really so bad? A comparison of positive and negative experiences in Antarctic winter stations. Environ. Behav. **32**, 84–110 (2000)
Young, L., Yajima, K.: Artificial gravity as a tool in biology & medicine, International Academy of Astronautics Study Group Report (in press)

Chapter 10
Flying with Strangers: Postmission Reflections of Multinational Space Crews

Peter Suedfeld, Kasia E. Wilk, and Lindi Cassel

Abstract After the Space Age began as part of the national rivalry between the USSR and the United States, space exploration gradually took on a multinational character as both countries included astronauts from their respective allies, and eventually from each other, in their missions. This trend became institutionalized in the Shuttle-*Mir* program and in the construction of the International Space Station (ISS). The latter is the first truly international, as opposed to multinational, space capsule, in that it does not belong to and was not built by one country. In previous cases, one national space agency was always the host and crewmembers from other nations were perceived and treated as guests. This "guest" status, which usually went with being a minority among a majority from the "host" nation, led to considerable dissatisfaction and frustration.

This chapter examines the archived reminiscences of both majority and minority astronauts and cosmonauts, relying primarily on the method of thematic content analysis (TCA). TCA is a set of techniques whereby trained scorers identify and quantify specific variables in narratives. In this study, TCA procedures were used to analyze how majority-minority status and other variables (e.g., gender, mission

---

This research was made possible by Contract No. 9F007-033006 with the Canadian Space Agency and is part of the project Long-term Effects After Prolonged Spaceflight (LEAPS). A briefer version of the chapter was presented at the meeting of the American Psychological Association in San Francisco, CA, in August 2007.

P. Suedfeld (✉)
Department of Psychology, University of British Columbia, Vancouver BC V6T 1Z4, Canada
e-mail: psuedfeld@psych.ubc.ca

K.E. Wilk
University of Manchester, W.J. Stelmaschuk and Associates Ltd, Manchester, UK
e-mail: katarzyna.wilk@postgrad.manchester.ac.uk

L. Cassel
Providence Health Care, Vancouver, BC, Canada
e-mail: lindi.cassel@gmail.com

D.A. Vakoch (ed.), *On Orbit and Beyond*, Space Technology Library 29,
DOI 10.1007/978-3-642-30583-2_10,

duration, and Space Age era) affected satisfaction, feelings about crewmates and home agencies, personal values, ways of coping with problems, and other psychosocial reactions of the mission participants. The study drew upon astronauts' and cosmonauts' memoirs, autobiographies, media interviews, and oral history interviews as the databases on which TCA scoring was performed.

10.1 Nationalistic Embodiments of a Universal Human Drive

The exploration of space may be attributed to two driving forces. One is an innate drive shared by many species but perhaps best exemplified by humanity: the urge to seek novelty, to enlarge the sphere of the known as we advance into the hitherto unknown, and to expand the habitat of humankind (Holquist et al. 1985). Long before technology made real space voyages possible, fictional explorations can be traced to the myth of Daedalus and Icarus and its counterparts in other traditions, to the writings of Cyrano de Bergerac, and eventually to the imaginations of Jules Verne and the multitude of early-twentieth-century science fiction writers.

The second motivator, which determined just when in our species' history space travel would move from fiction to reality, was international rivalry. Primitive military rocketry began centuries ago, accelerated and took the first large steps toward space during World War II, and was increasingly well supported and brought to eventual fruition as the "space race" component of the Cold War.

The first decades of human spaceflight were a series of competitions between the Soviet Union and the United States: who would be the first to launch an orbiting spacecraft, a piloted spacecraft, a space crew, a Moon rocket, a space station .... Flights were scheduled to preempt media publicity from the competition. Temporary victory veered from one bloc to the other, with each claiming—or at least implying – that being momentarily ahead in the race was proof of the superiority of its political and economic system, just as Olympic gold medals were (and are) risibly interpreted as markers of national quality.

In such a setting, it followed logically that cooperation between the two leading space nations would be unlikely. The original space travelers were exemplars of the virtues each country extolled: they were military pilots, the cream of that already hand-picked crop, who were used to flying experimental and operational aircraft at the very edge of new technology, individuals of demonstrated courage, coolness, and ability. The world was shown that they were all physically fit, psychologically stable, good husbands and fathers, modest, humorous, and loyal. They were patriotic citizens and, depending on which program they were in, strong supporters of either communism or democratic capitalism. Although these portraits omitted a number of what would have been more realistic, if less rosy, individual differences among these pioneers, both space agencies continued to paint such idealized pictures, and the spacemen did their best not to smear the paint (although later in the Space Age, revisionists have tried to rub off some of its luster by emphasizing the internal politics of the agencies, alleging arbitrary and biased decisions being made concerning the assignment of astronauts, and so on) (Burrough 1998). More recently, selection procedures have changed to reflect the expanded sources and duties of astronauts, to include civilians, nonpilots, women, and a variety of (mostly, but not entirely, technical and scientific) professionals; but there is a perception that some kinds of bias still exist – e.g., in favor of astronauts from the military (Mullane 2006).

It is worth remembering that the combination of the universal urge to explore and the particularistic urge to use exploration to exalt one's nation is neither new nor unique to space explorers. For centuries, it has been a prominent reason why terrestrial expeditions were funded and also a strong component of many expeditioners' motivation (Anderson 1970).

10.2 Guest Rooms in Space

In 1975, the two rivals cooperated to design a docking module that allowed spacecraft from each (Apollo and Soyuz) to join in space. Later, both superpowers began to offer room and board in their space capsules to citizens of their respective international blocs. The Soviet *Interkosmos* program made room for one cosmonaut from various Eastern bloc countries, as well as from France, Syria, and India; American crews have shared their spacecraft with colleagues from Canada, Western Europe, Australia, Japan, India, Israel, Brazil, and Saudi Arabia.

This trend was reinforced by the establishment of space agencies in countries that could select and train astronauts but had no independent crewed space vehicles. The most active among these are Canada, Germany, France (and eventually the European Union [EU]), and Japan. The People's Republic of China has since gone beyond such strategies to develop its own launch vehicles and begin an independent program of piloted spaceflight. Eventually, multinationality became routine, as did the inclusion of women and the broadening of selection to allow for the participation of people who were not military, not test pilots, and often not even pilots. The new participants were from a range of disciplines: engineers, scientists, physicians, politicians, and, most recently, private individuals who bought a brief stay on the ISS.

This major increase in the diversity of space voyagers sharpens a distinction that began when the USSR and the United States first added foreign crewmembers. Differences, sometimes invidious, were not only between nationalities per se, but also between the "host" crew of Americans or Soviets/Russians and the "visitors." At first, the inclusion of international crewmembers was primarily a propaganda move. It had relatively little beneficial effect on the missions themselves and angered both the established astronaut and cosmonaut corps by reducing the flight opportunities of their members (Furniss et al. 2007; Mullane 2006). Until the construction of the ISS, every capsule that carried human beings into space was either American or Soviet/Russian. Was it possible that mixed-nationality crews aboard felt equally at home and comfortable, or was a host-guest distinction unavoidable? Would the latter be strengthened by the fact that some of the “home” team inhabited the vehicle for a much longer period than did the foreign visitors? Could the distinction be eliminated, or at least minimized, by appropriate training and crew composition? And what did the answers to these questions imply for truly international efforts, such as building and working on the ISS, and perhaps the eventual exploration of space beyond low-Earth orbit and the Moon? This chapter presents data that address some, though not all, of these questions.

There has been considerable evidence that psychosocial stressors are among the most important impediments to optimal crew morale and performance (Kanas et al. 2003; Space Studies Board 1998; Stuster 1996; Suedfeld 2001; Suedfeld 2006). Positive reactions during and after spaceflight were relatively ignored as psychologists focused on problems that needed to be avoided or solved. After a somewhat slow start toward balancing the situation, attention to positive aspects has expanded in the past few years to look at eustress (positive stress), personal growth, excitement, enjoyment, feelings of satisfaction, camaraderie, and changes in values (Ihle et al. 2006; Kelly and Kanas 1993; Suedfeld 2005; Suedfeld 2006; Suedfeld and Steel 2000; Suedfeld et al. 2001).

It has been pointed out that “mixed” crews are mixed in many different ways. Intercultural issues can arise, and have arisen, not only between space voyagers of different nationalities, but also between those of different space agencies, sexes, and educational and professional backgrounds. Crewmembers who came to space with a military test pilot background and those with an academic science background may have problems understanding each other’s jargon and worldview (to say nothing of those of teachers and politicians). The same, to an even greater extent, is likely to be true in international crews that are not perfectly bilingual (Kelly and Kanas 1993; Kumar 2007). However, the current chapter focuses on only one kind of diversity, that based on nationality.

Whether the possible benefits of increased diversity in crew composition (such as reducing boredom, celebrating unaccustomed holidays, and becoming acquainted with new and useful approaches to interpersonal and operational problems) will outweigh the additional stresses that it generates, or vice versa, needs to be assessed through empirical data. To date, there have been three sources of relevant information. One advantage that they all share, which sets them off from simulation and analogue studies, is their high external validity: the information is produced by real participants in real space operations. This is the only kind of information that will be considered here.

The most colorful and memorable, but least generalizable and scientifically rigorous, source is the collection of anecdotes that has been generated by the space voyagers and others involved in the programs. Self-report studies using surveys and interviews have provided both qualitative and some quantitative information, usually from a relatively small number of crewmembers during a mission and occasionally from larger samples of ground staff personnel. Thematic content analyses applied to interviews, memoirs, and similar archival materials provide another form of quantitative analysis applied to qualitative materials. This is the method used in the current chapter.

10.3 "My House," or Joint Tenancy?

10.3.1 Anecdotal Evidence

As in much of space psychology, and more generally in the psychology of all extreme and unusual environments, the first bits of knowledge came from the anecdotes told and written down by participants. These stories have tended to emphasize the dramatic, and therefore mostly unpleasant, interactions between crewmembers of different demographic (including national/cultural) categories. It should be noted that most of them are "common currency" in the space community; the references given are only examples of several sources in which these stories have appeared.

The kind of diversity with which this chapter is concerned, that is, differences in national origin, has been the topic of many anecdotal reports. Some of the best known involve visitors to Soviet capsules. When the first *Interkosmos* cosmonaut, Vladimir Remek of Czechoslovakia, returned from space (Soyuz 28, 1978), the joke went around that he was suffering from "red hand syndrome": every time he reached for a switch or other control, a Russian crewmate would slap his hand and tell him not to touch it (Lebedev et al. 1990a; Zimmerman 2003).

Four years later, Jean-Loup Chrétien, a French air force officer and the first of a series of French cosmonauts, was likewise forbidden to touch anything during his crew training with two Russians; he not-so-subtly communicated his frustration (and annoyed his crew commander) by bringing a pillow and going to sleep during one training session. After the inhospitable commander was replaced and Chrétien reached the *Soyuz* space station for a 1-week visit, his expertise, good nature, and sophisticated equipment impressed the Russians – but one of them later expressed his relief at going back to black bread and borscht after a menu of canned French delicacies, including compote of pigeon with dates and dried raisins, duck with artichokes, boeuf bourguignon, and more (Lebedev et al. 1990b; Surprise! Astronauts Eat in Orbit 2008). Chrétien, in turn, criticized the excessive workload imposed on the crew (Hall et al. 2005).

The long-duration deployments to the *Salyut* and *Mir* space stations included the presence of mixed crews, and the Shuttle-*Mir* mission series was in fact designed for such crews. Each of the latter missions was constructed around an American-Russian team flying to *Mir* aboard a Space Shuttle orbiter and remaining on the station (with occasional crew changes and short-term visitors) for between 4 and 7 months.

The reluctance of Russian hosts to admit their guests to full coworker status persisted during this collaborative program. In 1995, Norman Thagard was the first American to be a long-term crewmember on the *Mir* space station. Despite his status as a full resident, rather than a short-term visitor like Remek and Chrétien, Thagard, like them, felt that he was left out of important and interesting activities on the aging and deteriorating spacecraft. He wound up doing crossword puzzles while his crewmates did the work. Shannon Lucid, who spent six seemingly happy months on *Mir* in 1996, was left "in command" of the station while her two Russian colleagues performed EVAs; however, she was told not to touch anything (Lucid 1988; Zimmerman 2003). In an oral history interview, one NASA psychologist said, "We were never able, I don't think, to have the American be on par with the Russian crew members . . . ." (Holland 1998). The problem may not be restricted to the astronauts. Several of the Shuttle-*Mir* astronauts indicated that more vigorous support from NASA ground personnel in Mission Control in Russia might have ameliorated their problems – but those personnel in turn felt themselves to be tense, unhappy, underutilized, and somewhat ignored by their own Russian counterparts (Linenger 2000; Kanas et al. 2000; Thagard 1998; Zimmerman 2003).

Even those astronauts who were given work to do could wind up with menial or routine jobs (Kanas et al. 2000a). David Wolf, a Shuttle-*Mir* resident astronaut, volunteered to clean "gooey, slimy, ice-cold fluid" from the station's walls, a job that then devolved on him for 4–8 h per day, almost every day, while his Russian colleagues performed sophisticated technical work. Wolf accepted this with equanimity: "that was the best thing I could come up with to free up their time for what they're better at and be part of the team" (David Wolf 1998).

The critical attitude toward people perceived to be not-quite-colleagues was not restricted to the Russian space program. Mike Mullane (2006, p. 197), referring to "part-time astronauts" – one-flight foreign visitors, payload specialists, politicians, and the like – asserts that their training had been cursory and superficial, that some of them exhibited psychological problems, and that "Mission commanders provided their own additional training in the form of the admonishment 'Don't touch any shuttle switches!'" (Mullane 2006) Obviously, "part-time astronauts" were seen as less expert and therefore undependable. J. M. Linenger, too, comments negatively on his and colleagues' attitude toward *American* "part-time astronauts" (Linenger 2000a).

Of course, this should not have applied to people such as Remek, Thagard, and Lucid. They and many others who flew as national minorities were in fact professional astronauts. They were just as well trained as the national majority with whom they flew, and in many cases, they trained together with the majority for a year or more. The comments of majority crewmembers are typically quite positive about their foreign colleagues' personality and ability to get along with the rest of the crew, but the distrust in their competence within the "home team's" spacecraft (and/or with the home team's language) persisted nonetheless (Thagard et al. 2007).

These and similar stories may not be representative of the general experiences of national minorities in a space crew. Many of these individuals' recollections were primarily positive. Nevertheless, the negatively toned anecdotes point out, even if they may overemphasize, problems of which planners should be aware.

To some extent, friction between majority and minority crewmembers may arise from differences in the home cultures of the two most populous groups, Russians (including citizens of the former USSR) and Americans. J.B. Ritsher, in an excellent summary of the relevant cultural differences between these two nations, cites research not only from space but also from aviation and from psychological, sociological, and anthropological studies more generally (Ritsher 2005). According to these studies, Russian culture values collectivism, hierarchical power, distance, and paternalism more than American culture and values individualism, egalitarianism, mastery, autonomy, and uncertainty less. In addition, the USSR was oriented more toward survival and less toward well-being.

On a number of dimensions, including the ones listed above, Russia is discrepant from all of the other nations involved in the International Space Station project. Russian and American approaches to spaceflight differ in significant ways, some of these reflecting the more general cultural differences discussed by Ritsher. Supporting the view that Russian culture is more hierarchical than American culture were perceptions that Russian mission commanders were more authoritarian, Russian communications were more structured and formal (at least in English translation), and Russians were more circumspect in criticizing mission control or the systems on the spacecraft (Kumar 2007).

Perhaps because of differences in the national economies and the funding of the space agencies, cosmonauts were more likely to feel that they had to try to repair malfunctioning systems, whereas astronauts tended to discard and replace them. Russians consequently were more satisfied with systems that worked adequately rather than demanding that they work perfectly. On a more personal level, cosmonauts (unlike astronauts) are paid a large spaceflight bonus, with deductions based on how many of the preset goals (experiments, repairs, extravehicular activities, etc.) they fail to complete successfully during the mission. As a result, their foreign colleagues sometimes considered them to be reluctant to do work that was not within the pay-for-action agreement and to do anything that might foil the completion of an agreed-upon paid task (Kumar et al. 2007).

Although it is certainly likely that cultural "root causes" (especially those stemming from differences between Russian and Western cultures) may underlie some of the frictions between majority and minority crewmembers, the host-guest dichotomy may have caused more problems than cultural or national diversity per se. If so, a completely different picture may emerge within a truly international facility such as the ISS when it becomes fully operational.

There is another possible explanation. Valentin Lebedev, a long-duration *Mir* cosmonaut, recognized a difference between his reactions to foreign and to compatriot visitors. Concerning one of the former, he wrote, "It's nice to have guests, but they make you tired" (Lebedev et al. 1990a, p. 101), even though most of his comments about his French colleague were positive; commenting on an upcoming visit by fellow Russian cosmonauts, he wrote, "I think it will be easier with this visiting crew; they won't disturb us as much . . ." (Lebedev et al. 1990a, p. 189). It may be that it is not nationality but familiarity that makes a visitor more welcome, so that more extensive pre-mission training and joint activities might erase or at least diminish the invidious difference.

10.3.2 Self-Report Studies

Kanas and Manzey summarized the few studies using self-report measures by space voyagers who had flown in foreign company (Kanas and Manzey 2003). Although there have been several simulation and analogue studies (respectively, group isolation experiments in specially designed settings and field studies in isolated areas such as the polar regions and undersea habitats), data from actual spaceflight are scarce. Participants have reported miscommunications due to both spoken and nonverbal interaction styles, abrasive differences in leadership decision-making, differences in work patterns, different standards of hygiene and food preparation, and personality clashes that may be related to cultural factors.

Kanas and his colleagues have conducted major studies of space crews in flight by administering standard questionnaires that crewmembers can complete on a computer while the mission is going on. One such study, of crewmembers on *Mir* (five Americans and eight Russians) and the ISS (eight and nine, respectively), found that cosmonauts on *Mir* experienced more direction, support from the leader, and self-discovery than astronauts; lower vigor and more tension and anxiety on the ISS; and less job pressure but higher task orientation and managerial control in both places (Kanas et al. 2005). Americans on the Russian station felt less comfortable and less well supported from the ground than did the "home team." In contrast, ISS procedures are more U.S.-influenced, which may have made the Russians feel that they were on unfamiliar territory. Another report on the same Shuttle-*Mir* crews found that during the second half of the mission, Russian crewmembers reported decreasing cohesion and work pressure compared to Americans (Kanas et al. 2001).

A more recent study reported that miscommunications abounded when members of international crews engaged in extravehicular activities, but not when all crewmembers were from the same country, and that besides the obvious language barrier, space fliers generally felt that coming from the same cultural background would also reduce interpersonal friction (Kumar et al. 2007). Most of those interviewed agreed that on long-duration missions, they would prefer to go with a homogeneous crew from their own culture. In fact, according to Linenger, many U.S. astronauts declined the opportunity to participate in the Shuttle-*Mir* program for reasons that included distrust of Russian technology and post–Cold War hostility toward Russians themselves. Some of those who did agree to join the program were so dissatisfied with the training they got in Russia that they threatened to quit – "a near mutiny" (Linenger 2000b, p. 45).

10.4 The Current Study: Thematic Content Analysis

Content analysis is a research method used in many disciplines to study narratives of interest. For example, anthropologists may content-analyze myths or folktales to identify important issues or beliefs of a culture; literary scholars may find in novels or plays the dominant patterns of social relations in a particular time or place, or hints as to the childhood experiences and personality of an author. Such qualitative or impressionistic methods are frequently used to explore hypotheses derived from a particular theory such as psychoanalysis, Marxism, or postmodernism. The scholar finds examples in the material that are relevant to the theory and uses those examples as evidence, as in Freud's inferences about Leonardo da Vinci's family background, childhood, and personality, based primarily on the analysis of a dream that Leonardo recorded in his diary (Freud 1961).

A purely quantitative counterpart is computerized content analysis, in which the occurrence of certain kinds of words or phrases is counted and used to infer either historical or personal characteristics. For example, a frequent appearance of the word “I” may indicate a degree of self-confidence, independence, or narcissism; hostile terminology (“enemy,” “threatening,” “evil”) reveals a bellicose emotional or cultural state: an increase in such words when referring to another person or country may be one indicator of a forthcoming confrontation (North et al. 1963).

The method used in our study, Thematic Content Analysis (TCA), is different from both the qualitative or impressionistic approaches and purely quantitative computerized ones. In TCA, either all available material or a randomly selected subset is used so that the researcher’s theoretical preconceptions cannot bias the selection of material to be analyzed; identifying information is removed as far as possible, as a safeguard against bias in the actual scoring; and the material is scored blindly by a qualified scorer using a detailed scoring manual to further reduce the chances of bias and of idiosyncratic scoring criteria. Generally, at least one other scorer analyzes a percentage of the same passages to ensure interscorer reliability. Thus, from a qualitative (e.g., narrative) database, quantitative data are generated in a scientifically rigorous way and statistical analyses are made possible (Smith et al. 1992).

The issue of accuracy always arises in retrospective materials, usually from two perspectives. One is the exactness of memory; the other is the possibility of impression management. In the current study, precision probably varied as a function of time since the experience (among other variables), which itself varied from very little, as in the *Life* magazine interviews of the first Mercury astronauts, to years in the case of book-length memoirs. In any case, the question of how precisely the narrators remembered events is not of critical importance to this study: we were not interested in compiling a history of their experiences, but rather in the emotions and motives that were associated with the events and that emerged during recall. As for impression management, although this is a likely mediating variable in any self-descriptive human narrative, the TCA scoring criteria are not very transparent, and the material includes a number of cross-checks (e.g., prepared versus spontaneous remarks). Many of the narratives included negative reflections on both other people and the narrator himself (or herself), and stories by several participants in the same event showed substantial differences, so at least the attempt to make oneself (or one’s colleagues or one’s agency) look good did not swamp all other considerations, and there was no evidence of externally imposed uniformity in the accounts.

10.4.1 Method

The current study applied TCA scoring methods to a collection of memoirs, interviews, and oral histories originated by 63 astronauts and cosmonauts.

The overwhelming majority of U.S. and Soviet/Russian participants were in the categories that NASA considers professional astronauts: pilots and mission specialists. The few exceptions were "spaceflight participants": individuals flown for some goal such as public relations. No "space tourists" (i.e., individuals who were allowed to fly for a brief visit to the ISS upon payment of a multi-million-dollar fee) were included; neither were payload specialists, who fly as members of nongovernmental institutions such as corporations or universities to carry out a specific task or experiment. Participants from countries other than the United States and the USSR/Russia were a more mixed group, which included both professional astronauts and others (many of them professional air force officers) who, after their one spaceflight, would have no long-term connection with a space program.

The collections covered the era of human spaceflight from the very first period through the construction of the ISS, but we omitted reports related to missions in which crews of only one nation were involved. Due to the extremely small sample available from ISS veterans, we also omitted those data from our analyses. With the increasing number of crewmembers who have served on the Station, this problem may be on the way to being solved.

Because the source materials of this study varied widely in length, all TCA results reported below are based on number of category mentions per page in the source. Not every subject had references to all of the dependent variables, so *n*'s (sample sizes within each subcategory) varied from measure to measure.

Anecdotes and a numerical content analysis software program were used as secondary data.

10.4.1.1 Independent Variables

Table 10.1 shows the breakdown of the subjects by relevant demographic and spaceflight categories, which served as the independent variables. "National origin" refers to the country with which the source was identified in the space program. For example, some "U.S.A." astronauts were originally immigrants from elsewhere; however, they were selected, trained, and chosen to fly by NASA. "USSR/Russia" includes cosmonauts whose citizenship was Soviet during the existence of the USSR and those who were Russian afterward. The "Other" category includes astronauts who had been recruited and selected by the established space programs of other nations (e.g., Canada or Japan, as well as France, Germany, or other EU nations) and who flew with either the U.S. or the Soviet/Russian space program. *Interkosmos* crewmembers and their equivalents flying with NASA are also classified as "Other".

In addition to these analyses, others were performed on disaggregations based on mission duration (2 weeks or less versus either 4 or 6 months or more) and mission phase (portions of the narrative referring to pre-, in-, or postflight periods). In some instances, the *n* within one of these cells was too small for analysis, and those scores are omitted from this report.

Table 10.1 Number of subjects by category

	Sex		Flew as		Flew with	
National origin	Male	Female	Minority	Majority	U.S.A.	USSR/Russia
U.S.A.	16	10	7	19	20	6
USSR/Russia	14	0	4	10	6	8
Other[a]	19	4	23	–	19	4
Total	49	14	34	29	43	18

[a]"Other" refers to crewmembers who are neither American nor Soviet/Russian. All of the subjects in this category flew as a minority with either American or Russian majorities

10.4.1.2 Dependent Variables

The scoring categories applied to the materials were as follows:

1. ***Value hierarchies:*** Schwartz defined values as having five major aspects. According to him, values

   (a) Are concepts or beliefs,
   (b) Pertain to desirable end states of behaviors,
   (c) Transcend specific situations,
   (d) Guide selection or evaluation of behavior and events, and
   (e) Are ordered by relative importance (Schwartz 1992).

   Eventually, Schwartz reported that 11 categories of values underlying stable, important life goals (see Table 10.2) had been empirically shown to have cross-cultural generality and high reliability.

   One way to group the values is to distinguish between those that serve individual interests and those that serve collective interests: Achievement, Hedonism, and Self-Direction versus Conformity, Security, and Tradition. The emphasis on individual versus collective cultural values is generally thought to separate American and Soviet cultures; a comparison between the two groups of values across the fliers representing the two national space agencies can be an interesting way to check on this widespread view.

   The value scores reported in the current chapter reflect the number of times the source person mentioned experiencing, advancing, or identifying with values in that category per page or section of text in the source material.
2. ***Social relations*** (two subcategories):

   (a) Affiliative Trust/Mistrust: positive, trusting relationships versus cynicism and negativity toward others (McKay 1992).
   (b) Intimacy: Positive Intimacy is a measure of readiness or preference for warm, close, and communicative interaction with others (McAdams 1984). In the current study, we supplemented this by scoring Negative Intimacy as well: negative affect in relationships, negative dialogue, rejection of commitment or concern for others, interpersonal disharmony, nonreciprocated friendliness, and escape from or avoidance of intimacy.

Table 10.2 Value categories and definitions (alphabetical order) (Adapted from Schwartz (1992))

Value	Brief definition
Achievement	Personal success through demonstrated competence according to social standards
Benevolence	Concern for close others in everyday interaction
Conformity	Inhibition of socially disruptive acts, impulses, or inclinations
Hedonism	Pleasure in satisfying organismic needs
Power	Social prestige/status, control over people and resources
Security	Safety, harmony, stability of society, relationships, and self
Self-direction	Independent thought and action: choosing, creating, exploring
Spirituality	Meaning and harmony by transcending everyday reality
Stimulation	Excitement, novelty, challenge
Tradition	Respect for one's cultural/religious customs and ideas
Universalism	Understanding, appreciation, tolerance, and protection of the welfare of all people and of nature

Each social relations measure was scored whenever the source mentioned that emotion in relation to the following:

- His or her crewmates,
- His or her own space agency,
- The space agency in charge of the mission,
- His or her own family, and
- People in general.

3. ***Coping strategies:*** A standard set of coping categories was used to analyze the source materials (Folkman et al. 1986). These include both problem-oriented and emotion-oriented strategies. "Supernatural Protection" was added in our previous studies of Holocaust survivors and was retained in the current analysis (Suedfeld et al. 1997). This is not a coping strategy per se, but rather an expression of the individual's invocation of spirituality, religion, mysticism, or fatalism in dealing with problems (see Table 10.3). The category was scored each time the narrative mentioned that the source had used that strategy in attempting to solve a problem.
4. ***LIWC computer analysis:*** In addition, material that was accessed through the Internet (oral histories and some interviews) was separately computer-analyzed by Linguistic Inquiry and Word Count (LIWC) (Pennebaker et al. 2001). LIWC is a word-count software application that identifies a variety of affective/emotional, cognitive, sensory/perceptual, and social processes, as well as references to personal space and orientation, motion, work, leisure, financial and metaphysical issues, and physical states. Because computer analysis is subject to many problems such as ignoring context and being restricted to those words and phrases that had been entered in the software dictionary, this was considered a secondary methodology in the current study.

Table 10.3 Coping categories and definitions

1.	Confrontation	Effort to resolve situation through assertive or aggressive interaction with another person
2.	Distancing	Effort to detach oneself emotionally from the situation
3.	Self-control	Effort to regulate one's own feelings or actions
4.	Accept responsibility	Acknowledging that one has a role in the problem
5.	Escape/avoidance	Efforts to escape or avoid the problem physically
6.	Planful problem-solving	Deliberate (rational, cognitively oriented) effort to change or escape the situation
7.	Positive reappraisal	Effort to see a positive meaning in the situation
8.	Seeking social support	Effort to obtain sympathy, help, information, or emotional support from another person or persons
9.	Endurance/obedience/effort	Trying to persevere, survive, submit, or comply with demands
10.	Compartmentalization	Encapsulating the problem psychologically so as to isolate it from other aspects of life
11.	Denial	Ignoring the problem, not believing in its reality
12.	Supernatural protection	Invocation of religious or superstitious practices; efforts to gain such protection (e.g., prayer, amulets); reliance on luck, fate

10.4.2 *Results*

We want to emphasize that the findings reported here concentrate on the impact of status as a member of the national majority or a minority within a space crew. Main effect differences as a function of minority-majority status based on characteristics other than nationality (e.g., gender, occupation, or job category on a space mission) are not reported.

In the findings described below, all cited differences were significant at $p = .05$ or better unless otherwise specified.

10.4.2.1 Value Hierarchies

1. ***Preflight differences:*** In references to their life before the mission, the 18 majority sources for whom we had complete data referred significantly more often to Achievement than did the 19 minorities.
2. ***In-flight differences:*** For the period of flight, internal analyses showed significant majority-minority differences on Achievement and Spirituality, with majority crewmembers higher on both.

Achievement and Conformity also showed interactions with nationality. Across all categories, Russians mentioned Achievement the most often. Americans ranked the highest on both values when flying in minority status, but Russians ranked highest when they were in the majority (see Fig. 10.1). With regard to Conformity, Americans were high when they were in the minority but low in the majority; Russians were the opposite (see Fig. 10.2).

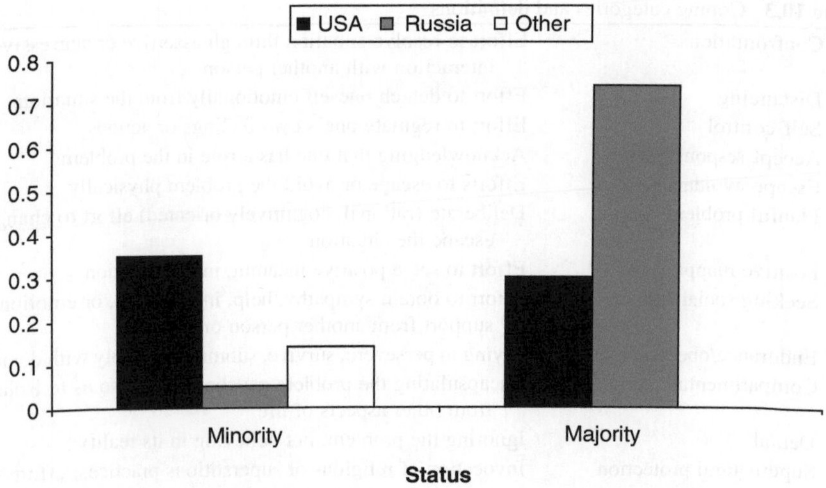

Fig. 10.1 Differences in achievement value during spaceflight

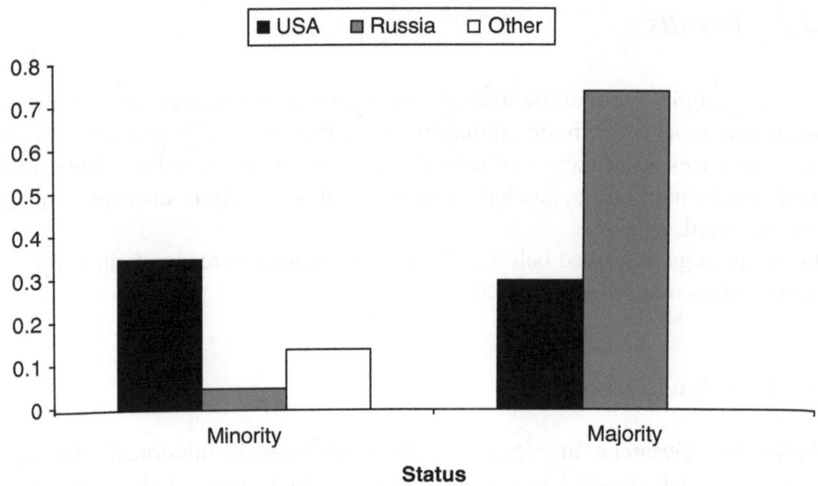

Fig. 10.2 Differences in conformity value during spaceflight

Minorities and majorities also differed significantly as a function of flight duration. When discussing in-flight periods lasting over 4 months, minorities emphasized Security more than majorities: the reverse was true for short (less than 2 weeks) missions.

Of the 18 minority astronauts who made references to values while they were in space, eight flew with predominantly American crews and ten with Russian crews. Those who flew with the Americans showed significantly less Hedonism, Self-Direction, Conformity, and Security than those who flew with the Russians. Comparing minorities and majorities flying in American or Russian crews, we found an

Table 10.4 Phase- and status-related changes (mean value scores)

Value	Status	Mission phase Preflight	In-flight	Postflight
Power	Majority	.08	.06	.23
	Minority	.11	.05	.29
Achievement	Majority	.76	.47	.47
	Minority	.46	.21	.29
Self-direction	Majority	.26	.17	.23
	Minority	.31	.10	.21

interaction: non-Americans flying with NASA mentioned Universalism more frequently than did their American colleagues; minorities flying with *Interkosmos* mentioned it less frequently than Russians in their own spacecraft.

10.4.2.2 Value Change

In general, all value references decrease when the source is discussing the period of his or her flight and mostly move back toward or above preflight baselines in descriptions of the postflight period. There were three significant phase-related value changes among the 17 minority and 15 majority crewmembers for whom we had complete data across the three mission phases (see Table 10.4).

For minority sources only, significant changes over the three phases were found: increasing references to Spirituality and Hedonism between the in-flight and postflight phases, as well as decreasing references to Stimulation and Conformity over the three phases (see Fig. 10.3). For majorities, none of the changes was significant.

There were no significant interactions of gender and status or nationality and status with regard to value change over time, nor were there significant national differences in individual- versus collective-oriented values.

10.4.2.3 Social Relations

There were four separate, although correlated, measures within this category: Trust, Mistrust, Positive Intimacy, and Negative Intimacy.

1. ***Majorities versus minorities:*** Minority or majority status made a significant difference in the social relations references: minorities were more mistrustful and more negative about intimacy than majorities. There were no reliable overall differences on the basis of gender or the nationality of crewmates, and no interaction effects. References to relations between the source and his or her family members were more positive in both subcategories when the source flew with a foreign majority. Minority astronauts showed more Negative Intimacy references toward their own space agency and showed more Mistrust toward the foreign home agency of their majority colleagues.

   There were no majority-minority differences in either positive or negative orientation toward astronauts' fellow crewmembers; these comments were

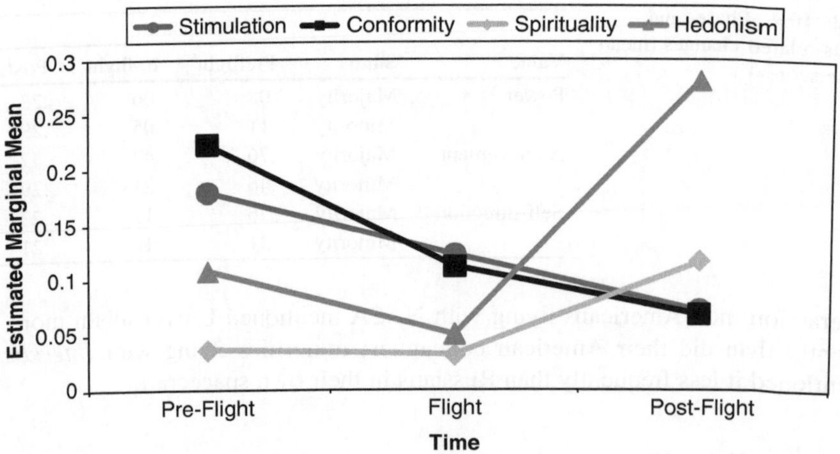

Fig. 10.3 Minority value changes by mission phase

predominantly positive (high Trust and Positive Intimacy). The same was true of references to other people in general. Minorities who flew in Russian spacecraft mentioned both positive and negative social relations (Trust and Mistrust, Positive and Negative Intimacy) more frequently than those who flew with American majorities, although only the difference in Trust reached the $p = .05$ level of significance.

2. ***Mission duration:*** Regardless of status, long-duration crewmembers made more references to Mistrust than those on shorter missions. Majorities had higher Mistrust and higher Negative Intimacy scores than minorities when discussing short missions, but minorities on long missions were more negative than their hosts.
3. ***Mission phase:*** Disaggregating minority and majority crewmembers, we found that both decreased in Trust from the pre- to the postflight portion of their narrative, but the drop was marginally ($p = .08$) steeper among minorities. On Mistrust, majorities remained stable while minorities showed a dramatic increase, especially from the in-flight to the postflight stages. There was also a marginal ($p = .07$) interaction effect on Negative Intimacy, with minorities showing a steep increase (again, especially from in- to postflight), while majorities started out much higher in the preflight stage and remained stable at that level.

10.4.2.4 Coping Strategies

Coping data were collected from 56 astronauts and cosmonauts. There were no significant overall differences for own nationality, gender, or nationality of crew colleagues, and there were no significant baseline (preflight) differences. Over all mission phases and nationalities, there were three statistically significant majority-minority differences (see Fig. 10.4).

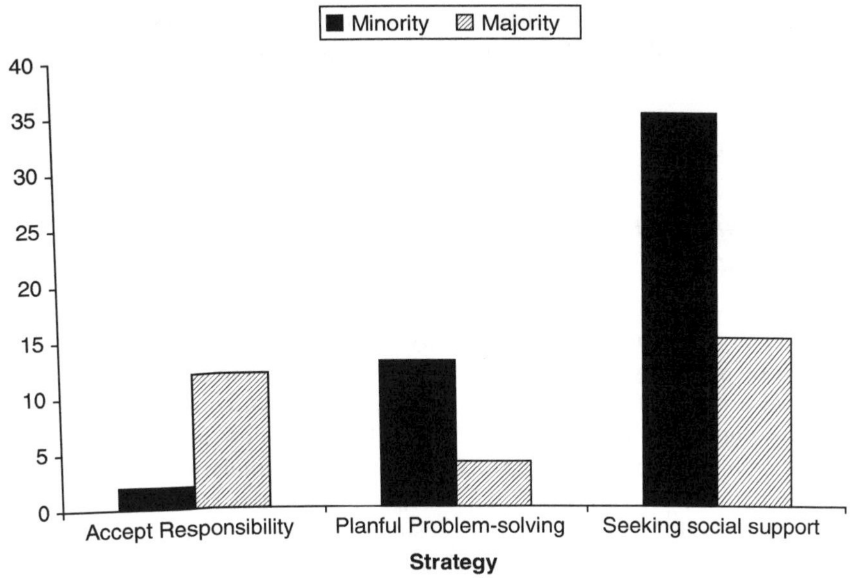

Fig. 10.4 Coping strategies by minority-majority status, measured in percentage of mentions

The remainder of the coping strategy analyses concentrated on descriptions of the mission phase where coping was most crucial – that is, during the flight. There was a significant majority-minority main effect for 3 of the 12 categories. Majority crewmembers were higher than minorities on Accepting Responsibility and lower on Planful Problem-Solving and Seeking Social Support.

Mission duration had significant effects on five coping strategies during flight, in each case with long-duration (4 months or more) fliers higher than the short (2 weeks or less). The strategies affected were Confrontation, Escape/Avoidance, Denial (all $p = .01$), Accepting Responsibility ($p = .05$), and Supernatural Protection ($p = .06$).

Still during the flight phase, duration also figured in four interactions with majority-minority status. In each case, the difference appeared in the long-duration group only: Accepting Responsibility, Denial, and Escape/Avoidance ($p = .07$), with majority fliers the highest on all three measures, and Supernatural Protection ($p = .06$), with minority crewmembers the highest.

10.4.2.5 LIWC

Analysis by nationality showed higher word count scores for Russians than Americans on references to affect in general, positive emotions, and optimism. Americans scored higher on references to social interaction. Word count differences as a function of majority-minority status showed that minorities used fewer words and phrases referring to social interaction, community, other individuals, and human beings as a group.

There were no significant differences on such LIWC categories as anxiety, anger, or sadness and no differences as a function of gender or mission duration.

10.5 Discussion

10.5.1 Flying with Strangers: The Influence of Minority-Majority Status

It seems clear that space voyagers who fly in a crew composed mostly of people from their own country have a different experience from those who are a minority flying with a mostly foreign crew. However, contrary to some assumptions that minority status would be generally aversive, the data show a mixed picture. For example, among minority participants, the value of Stimulation and Conformity decreases between the preflight and in-flight phases; this decrease presumably indicates that both boredom and the desire to submerge one's own culturally learned characteristics become less of a challenge over time. Simultaneously, Spirituality increases, indicating a growing internal recognition of transcendental values that is often found among astronauts and is apparently not thwarted – and may in fact be enhanced – by being the "odd person out" in the crew (Suedfeld 2006a). Hedonism also increases, implying a heightened concern with pleasure.

Being in the minority was associated with fewer references to social interaction, community, other specific individuals, and human beings in general. This datum emerged from the computerized frequency analysis and is difficult to interpret because LIWC merely counts words and phrases; it does not differentiate on the basis of context. More interesting is the fact that minority status also led to more positive comments about one's family, perhaps to compensate for some degree of social isolation; apparently, absence made the heart grow fonder (which was not found for majority crewmembers). One's own home organization evoked more negative references, confirming the complaints of inadequate preparation and support that characterize some anecdotal comments. The agency in charge of the mission – that is, a space agency foreign to the minority flier – was viewed with increasing Mistrust as the mission unfolded, perhaps with the recognition that its rules and procedures were alien and sometimes uncomfortable.

However, there was no evidence that bad feelings prevailed toward the majority crewmates, again despite conclusions sometimes drawn from selected anecdotal reports. In fact, the data showed a generally trustful and friendly attitude, compatible with such reports as that of Shannon Lucid, a long-duration *Mir* resident with two Russian crewmates: "Yes, we really had a good time together. We really enjoyed being there together. Yuri and Yuri were absolutely fantastic to work with. I mean, I could not have picked better people to spend a long period of time with. We just lived every day as it came. We enjoyed every day. We enjoyed working together and joking around together. It was just a very good experience, I think, for all of us" (Lucid 2012).

It should be noted that, in the same way, the majority crewmembers expressed trust and friendship toward their foreign colleagues – once again contradicting the negative picture drawn from selective quoting of particular complaints. However, comparisons of comments concerning the in-flight phase with those concerning the post-flight phase showed that these positive feelings did decline on both sides (and especially among minorities), and both majority and minority veterans of long-duration missions showed more Mistrust and Negative Intimacy than those who flew shorter missions. Growing interpersonal stress as a function of isolation and confinement with the same small group for over 4 months was thus confirmed by our data.

We expected to find changes in values, as in previous research (Suedfeld 2006). Among the most interesting changes was the drop in references to Power and Self-Direction, for both groups but especially for the minority, as the narratives moved from the pre- to the in-flight portion, followed by increases after the flight. The highly regimented aspects of the launch and the flight itself probably explain the general finding, and the somewhat tenuous and isolated role we have seen for many minority crewmembers, from which they were freed after the mission, explains their more dramatic changes. Minorities' position within the crew may also be implied by their higher scores on coping by Seeking Social Support and lower scores on Accepting Responsibility – which in many cases they were not permitted to do. However, they also used Planful Problem-Solving more frequently than their majority counterparts, perhaps because they had to face not only the problems of spaceflight, but also the problem of gaining full social equality.

Majority crewmembers' characterization of their pre-mission life was marked by more references to Achievement than that of the minority members; striving to become an astronaut may be a more vivid achievement goal for those hopefuls whose country has its own spaceflight capability (all majority subjects were from either the United States or the USSR/Russia). Achievement scores in general were high, compared to those for other values, as one would expect from a group with the high levels of achievement that spacefarers had reached even before becoming astronauts or cosmonauts (Suedfeld and Weiszbeck 2004).

We speculate that for most astronauts and cosmonauts, Achievement is a high-level background variable that tends to be taken for granted, not a primary concern, except at particular periods; preparing to become an astronaut and then to embark on a space mission may be such periods. This hypothesis is supported by the finding of overall decreases in Achievement references between the preflight phase and both subsequent phases as the successful mission and return reduce concern about the person's ability to function at the desired level. In fact, space agencies may want to provide Achievement opportunities for postflight astronauts to help them regain their interest in this value, as an astronaut for whom Achievement has become drastically less important than before may suffer serious adverse consequences in adjustment, health, and performance (Suedfeld and Weiszbeck 2004).

10.5.2 Status and Nationality

The often-cited cultural differences – especially between the two major spacefaring nations, the United States and the USSR/Russia – seemed to make no difference as to how positive and trusting relations among crewmembers were. Neither were they reflected in overall comparisons of the individual-oriented versus group-oriented values. However, Russian cosmonauts were higher in mentions of Achievement than were American or other astronauts.

In some of our data, nationality did interact with the majority or minority status of the crewmember. For example, it is intriguing to see Russians mentioning Achievement-related values especially often when they were in the majority during a space mission, but much less frequently when they were in the minority, whereas Americans did not change much as a function of their status. Achievement for the cosmonauts seems to be more closely linked to social approval from their compatriots than it is for astronauts, perhaps stemming from the collectivist-individualist difference.

Astronauts and cosmonauts varied in references to Conformity, but in opposite directions: the former were slightly higher when they were in the minority, and the latter were much higher when they were in the majority. It may be that Americans felt somewhat easier about being different from their compatriots but felt constrained to fit into their foreign crewmates' expectations. Russians, to the contrary, confirmed traditional mores when they formed the majority but were freer with dissent or individualism when flying alone with foreigners.

10.5.3 Status and Flight Duration

Discussing the spaceflight experience itself, majorities on short-duration missions referred to issues of Security more than minorities, perhaps because they were the hosts responsible for the safety of the capsule and the mission, and they may have felt responsible for the welfare of their guests as well as themselves. However, as missions lengthened, the pattern was reversed; the hosts may have come to feel more secure while the guests became more concerned, possibly because of problems that only the former knew how to solve (e.g., the gradual mechanical deterioration of *Mir*) or possibly because the latter were being prevented from full engagement in meaningful work dealing with those problems.

References to coping strategies in narratives of short flights showed no differences as a function of status. Long missions, however, evoked a majority-minority difference, with the majority higher on mentions of Accepting Responsibility, Denial, and Escape/Avoidance. It may be that these went together: as one accepted more responsibility for solving problems, it may also have become more important to use emotion-oriented means of coping to reduce psychological stress. Minorities mentioned Supernatural Protection more frequently, confirming their increased mention of spiritual values.

Minorities participating in short flights may have appreciated the experience so much that they let personal disharmony pass without comment, while the majority felt less compunction about mentioning interpersonal problems with the salient "other". In long flights, the guest may have both experienced more abrasiveness and become less reluctant to describe it later; the hosts may have become habituated to the strangeness of the visitor or (note the increased mentions of Escape/Avoidance and Denial, mentioned above) withdrawn from unpleasant interactions either physically or psychologically, or both.

10.5.4 Status and Host Nationality

Minority members also reacted differently depending on which nation constituted the majority. Minorities who flew in predominantly American crews cited pleasure and enjoyment, security, autonomy, and (paradoxically) conformity less frequently than those who flew with Russians and also made fewer valenced (positive and negative) references in either direction to social relationships. Whether these differences were the result of the minority member's changing to fit in with the majority (implying that the Russian colleagues themselves were more expressive than American crews, as is also indicated by the LIWC results) or of asserting their own cultural distinctiveness is impossible to tell.

Minorities who flew with *Interkosmos* expressed a feeling of global concern for Earth (universalism) less frequently than their Russian hosts, while the reverse was true for minorities who flew with NASA. However, this is a misleading datum: the two groups of minority fliers did not differ from each other; it was the hosts who differed, with higher scores among the Russian than the American majorities. Whether this is a function of the generally greater emotional expressivity of the Russians or is specific to the topic is not revealed by our data.

10.6 Conclusion

It appears that any problems related to mixed-nationality space crews may be more a function of the fact that space capsules have belonged to, and were predominantly operated according to the traditions and standard operating procedures of, one space agency. As a result, the minority "guest" tends to feel left out, unfamiliar with important matters that come naturally to his crewmates, and also feels neglected and let down by his own home organization. This may be an even more pressing problem for minorities who can fly only in that status – that is, all nationalities except Americans, Chinese, and Russians. The development and use of truly international missions, including international vehicles and common procedures, is a necessary countermeasure.

One possible way to reduce misunderstanding, miscommunication, and cultural friction would be for all mission participants, both space crews and ground staff, to have in situ language training and familiarization in each other's countries. When astronauts were being prepared for the Shuttle-*Mir* missions, they underwent extensive Russian language training and spent considerable time in Russia, both training and socializing with their future crewmates. Apparently it was not considered necessary for the cosmonauts in the crew to have equal exposure to American culture and folkways: cosmonauts did not have prolonged deployments to Houston to become linguistically and culturally adapted. This omission may have been economical in terms of money and time, but it was shortsighted in terms of smoothing performance and interpersonal relations in space, and the lack of similar provisions for mission controllers and staff exacerbated the problems.

We did not look at cultural or personal differences based on characteristics other than nationality, although they have also been thought of as causes of increased stress. However, there is no a priori reason to suspect that they would be any more important in that role than nationality itself, given its pervasive nature: it underlies language, values, history, traditions, child-rearing approaches, political ideologies, concepts of human nature and the individual-society relationship, and so on. We have found, as have other researchers, that differences within each group are greater than differences across groups; but the latter differences in any case were few except as they interacted with majority-minority (or host-guest) status.

As has been expected, based on anecdotes but without much empirical grounding, long-duration missions (4 months or more) reveal more abrasiveness and dissatisfaction (Lebedev et al. 1990a). Our data show that these negative tendencies also include more divergence in values and concerns, and on the part of the majority, emotion-oriented coping that does not really solve problems effectively. How this would develop further on voyages lasting several years is unknown, but is certainly something that space agencies need to think about.

Individual changes in astronaut personality – in values and social orientation – may be evanescent, persistent, or permanent. They may be particularly problematic for people who fly as minorities, especially on long missions. If they last into the postflight life of the crewmember, they may affect his or her family relationships, career progress, and physical and mental health. Again, it behooves the home organization to conduct nonthreatening and supportive post-return help where needed, both for astronauts and for their families.

10.7 Postscript

Much of the research concerning international crews has been based on the prospect that such crews will continue to be the norm, as they have been on the International Space Station. Some commentators have asserted that a project as massive and complex as the trip to Mars would have to be an international technical, scientific, and financial effort (although that does not necessarily imply a multinational crew). Politically, it seems that cooperation and collaboration have become the permanent hallmark of space exploration.

The old space race has been showing signs of reviving. NASA's three-stage plans, sparked by President George W. Bush, had turned the world's space agencies in new directions. Human return to the Moon, a Moon base, and the voyage to Mars, seemed to have become adopted as goals by all of the major space agencies. But by the same token, several agencies announced a hope that *their* astronauts would be among those who took these giant steps. Some countries (e.g., Canada) accepted that this will happen on an international vehicle, but others (e.g., Russia and China) indicated plans to go on their own. At that time, the Administrator of NASA deplored the possibility of Chinese "taikonauts" reaching the Moon before Americans return to it – an echo of the early years, when competition was the name of the game.

The Obama administration's 2010 decisions concerning the near future of NASA's human space exploration program – canceling the construction of new space vehicles, delaying if not abandoning a return to the Moon, delaying a voyage to Mars, and introducing the novel possibility of using an asteroid as the next new destination – may have put the United States on the sidelines in any such renewed space race. The impact of these changes for the future of multinational space crews remains to be seen.

References

Anderson, J.R.L.: The Ulysses Factor: The Exploring Instinct in Man. Harcourt Brace Jovanovich, New York (1970)

Burrough, B.: Dragonfly: NASA and the Crisis Aboard Mir. HarperCollins, New York (1998)

Folkman, S., Lazarus, R.S., Dunkel-Schetter, C., DeLongis, A., Gruen, R.: Dynamics of a stressful encounter: cognitive appraisal, coping, and encounter outcomes. J. Pers. Soc. Psychol. **50**, 992–1003 (1986)

Freud, S.: Leonardo da Vinci and a Memory of his Childhood. Norton, New York (1961) (original publication, 1910)

Furniss, T., Shayler, D.J., Shayler, M.D.: Praxis Manned Spaceflight Log, 1961–2006. Springer Praxis, Chichester (2007)

Hall, R.D., Shayler, D.J., Vis, B.: Russia's Cosmonauts: Inside the Yuri Gagarin Training Center, pp. 235–236. Springer Praxis, Chichester (2005)

Holland, A.: Interview by Rebecca Wright, Frank Tarazona, and Summer Bergen, 13 August 1998, published through, Shuttle-Mir Oral History Project, Johnson Space Center History Portal, http://www.jsc.nasa.gov/history/oral_histories/participants.htm (1998). Accessed 25 Mar 2012

Holquist, M.: The Philosophical Bases of Soviet Space Exploration. The Key Reporter **51**(2), 2–4 (winter 1985–1986)

Ihle, E.C., Ritsher, J.B., Kanas, N.: Positive psychological outcomes of spaceflight: an empirical study. Aviat. Space Environ. Med. **77**, 93–102 (2006)

Kanas, N., Manzey, D.: Space Psychology and Psychiatry. Kluwer, Dordrecht (2003)

Kanas, N., Salnitskiy, V.P., Grund, E.M., Gushin, V.I., Weiss, D.S., Kozerenko, O., Sled, A., Marmar, C.R.: Social and cultural issues during space missions. Acta Astronaut. **47**, 647–655 (2000a)

Kanas, N., Salnitskiy, V., Grund, E.M., et al.: Interpersonal and cultural issues and ground personnel during Shuttle/MIR space missions. Aviat. Space Environ. Med. **71**(9), A11–A16 (2000b)

Kanas, N., Salnitskiy, V.P., Weiss, D.S., Grund, E.M., Gushin, V.I., Kozerenko, O., Sled, A., Bosrom, A., Marmar, C.R.: Crewmember and ground personnel interactions over time during Shuttle/Mir space missions. Aviat. Space Environ. Med. **72**, 453–461 (2001)

Kanas, N., Salnitskiy, V.P., Ritsher, J.B., Gushin, V.I., Weiss, D.S., Saylor, S.A., Kozerenko, O.P., Marmar, C.R.: Human interactions in space: ISS vs. Shuttle/Mir. Paper IAC-05-A1.5.02, presented at the International Astronautical Congress, Fukuoka, Oct 2005

Kelly, A.D., Kanas, N.: Communication between space crews and ground personnel: a survey of astronauts and cosmonauts. Aviat. Space Environ. Med. **9**, 795–800 (1993)

Kumar, P.: Intercultural interactions among long-duration spaceflight crew (LDSF). Paper presented at the International Astronautical Congress, Hyderabad, Sept 2007

Lebedev, V.: Diary of a Cosmonaut: 211 Days in Space. Bantam Books, New York (1990a) (original publication, 1983); cited in Zimmerman, R.: Leaving Earth: Space Stations, Rival Superpowers, and the Quest for Interplanetary Travel. p. 134. Joseph Henry Press, Washington, DC (2003)

Lebedev, V.: Diary of a Cosmonaut: 211 Days in Space. Bantam Books, New York (1990b) (original publication, 1983)

Linenger, J.M.: Off the Planet: Surviving Five Perilous Months Aboard the Space Station Mir. McGraw-Hill, New York (2000)

Lucid, S.: Interview by Mark Davison, Rebecca Wright, and Paul Rollins, 17 June 1998, published through "Shuttle-Mir Oral History Project," Johnson Space Center History Portal, http://www.jsc.nasa.gov/history/oral_histories/participants.htm (2012). Accessed 25 March 2012

Lucid, S.: Six Months on *Mir*. Scientific American. pp. 46–55 (1988)

McAdams, D.P.: "Scoring Manual for the Intimacy Motive", Psychological Documents, vol. 2613. Select Press, San Rafael (1984)

McKay, J.R.: Affiliative trust-mistrust. In: Smith, C.P., Atkinson, J.W., McClelland, D.C. (eds.) Motivation and Personality: Handbook of Thematic Content Analysis, pp. 254–265. Cambridge University Press, Cambridge (1992)

Mullane, M.: Riding Rockets: The Outrageous Tales of a Space Shuttle Astronaut. Scribner, Mullane, Riding Rockets New York (2006)

North, R.C., Holsti, O.R., Zaninovich, M.G., Zinnes, D.A.: Content Analysis: A Handbook with Applications for the Study of International Crisis. Northwestern University Press, Evanston (1963)

Pennebaker, J.W., Francis, M.E., Booth, R.J.: Linguistic Inquiry and Word Count (LIWC): LIWC 2001. Erlbaum, Mahwah (2001)

Ritsher, J.B.: Cultural factors and the international space station. Aviat. Space Environ. Med. **76**, 135–144 (2005)

Schwartz, S.H.: Universals in the content and structure of values: theoretical advances and empirical tests in 20 countries. Adv. Exp. Soc. Psychol. **25**, 4 (1992)

Smith, C.P., Atkinson, J.W., McClelland, D.C. (eds.): Motivation and Personality: Handbook of Thematic Content Analysis. Cambridge University Press, Cambridge (1992)

Space Studies Board: A Strategy for Research in Space Biology and Medicine in the New Century. National Research Council, Washington, DC (1998)

Stuster, J.: Bold Endeavors: Lessons from Space and Polar Exploration. Naval Institute Press, Annapolis (1996)

Suedfeld, P.: Applying positive psychology in the study of extreme environments. J. Hum. Perform. Extreme Environ. **6**, 21–25 (2001)

Suedfeld, P.: Invulnerability, coping, salutogenesis, integration: four phases of space psychology. Aviat. Space Environ. Med. **76**, B61–B66 (2005)

Suedfeld, P.: Space memoirs: value hierarchies before and after missions – a pilot study. Acta Astronaut. **58**, 583–586 (2006)

Suedfeld, P., Steel, G.D.: The environmental psychology of capsule habitats. Annu. Rev. Psychol. **51**, 227–253 (2000)

Suedfeld, P., Weiszbeck, T.: The impact of outer space on inner space. Aviat. Space Environ. Med. **75**(7), C6–C9 (2004)

Suedfeld, P., Krell, R., Wiebe, R., Steel, G.D.: Coping strategies in the narratives of Holocaust survivors. Anxiety Stress Coping **10**, 153–179 (1997)

Surprise! Astronauts Eat in Orbit. Space Today Online. http://www.spacetoday.org/SpcShtls/AstronautsEat.html. Accessed 25 March 2012

Thagard, N.E.: Interview by Rebecca Wright, Paul Rollins, and Carol Butler, 16 Sept 1998, published through, Shuttle-Mir Oral History Project, Johnson Space Center History Portal, http://www.jsc.nasa.gov/history/oral_histories/participants.htm (1998). Accessed 25 May 2012

Thagard, N.: Interview with the Panel on Human Behavior, Space Studies Board, National Research Council, Washington, DC, 2 May 2007

Wolf, D.: Interview by Rebecca Wright, Paul Rollins, and Mark Davison, 23 June 1998, Shuttle-Mir Oral History Project, Johnson Space Center History Portal, http://www.jsc.nasa.gov/history/oral_histories/participants.htm (1998). Accessed 25 Mar 2012

Zimmerman, R.: Leaving Earth: Space Stations, Rival Superpowers, and the Quest for Interplanetary Travel. Joseph Henry Press, Washington, DC (2003)

Chapter 11
Cross-Cultural and Spaceflight Psychology: Arenas for Synergistic Research

Juris G. Draguns and Albert A. Harrison

Abstract Shortly after the competitive Moon Race the United States and the Soviet Union joined together for the Apollo-Soyuz Test Project, an activity that drew attention to cross-cultural issues in space missions. In 1978, the Soviets began their *Interkosmos*, or "guest cosmonaut," program, whereby non-Soviet cosmonauts joined Soviet crews on space stations, while Europeans developed Spacelab which flew with the U.S. Space Shuttle Columbia in 1983. Over the years, the largely symbolic *Interkosmos* program grew into flights involving true international partnerships, and NASA drew payload specialists from an increasing array of nations. In the 1990s, astronauts joined cosmonauts on *Mir*, and today the International Space Station routinely hosts international crews. Over the past 50 years, both cross-cultural and spaceflight psychology have matured, and we develop the thesis that each offers opportunities and benefits to the other. After sampling relevant research on cultural dimensions, values, and social axioms, we present the culture assimilator and other methods as potential aids for preparing personnel for future missions. We stress the need for additional research on other cultures, especially China, whose representatives are likely to become increasingly prominent in mission planning, management, and execution. We end with a brief discussion of cautions and limitations and conclude that even as spaceflight psychology will benefit from cross-cultural psychology, cross-cultural psychology will benefit from behavioral research pertaining to space.

J.G. Draguns (✉)
Pennsylvania State University, University Park, PA, USA
e-mail: jgd1@psu.edu

A.A. Harrison
University of California, Davis, CA, USA

D.A. Vakoch (ed.), *On Orbit and Beyond*, Space Technology Library 29,
DOI 10.1007/978-3-642-30583-2_11,

11.1 Space Exploration and Culturally Shaped Behavior: Anticipations and Preparations

Both spaceflight and cross-cultural psychology are young endeavors. Like spaceflight, systematic and continuous pursuit of research in cross-cultural psychology has a lengthy set of scattered antecedents going back to the first decades of the twentieth century (Jahoda 1990). The inaugural meeting of the International Association for Cross-Cultural Psychology took place in Hong Kong in 1972, and this event marked the launching of cross-cultural psychology as a continuous, systematic, and cumulative endeavor 1972 (Dawson and Lonner 1974).

The concerns of these two ventures began to coalesce in 1975, when the Cold War defrosted a bit and the United States and Soviet Russia worked together on the Apollo-Soyuz Test Project. This required substantial cooperation among project managers, mission control personnel, and engineers to ensure compatibility of the U.S. and Soviet spacecraft as well as the collaborative training of astronauts and cosmonauts. Participants grappled with each other's languages and customs, as well as each other's technology. Occasionally, astronauts and cosmonauts would slip away from program officials for hunting trips and other enjoyable, morale-building activities (Slayton and Casutt 1995). As NASA historians point out, "The flight was more a symbol of the lessening of tensions between the two superpowers than a significant scientific endeavor, a sharp contrast with the competition for international prestige that had fueled much of the space activities of both nations since the late 1950s" (NASA).

The initially daring and innovative enterprise of international missions has expanded greatly over the years, although approximately 20 years would pass before Americans and Russians again flew together. However, other international missions appeared early on. In 1976, the Soviet Union announced its *Interkosmos* or "guest cosmonaut" program. As James Oberg explains, the first, 3-week flights in March, June, and August 1978 included Czech Air Force Pilot Vladimir Remek, Polish pilot Miroslaw Hewrmaszewski, and East German Air Force officer Sigmund Jahn (Oberg 1978). Actual flying was done by Soviet cosmonauts, who had many more years of training than their guests. Oberg (1978, p. 1) recounts a joke about Remek returning to Earth with "red hands" disease. When flight surgeons asked how he had acquired this malady, Remek explained, "Well, in space, whenever I reached for this or that switch, the Russians cried 'Don't touch that!' and slapped me on my hands" (Oberg 1978). Other early guest cosmonauts – each with a specific assignment – were recruited from Cuba, Bulgaria, Mongolia, and Romania.

Although Oberg portrays *Interkosmos* as largely symbolic, he also predicted – correctly – that it would give rise to missions with more active participation on the part of the international partners. More recently, in a report for NASA, Oberg pointed out that Russia has flown cosmonauts from many countries including Germany, India, Italy, Japan, South Africa, Syria, the United Kingdom, the United States, and Vietnam (Astronaut 2010). The first Western European to fly, Frenchman Jean-Loup Chrétien, did so aboard Russian craft in 1982.

In the late 1970s, while the *Interkosmos* program flourished, the last Apollo rocket had been launched, but the Space Shuttle had not yet completed the transition from drawing board to orbit. In 1969, NASA had invited Europeans to participate in post-Apollo missions, and in the 1970s, while the Europeans were hard at work designing Spacelab, Ulf Merbold from West Germany, Wubbo Oeckels from Holland, and Claude Nicollier from Switzerland were training for the Shuttle. Merbold, flying aboard *Columbia*, was the first to reach orbit and did so in November 1983. By 2003, the 25th anniversary of Remek's historic flight, 30 European astronauts had participated in 44 missions: 26 in collaboration with NASA and 18 aboard Russian spacecraft (Feustl-Beuchl 2003). As of early 2012, Canadian astronauts had participated in 1 Soyuz and 14 Space Shuttle flights (Canadian Astronaut Missions 2012). By the same time eight Japanese astronauts had successfully completed their missions, and as early as April 2008, the first South Korean astronaut, Yi So-Yeon, blasted off with two cosmonauts to head to the ISS (JAXA).

According to Peter Suedfeld, Kasia E. Wilk, and Lindi Cassel, the success of the ISS portends a transition from mislabeled multinational to truly international missions (Suedfeld et al. 2011). Many early missions described as "international" should have been called "multinational" because of a landlord tenant type of relationship between the Soviets, Russians and Americans who controlled access to space and the foreign spacefarers who were given limited access to their spacecraft. The ISS, a tangible product of stakeholders from many different lands, may erase the landlord-tenant distinction and initiate a transition from nominal to truly international crews.

International flights make tremendous sense. First, by drawing from an international pool, managers can tap a broader range of interests and skills. Second, space missions are, in effect, overarching or superordinate goals that encourage different nations to work together and may serve as a prototype for other collaborative ventures. Third, the more nations that are involved in a mission, the greater the number of people who can identify (if only vicariously) with the challenge and triumphs of spaceflight. Space belongs to everyone, and international missions underscore this point. International missions reduce duplication of effort. And, most obviously, international cooperation defrays the enormous expense, increasing the palatability of flight for large, wealthy nations and enabling nations with fewer resources to participate in space.

Since the United States and Russia control access to space, relations between U.S. astronauts and Russian cosmonauts have gained the most attention (Ritsher 2005). There have been too few astronauts from other cultures for meaningful inclusion in systematic statistical studies. But this is likely to change as more and more countries enter into collaborative space ventures, especially piloted ventures that send more representatives into orbit and beyond. The success of the Europeans with ESA and its coordination with Russian and American programs has yet to be achieved in Asia, where there is little in the way of regional coordination (Moltz 2011).

We found little discussion of cultural factors in early space psychology papers, perhaps because during the tensions of the 1960s it was all but impossible to imagine astronaut-cosmonaut collaboration. By the mid-1980s, in recognition of the Apollo-Soyuz Test Project, *Interkosmos*, and Spacelab, cultural factors had joined the list of behavioral concerns (Connors et al. 1985). A 1993 study based on debriefing American astronauts who flew with international crewmembers found 9 preflight, 26 in-flight, and 7 postflight incidents of misunderstanding, miscommunication and interpersonal conflict (Santy et al. 1993). Interest was piqued further in the mid to late 1990s when astronauts joined cosmonauts on *Mir*. Journalist Bryan Burrough described (and perhaps slightly sensationalized) several instances where cultural differences influenced astronaut performance and morale (Burrough 1998). Cultural problems that the astronauts reported pertained to personal hygiene, food preferences, and chosen activities as well as to interpersonal distance, privacy, and work styles (Ritsher 2005). Jason Kring summarized experience with international crews prior to the ISS as follows:

> Apollo-Soyuz, Shuttle, and Shuttle-Mir proved, for the most part, that international cooperation was feasible and rewarding. During Shuttle-Mir, the U.S. and Russia gained from each other's strengths and weaknesses, learned how to work together, and gathered insights that directly benefitted the ISS. There were, however, moments of confusion and disagreement between the two countries, incidents that affected crew performance and mission activities. Differences in management, training, decision-making and problem resolution, for example, were tied to differences in national culture and the backgrounds of the U.S. and Russian crewmembers and personnel (Kring 2001, p. 11).

A recent International Academy of Astronautics study group chaired by Nick Kanas pointed out that one of the reasons that culture is so important is because it influences almost all forms of behavior (Kanas et al. 2013). In addition to contributing to the problems noted above, culture influences psychological reactions to stress, shapes social behavior, and molds preferences in such areas as recreational activities, humor, and privacy. As Jennifer Ritsher has shown, culture affects the likelihood and kinds of mental illness that arise, as well as diagnosis and treatment. Forms of therapy that work in one culture may be ineffective in another (Ritsher 2005).

As early as the 1970s, then, NASA and its Soviet, later Russian, counterpart agency were faced with the challenge of ensuring that astronauts and cosmonauts would work together smoothly and effectively. To this end, it was imperative to identify and overcome obstacles rooted in socialization within different cultures and contrasting sociopolitical systems. They had to recognize and accommodate the weight of lifelong training and indoctrination, inextricably tied to the divergent traditions and ideologies of the two countries.

When initially confronting cultural issues, space agencies had to rely on intuition, trial and error, and common sense. Now they can draw on the results of several decades of research on interaction in small groups, including interviews of astronauts and support personnel, to better understand the social realities within the space habitat. Researchers and operational personnel now build on the store of information accumulated in preparing sojourners, visitors, and immigrants for encounters with a new culture. Contemporary researchers both utilize and contribute to the theories and tools of emerging academic cross-cultural psychology. Thus, we see a shrinking gulf between cross-cultural psychology and efforts to prepare personnel for international missions. We turn now to some of the theories and principles that help reduce this gulf.

11.2 The Domain of Cross-Cultural Psychology: Cultural Characteristics and Intercultural Interaction

Cross-cultural psychology is concerned with "the study of similarities and differences in individual psychological functioning in various cultural and ethnic groups; of the relationships between these variables and socio-cultural, ecological, and biological variables, and of ongoing changes in these variables" (Berry et al. 2002, p. 3). Within this broad field, intercultural research focuses on the study of contacts and interactions between persons from different cultures (Brislin 2000). In his discussion of international crews, Kring (2001, p. 12) described culture as "an individual's values, beliefs, behavior patterns and language that are directly linked to his or her national and ethnic background" (Kring 2001). Other researchers have favored a broader definition that views culture as multi-leveled (Boyd et al. 2009). Advanced by Helmreich and Merritt (1988, p. 5) it states that "culture fashions a complex framework of national, organizational, and professional attitudes and values within which groups and individuals function. The power of culture often goes unrecognized since it represents 'the way we do things *here*' – the natural and unquestioned mode of viewing the world" (Helmreich and Merritt 1988). This definition acknowledges that culture is found in national, organizational, and occupational differences and each requires study. For example, while national differences (US and Russia) and organizational differences (NASA and the RSA) might separate astronauts and cosmonauts, occupational similarities (along with other factõrs traditions and rituals that evolve in the course of specific missions) draw them together. Thus, cross-cultural research extends beyond flight crews. It includes space agencies, program managers, and mission control, and differences that affect how mission support personnel get along with one another and interact with the space crew (Boyd et al. 2009). There has been considerable research on cultural dimensions and values, and social axioms may be a fruitful area for future study.

11.2.1 Cultural Dimensions

On the basis of multistage factor analyses conducted on a huge store of data from 72 countries, G. Hofstede identified four relatively independent cultural dimensions (Hofstede 1980). Originally based on questionnaire responses on values in the workplace, Hofstede's four factors have been studied around the world in educational, social, mental health, and many other settings. The second edition of his monograph and more recent publications, for example by G. Hofstede and G. J. Hofstede, document a prodigious amount of research-based findings (Hofstede 2001; Hofstede and Hofstede 2005). It is no exaggeration to say that Hofstede's dimensions are among the most intensively investigated psychological constructs in the last four decades.

As labeled by Hofstede, the four dimensions pertain to power distance, individualism-collectivism, masculinity-femininity, and uncertainty avoidance, respectively. Subsequent to the completion of Hofstede's original data collection, a fifth dimension was added on the basis of a somewhat different methodology. It refers to short- versus long-term orientation and was originally designed to span the continuum between Confucian and Western dynamism. Power distance was defined by Hofstede and Hofstede as the extent to which unequal distribution of power is accepted as a normal or natural state of affairs. Individualism-collectivism is a bipolar dimension, bounded by the exaltation of the self at one extreme and the overriding prominence of the group or collectivity at the other. Masculinity-femininity refers to the characteristic overlap between gender roles that is encountered within a culture. Such overlap is less extensive in masculine and more extensive in feminine cultures. Moreover, masculine cultures are tilted toward achievement and performance, and feminine cultures favor caring. Uncertainty avoidance is high in those cultures where situations that lack clarity or structure are experienced as unpleasant or noxious. Low uncertainty-avoidance settings thrive on ambiguity, unpredictability, and improvisation. Long-term orientation differs from short-term orientation in the degree to which a culture's members are willing to postpone immediate reinforcement in favor of delayed rewards.

In the present context, it is worth noting that, as reported by Hofstede and Hofstede, the U.S. culture has been found to be the highest in individualism among the 72 cultures for which such indicators are available and that Russia's rank on this dimension is 37th and Japan's, 46th (Hofstede and Hofstede 2005). On the other hand, Russia clearly exceeds both Japan and the United States in power distance, in which it ranks 6th while Japanese and American ranks are 50th and 58th, respectively. In uncertainty avoidance, the ranks are 7th for Russia; for Japan, 12th; and for the United States, 63rd. In masculinity-femininity, Japan is decidedly more masculine, ranking 2nd, while the United States and Russia placed 19th and 63rd, respectively. Japan is also substantially higher in long-term orientation, placing 4th on this dimension, with United States coming in 29th out of 38 countries. Russian university student samples were higher in long-term orientation than those of their counterparts in the United States, a finding that indicates persistent paternalism in the Russian culture (Naumov and Puffer 2000).

To what extent (if any) and in what specific situations are these cultural value dimensions practically relevant to international spaceflight? We might expect that the demanding characteristics of spaceflight and the rigorous training that all astronauts undergo would override any cultural differences in Hofstede's dimensions. Moreover, we admit freely that the available cultural ranks for these traits may not be predictive of a specific individual's value pattern or of his or her performance. However, as Jennifer Ritsher and Pascale Ehrenfreud and associates point out, observations over several decades of international spaceflights forcefully show that cultural differences do matter. Extrapolation from the rapidly growing body of findings pertaining to personality differences between Russians and Americans strongly suggests that Hofstede's dimensions tie together many results (Ritsher 2005; Ehrenfreund et al. 2010). In particular, the importance of collectivism-individualism has been widely recognized. According to Ritsher (2005: B135), Yuri Gagarin anticipated Hofstede's findings and formulations by stating in 1968 that "in our country, it is much easier to form a crew for a long-duration space mission than in capitalist countries [because we] are collectivist by nature" (Ritsher 2005).

Merritt's comparison of airline pilots from 19 nations on Hofstede's dimensions found a significant effect of these variables on attitudes and behaviors, even though there was a detectable commonality across countries based on pilots' occupational culture. Differences across cultures were more pronounced when the indicators were made more relevant to the pilots' occupational context. Merritt (2000, p. 299) concluded that these results "may extend to any population that is hierarchical in nature and involves teams of individuals interacting in high stress, high technology environments."

Relevant as the differences in collectivism-individualism are, they should not be the sole or principal focus of investigation. As already noted, even greater contrasts have been uncovered in power distance and uncertainty avoidance, as have substantial differences in masculinity-femininity and long-term orientation. Recent historical and sociological analyses of Russian society and polity emphasize the role of vertical and hierarchical, dominant-submissive relationships within the Russian social structure (Afanas'ev 2001; Shlapentokh 2006, 2007). Their reverberations in human interaction, especially in the context of teamwork under demanding and stressful conditions, do not appear to have been studied systematically or intensively. Similarly, it is not yet clear in what ways, if at all, the substantial Russian-American differences in uncertainty avoidance may be manifested in space. The discrepancies in ratings in masculinity-femininity and long-term versus short-term orientation appear to be baffling or even counterintuitive and call for open-ended exploration through basic and applied research. In what sense are the Russian society and its members committed to long-term goals, and what form does the feminine orientation take in social interaction, attitudes, and behavior? The initial impression is that of a conservative, authoritarian, though more matriarchal than patriarchal, society, which appears paradoxical in light of Russia's radical and revolutionary history through much of the twentieth century.

11.2.2 Values

A second line of cross-cultural research comes from Shalom Schwartz's and A. Bardi's assertion that values constitute the central, nodal features of cultures that inevitably radiate to and affect all aspects of human behavior, conduct, and performance (Schwartz and Bardi 1997). Schwartz has developed and extensively investigated in 20 nations a set of 11 values based on important and stable life goals (Schwartz 1992). Schwartz's value categories have been studied in relation to national differences (American versus Russian), majority versus minority status, and host versus guest status in multinational space crews. On Earth, Russian teachers endorsed hierarchy and conservatism to a greater degree and intellectual and affective autonomy to a lesser degree than did their colleagues in several countries of Western Europe.

11.2.3 Social Axioms

The relatively new concept of social axioms was introduced in Hong Kong by Leung and Bond (Bond and Leung 2004; Leung and Bond 2004). It refers to general, context-free beliefs that a great many people at a point in time and space hold, presumably on the basis of their socialization and lifetime personal experience. Collectively, a person's social axioms represent his or her convictions of how the world functions. As is true of assumptions in general, social axioms are accepted as given and are not questioned. As a rule, they are encoded in the form of assertions about the relationship between two concepts or phenomena. Thus, "wars do not resolve the problems that cause them" is a social axiom, but "wars are inherently bad" is a value judgment. Major international surveys have been completed to ascertain the factorial structure of the composite of social axioms at the individual and national levels and to gather comparative data across countries. On the country level, a two-factor solution composed of social cynicism and dynamic externality has emerged (Bond et al. 2004). The former factor encompasses a negative view of human nature and social structure and is characterized by distrust and skepticism about ethical rules and principles; the latter factor includes and juxtaposes the interplay of religion and fate with personal effort and control of one's destiny.

It is implausible to expect that space explorers of any nationality would be high on the pessimistic and misanthropic social cynicism dimension. However, it would be exceedingly interesting to learn more about astronauts' scores on dynamic externality, a dimension that is germane to plunging into the unknown, tapping the limits of personal control over the environment, and striking a balance between fate and effort. Are there any national differences and, if so, how do they manifest themselves in space flight?

Are they related to any of Hofstede's dimensions or Schwartz's and Bardi's values? With the instruments developed by the Hong Kong researchers and their international framework of collaborators these questions should be relatively easy to answer. And will the commonalities that the highly atypical and rigorously selected spaceflight personnel may demonstrate, across nationality, gender, and other social parameters, be reflected in convergent responses on the social axiom scales, perhaps pointing to the development of a discernible culture of space explorers?

11.3 The Culture Assimilator

The culture assimilator, developed by F. E. Fiedler, T. Mitchell, and H. C. Triandis, is an extensively used and systematically validated programmed learning approach to cross-cultural training (Fiedler et al. 1971). As described by Triandis (1994, p. 278), "it consists of 100–200 episodes, i.e. scenarios where people from two cultures interact. Each episode is followed by four or five explanations of why the member of the other culture has acted in a specific way. The trainee selects one of the explanations and is asked to turn to another page (or computer screen) where feedback is provided concerning the chosen explanation." The objective is not only to help people anticipate concrete situations that are construed and reacted to differently in another culture, but to help them gradually grasp the rules that underlie the various expected and "normal" courses of action in the culture in question. Controlled studies of training by means of the culture assimilator demonstrate that it results in trainees' selecting explanations of others' behavior that are closer to those chosen by the members of the target culture. Moreover, the learners' attributions tend to become more specific and complex and less ethnocentric, and they become better at estimating what members of the target culture will do (Triandis 1994). Thus, a great deal of anticipatory culture learning takes place in a relatively short period of time.

Conceptually, the development of the culture assimilator was linked with basic research in Greece, India, and Japan, which Triandis conducted on subjective culture (Triandis 1972). Subjective culture encompasses the characteristic ways of viewing the human made environment (Triandis 1972). More specifically, it comprises the standards by which events and actions are evaluated in a culture as well as the ideas, theories, and explanations shared by that culture. The investigation of subjective culture focuses on categorizations, associations, and beliefs, and on their interrelationships. In this manner, it becomes possible to ascertain the inner coherence of culture as it is experienced by its members. Subjective culture is implicitly known to its members, even though this knowledge is rarely articulated. Somewhat like a speaker's first language, it is used effectively and correctly by the members of the culture, but its rules and practices must be made explicit for an outsider.

Triandis developed a number of complex methods for bringing facets of subjective culture to the fore (Triandis 1994). Thus, beliefs, behavior patterns, and social rituals are related to the other features of the culture. In this manner, the culturally characteristic meanings of "happiness," "risk," "science," "family," and other terms and concepts are established empirically and related in pinpoint fashion to behaviors, practices, and convictions.

Subjective culture allows investigators to peer beneath the behavioral surface and to uncover the implicit, yet socially shared levels of consciousness in their relation to characteristic behavior. Just as no two persons speak their first language alike or perfectly, each person's subjective culture is the best approximation of his or her culture based on the individual's lifetime of experience. To further optimize the intense and demanding interaction in the space capsule, we can ascertain the subjective cultures of specific individuals in their personal uniqueness, cultural distinctiveness, and human universality.

K. Cushner and R. W. Brislin extended culture assimilator methodology to a set of 100 sketches pertaining to 18 themes representative of various domains of intercultural experience (Cushner and Brislin 1995). These vignettes represent critical incidents frequently encountered by newcomers to a culture other than their own. Each vignette is accompanied by four options, one of which, through thorough pretesting, has been deemed to be the most culturally appropriate. Critical incidents can serve as points of departure for the development of more specific, yet critically important, items that would sensitize astronauts to crucial, yet culturally divergent, aspects of interaction and cooperation during space missions. Extending these methods to space, researchers would pool knowledge of participants' cultures with information on the personal, social, and technical challenges of flight. That is, previously encountered problems, conflicts, and crises are the requisite "raw material" for the construction of the vignettes to make them more relevant and applicable to spaceflight and its cultural aspects. The culture assimilator thus represents a vicarious enculturation experience preparatory to the actual encounter with a new and different culture. We see at least five advantages that the assimilator may provide.

First, the rationale, if not the actual procedure, of the culture assimilator pinpoints and makes explicit the themes and features of the other culture. This is particularly germane to the debriefing phase as specific incidents and encounters come to be imbued with more general meaning and patterns are established.

Second, properly constructed, the culture assimilator increases the range of situations that people are likely to encounter. This helps people "fill in the gaps" and respond appropriately to spontaneous and somewhat unpredictable instances of unfamiliar cultural realities.

Third, as well illustrated by prior research, on the basis of the accumulated experience with mixed American-Russian space crews over three decades, we can anticipate certain problems (Ritsher 2005; Kanas et al. 2013; Boyd et al. 2009; Kring 2001). We can then devise specific items for incorporation into custom-made versions of the assimilator. That is, the assimilator can be adapted to cover the special circumstances of mission managers, control personnel, and flight crews.

Fourth, the culture assimilator methodology is useful in both its specific and its general aspects. Given the increasing range of nations from which spaceflight participants are increasingly drawn, it may be possible to develop a general culture assimilator for increasing generic awareness of sensitivity to different cultures, especially into the early phases of astronaut training. This would involve promoting a specific aspect of the acceptance of uncertainty, especially in the context of intensive interpersonal, task-oriented interaction in mission control or the space capsule. To this end, traditional verbal modes of the culture assimilator may be creatively and innovatively supplemented by the filmed or other illustrations or simulations of particularly relevant or crucial intercultural interactions.

Fifth, since assimilators can be geared to many different levels: national, organizational, and occupational, they can be applied not only to different nations, but also different organizations (such ESA, NASA, and RSA) and subcultures within these organizations. Eventually the culture assimilator methodology may be applied to new groups – such as tourists, industrial workers, and migrants and others as well as highly trained professional astronauts who may eventually fly to new destinations such as space hotels, Mars, and the outer reaches of the solar system and beyond. This may make it easier to understand and ease the tensions that are likely to arise between the professionals and the various classes of passengers that may someday accompany them (Harrison 2010).

11.4 Expanding the Range of Cultures

As space exploration becomes ever more international, the approaches and findings of cross-cultural psychology may be extended to other cultures. Japanese astronauts have flown with both Americans and Russians. Observations and experiences from these flights, from both the hosts' and the Japanese participants' points of view, would provide valuable guidance in planning future spaceflights. There is substantial and diverse research-based literature on cultural characteristics in personality, emotional expression, and social behavior in Japan. Thus, in addition to Hofstede's dimensions in Japan, described earlier, one can draw upon studies of differences between Japanese and Americans and ways to bridge them (Goodykunst and Nishida 1994); characteristics of Japanese selves (Rosenberger 1992), modes of interpersonal relating (Doi 1973) and socialization within the family and at school (Shwalb and Shwalb 1996); indirect control, empathic nonverbal communication (Ehrenfreund et al. 2010), and many other themes. Collectively, these contributions point to the prominence of group orientation in Japan that is reflected in self-experience, interpersonal relations, and styles of adaptation. But it would be an oversimplification to describe the Japanese self as sociocentric or the Japanese social orientation as collectivistic. Instead, over more than five decades of systematic research reveals a more complex picture pointing to an intricate intertwining of collectivistic attitudes with individualistic striving. Japanese and Americans both strive toward attaining and maintaining a more positive sense of self. However, Heine et al. (2000, p. 71) have concluded that "Japanese self seems better characterized by a need to secure a positive view from others rather than from oneself and this securing of others' approval seems to be better served by self-improvement rather than self-enhancement." Performance and well-being of the Japanese tends to be enhanced in groups composed of their compatriots; subjective and output effects of being a minority of one among strangers in mission control or space awaits systematic investigation.

China is a new and dynamic participant in space exploration, with plans for extended, and possibly international, piloted missions. From the Chinese perspective, space exploration would a natural extension of a long history of intermittent peaceful exploration. China has entered into cooperative space ventures with other countries (including both Russia and the US) and a crew member identified as Chinese took part in the Mars500 simulation, noting that one of the things he missed most was his mother's Chinese cooking. (Apparently, a bamboo-shoot pizza that was served one evening was not completely satisfying.) However, whether or not there is an informal "ban," taikonauts have yet to be prominent on the ISS. A national space program guarantees access to space. Over time, the Chinese space program may become regional (something like ESA) or global. There are many complexities here, and only time will tell.

Perhaps reflective of its long-term perspective, China has maintained a clear vision for space, in sharp contrast to the ever-shifting American plans (Wong and Chang 2011). Many countries have either abated their appetites for piloted missions or have joined other nations that can already provide humans access to space. Why, 30 years after the Moon Race, would China develop its own program to send taikonauts into orbit despite knowing that the American and Russian programs were enormously expensive and yielded little of material value? Fiona Cunningham writes that, in the eyes of Chinese political leaders who are responsible for China's international standing, status is the primary motivator. "China" she writes, "has possessed a manned space program because it understands manned spaceflight as a status marker for the superpower status group, and, desiring membership to that group, has attained one of its status markers" (Cunningham 2009, pp. 73–74). This is not the sole explanation, she adds, as orbiting a manned spacecraft requires a coalition of political, military, scientific and engineering interests, each of which benefits from the achievement.

Two recent resources may help us in bringing the findings of cross-cultural psychology to bear upon the anticipated joint space flights with Chinese astronauts. The 31 chapters of *The Oxford Handbook of Chinese Psychology* edited by Michael Bond cover and integrate a wealth of research-based findings on virtually all domains of behavior and experience of the Chinese people (Bond 2010). An article by Ehrenfreund and her coworkers is focused on cross-cultural management in support of global space exploration and prominently includes relevant empirically-grounded information on pertinent Chinese characteristics (Ehrenfreund et al. 2010).

In light of Hofstede's global aggregate of data, China is characterized by high ranks in three out of five dimensions: long-term orientation, in which it ranks first among the 39 countries for which scores were available (accounting, perhaps, for the stability of China's goals in space), power distance, wherein it shares the score of 12th–14th, and masculinity with a rank of 10–11, all in a sample of 74 countries (Hofstede and Hofstede 2005). Within the same sample, China's rank in individualism is between 56th and 61st, consequently placing it toward the high end of the distribution in collectivism. In uncertainty avoidance, China's rank, perhaps surprisingly, is near the lower end of the distribution, 68th–69th of 74, in contrast to Japan's high score on the same dimension, which was presented earlier in this chapter.

Two general impressions emerge from these findings. First, on all five dimensions, Chinese ranks tend toward the extreme; they are either very high or very low. It is not clear what inference can be drawn from this mix of polarities, but it is worth noting. Second, the composite of these five characteristics does not reflect the cataclysmic and convulsive transformations that the Chinese society and polity have experienced since 1949. In fact, the scores for the People's Republic of China on all five of Hofstede's dimensions are remarkably similar to the scores for the other three entities with a predominantly if not wholly Chinese population: Taiwan, Hong Kong, and Singapore.

Consistent with Ehrenfreund et al. and Smith we would like to characterize the Chinese culture as hierarchical and collectivist with a Confucian long-term orientation and a quest for relatedness and interpersonal harmony as fundamental goals (Ehrenfreund et al. 2010; Smith 2010). Despite upheavals and momentous changes, Confucian values remain important, especially in respect for age and status, and expectations of benevolent paternalistic leadership (Hwang and Han 2010; Chen and Farh 2010; Liu et al. 2010). Long-term relationships are emphasized for the accomplishment of joint objectives and trust is important or even essential for cooperation. At the same time, the Chinese have shown themselves to be remarkably adaptable to external change and resilient to stress, perhaps on the basis of their low uncertainty avoidance. In intercultural encounters, the Chinese eschew confrontation in favor of compromise (Leung and Au 2010).

A characteristic Chinese cultural feature is the concern with maintaining and avoiding loss of face (Smith 2010). Face can be described as the concern with maintaining one's reputation and dignity that accrues with status and age, and the status of a piloted space program mentioned earlier in this section may help account for the Chinese space program. Loss of face occurs when there is a discrepancy between one's reputation and role-based behaviors and actions. Loss of face can be precipitated by situational circumstances or personal actions and deficiencies. Incompetence in performing professional tasks is definitely one of the conditions that may trigger loss of face. Although loss of face is a traditional concept rooted in Confucianism, contemporary young adults in China are no strangers to this phenomenon. In social interactions across cultures, inadvertently inflicting loss of face upon the Chinese partner should be avoided, which may be best accomplished by following the expected rules of behavior, avoiding excessive informality and spontaneity, and anticipating Chinese vulnerabilities and sensitivities to specific topics. Again, immersion in the other culture, this time by the non-Chinese participants in joint programs, may reduce these risks, perhaps with the help of culture assimilators and other relevant approaches.

James R. Hansen points out that like astronauts and cosmonauts, the first taikonauts may be cast as cultural icons (Hansen 2007). Like Yuri Gagarin and John Glenn they are highly fit, talented, and skilled individuals who have survived rigorous selection and training procedures. Like Gagarin and Glenn taikonauts are effusively patriotic, expressing great love for their country. However, in comparison to some of their western counterparts they express humility. They do not swagger, revel in partying, or use "macho" language. Rather, they have a strong interest in relationships, with people in general, especially for parents, spouses and children. The combined task and relationship orientation may serve them well on missions that require a group of people to work and live together on an extended-duration mission. Hansen points out that developing publicity to make the first taikonaut (and his son) heroes represented a sea-change in a society whose existence depended on being part of a group and typically down-playing the significance of the individual. Whereas the Chinese program is a symbol of achievement and status it is also a source of national pride.

11.5 Concluding Comments

In this chapter, we have noted that over the past few decades both cross-cultural and spaceflight psychology have moved towards maturity. The interests of the two endeavors began to merge with international missions in the 1970s. Since that time, psychologists have shown growing interest in cross-cultural differences, primarily at the national but also at the organizational occupational levels. Researchers cited throughout this chapter have already demonstrated the utility of Hofstede's cultural dimensions and Schwartz's and Bardi's values. Social axioms may be another fruitful area for future research.

We suggest that the culture assimilator is a flexible and useful tool for preparing international partners for future missions. We propose this as an adjunct to in-depth language training, immersion in different cultures, cultural sensitivity training, debriefings, and other established procedures. Preparation has didactic and experiential components, and no single research or training strategy can resolve all cross-cultural issues (Hysong et al. 2007).

Even as cross-cultural psychology has offered some leads to spaceflight psychology, studies of intercultural interaction in space contribute to the store of knowledge of cross-cultural psychology. Oftentimes observations from space are thoroughly documented. Given the singularity of the situations involved, qualitative observations of interactions and encounters have led the way. But slowly, as the number of people who fly in space increases and attention extends to flight support personnel (who often exist in large numbers) researchers are obtaining quantitative findings. Of course, culture itself does not stand still, so cross-cultural research must be on-going. Consider how national, organizational, and personal goals been transformed in the course of the last 50 years. Cultures of both the United States and Russia have undergone various degrees of change, from the social upheaval in America in the 1960s and 1970s to the demise of the Soviet political and economic system.

With only slight exaggeration, adaptation to space missions, especially long-duration missions, can be likened to adaptation to a small society or a mini-culture, with its own internal rules and evolving customs as well as externally imposed imperatives. Contemporary researchers are exploring the characteristics of the communities of space exploration on the ground and in space, their commonalities across nations, and their culturally specific hallmarks.

Perhaps future researchers will conduct cross-cultural studies of different space settlements. Presumably, if and when colonists set forth on lifetime missions, pre-existing national, organizational and occupational norms will hold sway at first, but with the passage of time each settlement will begin to develop its own constellation of cultural attributes (Harrison 2001). Terrestrial ideology and ritual may persist, but increasingly in a skeletal or stripped down form. Meanwhile, developments and cultural changes on Earth may have very little impact on groups that departed their planet of origin years or generations ago. Local myths, heroes and villains, triumphs and tragedies, and chance events will gain ascendance in off-world cultures. At some point, there may be many different human cultures scattered throughout the solar system – or beyond – each with its own identifiable culture, but with many of the cultures differing substantially from each other. On Earth, and possible elsewhere, cultures rise, change, and fall, and the task of the cross-cultural psychologists – by that name or any other – will never end.

We conclude with some notes of caution. Let us forever keep in mind that differences between individuals who come from different cultures are not automatically traceable to culture. In complex situations encountered in the course of international spaceflight, national, organizational and occupational culture are likely to interact with other factors in affecting behavior and well-being, and in some instances, national culture may have had a minor or negligible influence. Moreover, astronauts and cosmonauts are of necessity exceptional individuals, certainly shaped by their respective cultures, but also statistically discrepant from most of their cultural peers. We face a formidable undertaking to disentangle the cultural threads operating in any concrete instance. It is a challenge that requires persistence, patience, and creativity to formulate general principles of culture's influence in interaction with the person and his or her context.

In this chapter we have said nothing about gender-by-culture interaction in space flight. The principal reason for this omission is the small number of women who have participated in international space missions. In the future, this situation is expected to change. As more women of various nationalities take part in spaceflight, instances of culture's interplay with gender will invariably accumulate and provide bases for generalizations.

In past decades, cultural factors in behavior and experience tended to be slighted or overlooked; with the current emphasis on cultural awareness, there is an emergent danger of exaggerating the importance of culture. The former tendency leads to denial of culture's reality and consequently to cultural insensitivity; the latter trend fosters stereotyping and thereby reduces sensitivity to personal and unique characteristics.

Space exploration has been both a challenge and a boon to cross-cultural psychology: a challenge by demonstrating that, even though it is a young branch of its discipline, it has methods and findings to contribute to the continued progress of space exploration, and a boon by making it possible to collect a wealth of unique data of cultural relevance. These data remain to be sifted, categorized, and incorporated into a set of general principles. Potentially, cross-cultural psychology stands to be tremendously enriched as this happens. We hope that cross-cultural psychology's contribution to future missions will constitute a partial repayment.

References

Afanas'ev, Y.: Opasnaya Rossia (Dangerous Russia). Russian State Humanities University Press, Moscow (2001)

Astronaut.: World Book at NASA, http://www.nasa.gov/worldbook/astronaut_worldbook.html. Accessed 23 May 2010

Berry, J.W., Poortinga, Y.H., Segall, M.H., Dasen, P.R.: Cross-Cultural Psychology: Research and Applications, 2nd edn. Cambridge University Press, New York (2002)

Bond, M.H. (ed.): The Oxford Handbook of Chinese Psychology. Oxford University Press, Oxford (2010)

Bond, M.H., Leung, K., et al.: Culture-level axioms and their correlates across 41 cultures. J. Cross-Cult. Psychol. **35**, 548–570 (2004)

Boyd, J.E., Kanas, N.A., Salnitskiy, V.P., Gushin, V.I., Saylor, S.A., Weiss, D.S., Marmar, C.R.: Cultural differences in crewmembers and mission control personnel during two space station programs. Aviat. Space Environ. Med. **80**(6), 532–540 (2009)

Brislin, R.W.: Understanding Culture's Influence on Behavior, 2nd edn. Harcourt College Publishers, Fort Worth (2000)

Burrough, B.: Dragonfly: NASA and the Crisis aboard Mir. Harper Collins, New York (1998)

Canadian Astronaut Missions.: http://www.asc-csa.gc.ca/eng/astronauts/missions.asp. Accessed 25 Mar 2012

Chen, C.C., Farh, J.L.: Developments in understanding Chinese leadership: paternalism and its elaborations, moderations, and alternatives. In: Bond, M.H. (ed.) The Oxford Handbook of Chinese Psychology, pp. 599–622. Oxford University Press, Oxford (2010)

Connors, M.M., Harrison, A.A., Aikens, F.R.: Living Aloft: Human Requirements for Extended Spaceflight. National Aeronautic and Space Administration, Washington, DC (1985). SP-483

Cunningham, F.: The stellar status symbol: true motives for China's manned space program. China Secur. **5**(3), 73–74 (2009)

Cushner, K., Brislin, R.W.: Intercultural Interactions: A Practical Guide, 2nd edn. Sage, Thousand Oaks (1995)

Dawson, J.L.M., Lonner, W.J. (eds.): Readings in Cross-Cultural Psychology. University of Hong Kong Press, Hong Kong (1974)

Doi, L.T.: Anatomy of Dependence. Kodansha, Tokyo (1973)

Ehrenfreund, P., Peter, N., Schrogl, K.U., Logsdon, J.M.: Cross-cultural management supporting space exploration. Acta Astronaut. **66**, 245–256 (2010)

Feustl-Beuchl, J.: 25 years of European human spaceflight. ESA Bulletin **116**, 6–16 (2003)

Fiedler, F.E., Mitchell, T., Triandis, H.C.: The culture assimilator: an approach to cross-cultural training. J. Appl. Psychol. **55**, 95–102 (1971)

Goodykunst, W., Nishida, T.: Bridging Japanese/North American Differences. Sage, Thousand Oaks (1994)

Hansen, J.R.: The Taikonaut as cultural icon: the cultural and political significance of Yang Liwei, China's first space traveler. In: Dick, S.J., Launius, R.L. (eds.) Societal Impact of Spaceflight, pp. 103–117. National Aeronautics and Space Administration, Washington, DC (2007)

Harrison, A.A.: Spacefaring: The Human Dimension. University of California Press, Berkeley (2001)

Harrison, A.A.: Humanizing space. Acta Astronaut. **66**, 890–896 (2010)
Heine, S.J., Takata, T., Lehman, D.R.: Beyond self-presentation. Evidence for self-criticism among Japanese. Pers. Soc. Psychol. Bull. **26**, 71–78 (2000)
Helmreich, R.L., Merritt, A.: Culture at Work in Aviation and Medicine: National, Organizational and Professional Influences. Ashgate, Aldershot (1988)
Hofstede, G.: Culture's Consequences: International Differences in Work-Related Values. Sage, Beverly Hills (1980)
Hofstede, G.: Culture's Consequences: Comparing Values, Behaviors, Institutions and Organizations across Nations, 2nd edn. Sage, Thousand Oaks (2001)
Hofstede, G., Hofstede, G.J.: Cultures and Organizations: Software of the Mind, 2nd edn. McGraw-Hill, New York (2005)
Hwang, K.K., Han, K.J.: Face and morality in confucian society. In: Bond, M.H. (ed.) The Oxford Handbook of Chinese Psychology, pp. 479–498. Oxford University Press, Oxford (2010)
Hysong, S.J., Galarza, L., Holland, A.W.: A review of training methods and instructional techniques: implications for behavioral skill training in U.S. astronauts. (NASA TP-2007-213726, May 2007)
Jahoda, G.: Our forgotten ancestors. In: Berman, J.J. (ed.) Cultural Perspectives. Nebraska Symposium on Motivation, vol. 37, pp. 1–40. University of Nebraska Press, Lincoln (1990)
JAXA.: JAXA's Astronauts, Human Space Activities, http://www.jaxa.jp/projects/iss_human/astro/index_e.html. Accessed 25 Mar 2012
Kanas, N., Sandal, G.M., Boyd, J.E., Gushin, V.I., Manzey, D., North, R., Leon, G.R., Suedfeld, P., Bishop, S.L., Fiedler, E.R., Inoue, N., Johannes, B., Kealey, D.J., Kraft, N.O., Matsuzaki, I., Musson, D., Palinkas, L.A., Salnitskiy, V.P., Sipes, W., Stuster, J., Wang, J.: Psychology and culture during long-duration space missions. In: Vakoch, D.A. (ed.) On Orbit and Beyond: Psychological Perspectives on Human Spaceflight. Springer, Heidelberg (2013)
Kring, J.: Multicultural factors for international spaceflight. J. Hum. Perform. Extreme Environ. **5**(3), 11–32 (2001)
Leung, I., Au, W.-I.: Chinese cooperation and competition. In: Bond, M.H. (ed.) The Oxford Handbook of Chinese Psychology, pp. 499–514. Oxford University Press, Oxford (2010)
Leung, K., Bond, M.H.: Social axioms: a model for social beliefs in multicultural perspective. Adv. Exp. Soc. Psychol. **36**, 119–197 (2004)
Liu, J.H., Li, M.C., Yue, X.: Chinese social identity and inter-group relations: the influence of benevolent authority. In: Bond, M.H. (ed.) The Oxford Handbook of Chinese Psychology, pp. 579–598. Oxford University Press, Oxford (2010)
Merritt, A.: Culture in the cockpit: do Hofstede's dimensions replicate? J. Cross-Cultur. Psychol. **31**, 283–301 (2000)
Moltz, J.C.: China, the United States, and prospects for Asian Space Cooperation. J. Contem. China **20**, 69–87 (2011)
NASA.: Chronology of selected highlights in the first 100 American spaceflights, 1961–1995. http://history.nasa.gov/Timeline/100flt.html. Accessed 25 Mar 2012
Naumov, N., Puffer, S.M.: Measuring Russian culture using Hofstede's dimensions. Appl. Psychol. **49**(4), 709–718 (2000)
Oberg, J.: Russia's 'guest cosmonaut' program, a commentary. L-5 99 News **3**(11), 1–2 (1978)
Ritsher, J.B.: Cultural factors and the International Space Station. Aviat. Space Environ. Med. **76**(6), B135–B144 (2005). sect. II
Rosenberger, N. (ed.): Japanese Sense of Self. Cambridge University Press, Cambridge (1992)
Santy, P.A., Holland, A.W., Looper, L., Macondes-North, R.: Multicultural factors in an international crew debrief. Aviat. Space Environ. Med. **64**, 196–200 (1993)
Schwartz, S.H.: Universals in the content and structure of values: theoretical advances and empirical tests in 20 countries. Adv. Exp. Soc. Psychol. **25**, 1–65 (1992)
Schwartz, S.H., Bardi, A.: Influences of adaptation to communist rule on value priorities in Eastern Europe. Polit. Psychol. **18**, 385–410 (1997)

Shlapentokh, V.: Fear in Contemporary Society: Its Negative and Positive Effects. Palgrave Macmillan, New York (2006)
Shlapentokh, V.: Contemporary Russia as a Feudal Society. Palgrave Macmillan, New York (2007)
Shwalb, D.W., Shwalb, B.J. (eds.): Japanese Childrearing: Two Generations of Scholarship. Guilford, New York (1996)
Slayton, D.K., Casutt, M.: Deke! An Autobiography. Forge Books, New York (1995)
Smith, P.B.: On the distinctiveness of Chinese psychology, or: are we all Chinese? In: Bond, M.H. (ed.) The Oxford Handbook of Chinese Psychology, pp. 699–710. Oxford University Press, Oxford (2010)
Suedfeld, P., Wilk, K.E., Cassel, L.: Flying with strangers: postmission reflections of multinational space crews. In: Vakoch, D.A. (ed.) Psychology of Space Exploration: Contemporary Research in Historical Perspective, pp. 143–175. National Aeronautics and Space Administration, Washington, DC (2011)
Triandis, H.C.: Subjective Culture. Wiley, New York (1972)
Triandis, H.C.: Culture and Social Behavior. McGraw-Hill, New York (1994)
Wong, E., Chang, K.: Space plan from China broadens challenge to US. New York Times, 29 Dec 2011. http://www.nytimes.com/2011/12/30/world/asia/china-unveils-ambitious-plan-to-explore-space. Accessed 1 Jan 2012

Part IV
Autonomy in Future Space Missions

Chapter 12
High Versus Low Crewmember Autonomy in Space Simulation Environments

Nick Kanas, Stephanie Saylor, Matthew Harris, Thomas Neylan, Jennifer Boyd, Daniel S. Weiss, Pamela Baskin, Colleen Cook, and Charles Marmar

Abstract Given the long distances involved and the kinds of activities planned, crewmembers participating in long-duration exploratory space missions such as an expedition to Mars will have more autonomy than in previous space missions. In order to study the impact of high versus low crew autonomy on crewmembers and the crew–mission control interaction, we conducted a series of pilot studies involving three space simulation settings: NASA Extreme Environment Mission Operations (NEEMO) missions, the Haughton-Mars Project, and the pilot phase of the Mars500 Program. As in our previous on-orbit studies on the Mir and International Space Station, crew and mission control subjects working in missions involving these three settings completed a weekly study questionnaire that assessed mood and interpersonal interactions using the Profile of Mood States, the Group Environment Scale, and the Work Environment Scale. The Mars500 pilot study also directly assessed individual and group autonomy. In these studies, high autonomy periods were those where crewmembers planned much of their work schedule, whereas low autonomy periods were those where mission control personnel developed the schedule, much as happens now during actual space flight conditions. Our results suggested that high work autonomy was well-received by the crews, mission goals were accomplished, and there were no adverse effects. During high autonomy periods, crewmember mood

---

Adaptation of Kanas et al. (2010), reprinted with permission from Elsevier. Parts of this chapter were presented at the 60th International Astronautical Congress, Daejeon, South Korea, October 12–16, 2009.

N. Kanas (✉) • S. Saylor • M. Harris • T. Neylan • J. Boyd • D.S. Weiss • C. Cook
University of California, San Francisco, and Veterans Affairs Medical Center, San Francisco, CA, USA
e-mail: nick.kanas@ucsf.edu

P. Baskin
Wyle, Integrated Science and Engineering Group, Houston, TX, USA

C. Marmar
New York University Medical Center, New York, NY, USA

D.A. Vakoch (ed.), *On Orbit and Beyond*, Space Technology Library 29,
DOI 10.1007/978-3-642-30583-2_12, © Springer-Verlag Berlin Heidelberg 2013

was generally reported as being better and creativity was higher, but mission control personnel reported some confusion about their work role. The crewmember group environment in the Mars500 pilot study was dependent on the nationality mix. Despite scoring lower in work pressure overall, the four Russian crewmembers reported a greater rise in work pressure from low to high autonomy than the two Europeans. In contrast, the European crewmembers reported a greater rise in dysphoric mood in going from low to high autonomy, whereas the Russians' emotional state remained the same or slightly improved. It is time to study the effects of high autonomy with larger subject samples during actual space missions on-orbit in preparation for future exploratory missions, where high autonomy will be the norm.

12.1 Background

During future long-duration exploratory space missions such as an expedition to Mars, the long distances involved, the delays in communicating with the Earth, and the sense of extreme isolation will cause the crewmembers to experience more autonomy and possible breakdowns in interpersonal interactions, both within the crew and with mission control. Space agencies have recognized these issues as risk factors that could threaten crew safety and performance. For example, in NASA's Bioastronautics Roadmap, Risk 24 titled "Human Performance Failure Due to Poor Psychosocial Adaptation" states: "The isolated and confined nature of space flight, along with its potential hazards, pose human performance related challenges. This risk may be influenced by boredom with available food, crew autonomy and increased reliance on each other, crowding, distance from family and friends, duration of flight, incompatible crewmembers, interpersonal tensions, mechanical break-downs, poor communication, scheduling constraints and requirements, sleep disturbance, or social isolation" (Bioastronautics Roadmap 2010).

Many of these factors have been studied in space. For example, Kanas and colleagues have conducted two major on-orbit studies of psychological and interpersonal issues affecting crewmembers and mission control personnel during missions to the Mir and the International Space Station (ISS). Issues examined included changes in mood and group social climate over time, displacement of tension to individuals outside the group, cultural differences, and leadership roles. The Mir study sample consisted of 5 American astronauts, 8 Russian cosmonauts, and 42 American and 16 Russian mission control personnel (Kanas and Manzey 2008; Kanas et al. 2000a, b, 2001a, b, c, d, 2002). The ISS study sample consisted of 8 American astronauts, 9 Russian cosmonauts, and 108 American and 20 Russian mission control personnel (Kanas and Manzey 2008; Kanas et al. 2006, 2007a, b; Boyd et al. 2009). Subjects completed a weekly study questionnaire that included items from the Profile of Mood States (POMS) (McNair et al. 1992) and two group climate measures, the Group Environment Scale (GES) (Moos 1994a) and the Work Environment Scale (WES) (Moos 1994b).

Most findings from the two studies were similar (Kanas and Manzey 2008; Kanas et al. 2007b). There was little support for the presence of changes in mood and social climate over time, and no evidence for a 2nd half decrement or "3rd quarter phenomenon." There was evidence for displacement of negative dysphoric emotions to outside personnel by both crewmembers and mission controllers. The support role of the leader was related to group cohesion among crewmembers, and both the task and support roles of the leader were related to cohesion among mission control personnel.

There was also evidence for national (American versus Russian), occupational (crewmember versus mission control), and organizational (NASA versus Russian Space Agency) cultural differences (Kanas and Manzey 2008; Boyd et al. 2009). Crewmembers had higher scores on a measure of cultural sophistication than mission control personnel, especially American mission control, but cultural sophistication was not related to mood or social climate. Russians reported greater language flexibility than Americans. Although crewmembers reported more positive mood states than mission control, both groups scored in the healthy range. Analyses showed a few Mir/ISS differences in social climate among crewmembers, but these disappeared when the full sample (crewmembers and mission control subjects) was considered. There were several Russian/American differences in social climate in each setting, but the most robust difference in both the Mir and ISS studies was a high score in Work Pressure among the Americans and a low score among the Russians. For the crewmembers, the astronauts scored one standard deviation above published norms for this measure, whereas the cosmonauts scored below the norm. Both national and organizational factors were implicated in accounting for this difference, possibly reflecting collectivist characteristics of Russian culture (Boyd et al. 2009; Ritsher 2005).

The Mir and ISS studies were both done under conditions of low crewmember autonomy, where mission control timelined most of the crew activities and there were frequent crew–ground interactions in real time. There have been no studies of high autonomy in space, and until recently none in a space simulation environment.

An example of the latter was a study reported by Roma et al. (2009). They conducted a distributed interactive computer space mission simulation where three-person groups participated in a variety of tasks aimed at collecting geological specimens on a planetary body under both autonomous and highly scheduled conditions. The research team found that performance was consistently higher during the autonomous missions relative to the scheduled missions. High autonomy also resulted in increased positive and decreased negative self-report ratings and decreased linguistic expressions of negative emotion. Salivary cortisol was lower during autonomous missions. The authors concluded that autonomous mission management strategies resulted in improved performance, better group cohesion, and reduced stress-induced risks.

Since 2006, our group has been directed by NASA/Johnson Space Center management to test the feasibility and safety of high autonomy periods during pilot studies involving different space simulation environments on the Earth. In these studies, crewmember and mission control personnel who agreed to be subjects after being given informed consent completed the POMS, GES, and WES on a weekly basis. Both low and high autonomy conditions were employed, and three separate simulation environments were studied (see below). Although the findings from two of these were presented earlier in scientific conferences (Saylor et al. 2008; Kanas et al. 2009), only one resulted in a written report (Saylor et al. 2008). In this chapter, we will describe the results from all three settings, integrate the findings, and propose some conclusions that will have relevance for a hypothesis-testing study on-orbit using the ISS.

12.2 NEEMO 12 and 13

Our first project involved the NASA Extreme Environment Mission Operations (NEEMO) program. The purpose of this program is to study human survival in the Aquarius underwater laboratory in preparation for future space missions. The Aquarius is located near Key Largo, Florida. Missions typically last 10–14 days and involve research and other activities conducted by four or more "aquanauts," who are usually astronauts or other NASA personnel. These activities are conducted within the habitat or during simulated EVAs in the ocean. "Mission control" personnel located topside monitor crew activities and timeline the schedule.

We studied two NEEMO missions: 12 and 13. NEEMO 12 consisted of 12 days of low autonomy, where the crewmember work schedules were timelined by the topside mission control. Subjects included three NASA crewmembers and six topside personnel. NEEMO 13 comprised a total of 5 days of low autonomy and 5 days of high autonomy (where the crewmembers had more flexibility to plan their own work schedule and experienced a 40-min two-way communication delay with mission control, such as might be expected from the surface of Mars). Subjects included four NASA crewmembers and eight topside personnel. At the middle and end of each mission, subjects completed a questionnaire containing many subscales from the POMS, GES, and WES.

Due to the small number of data points, the data could only be analyzed descriptively by examining change scores of the subscales of our measures from the first and second half, separately for the two missions. Our analyses suggested that the autonomy condition was successfully employed in NEEMO 13 with no adverse results. There seemed to be clear differences between the low and high autonomy periods on some subscales that tended to favor the autonomy condition. For example, as suggested in Fig. 12.1, the crewmembers reported more direction from the mission commander (reflected by an increase in Leader Control) as well as a decrease in fatigue from the low to the high autonomy period during NEEMO 13 as compared with NEEMO 12, which was conducted under low autonomy. For the topside subjects, work pressure was stable in NEEMO 13, whereas it rose in the second half during NEEMO 12. However, confusion increased during the high autonomy period of NEEMO 13 as compared with the second half of NEEMO 12 (where it declined), suggesting that the topside personnel were not as clear in their roles when the crew took over more of the schedule.

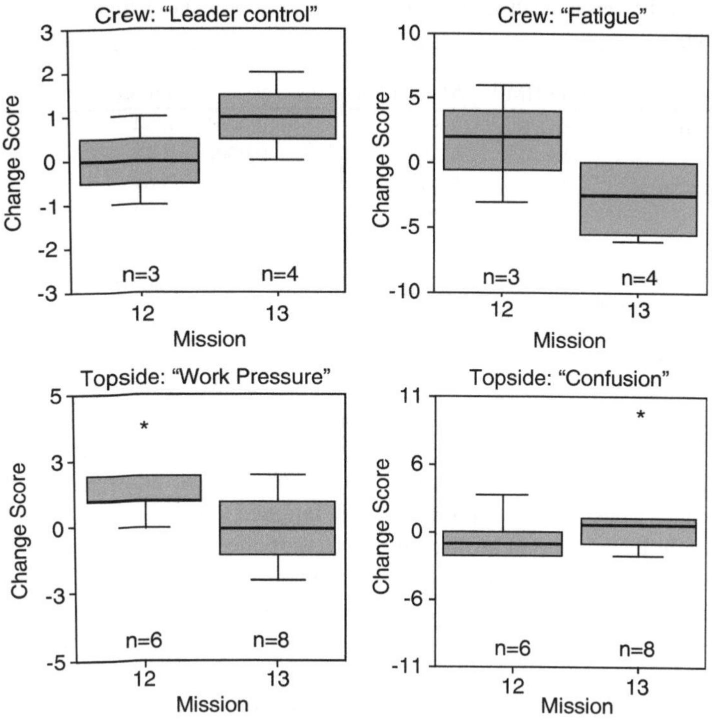

Fig. 12.1 Effects of autonomy in NEEMO 12 (low autonomy throughout) versus NEEMO 13 (one low and one high autonomy period)

12.3 HMP 2008

Our second study involved some of the crewmembers working on the Haughton-Mars Project (HMP) on Devon Island in Canada. The location is an old, large impact crater that has similarities to the Martian surface in appearance and geological characteristics. People participating in this program dress up in space suits, travel in a Mars rover, and conduct tasks similar to those anticipated during an actual expedition to Mars. There is no formal mission control, and the participants have great freedom to plan their work and leisure activities each day. Hence, there is a high level of autonomy throughout the day.

We studied the last 3 weeks of the 2008 summer mission. Eight non-NASA scientists and graduate students served as subjects and completed weekly POMS, GES, and WES measures. Since there was no contrasting low autonomy period with which to compare the results, we decided to compare the subscale scores of the HMP subjects with those from published adult norms in the POMS, GES, and WES manuals, and with subscale scores from mission control subjects who participated in our previous on-orbit studies. Significant findings are shown in Table 12.1. Note that corrections have been made to reduce the possibility of Type I errors due to multiple comparisons using the procedure of Benjamini and Hochberg (1995).

Table 12.1 HMP versus adult norms and Mir/ISS mission control mean scores (significant subscales only)

Subscale	Mean HMP	Mean adult norm	p*	Mean Mir/ISS mission control	p*
Leader control	6.88	4.55	0.0001	6.44	NS
Innovation	6.38	4.71	0.0006	3.45	<0.0001
Managerial control	4.25	5.26	0.003	5.67	NS
Cohesion	8.15	6.66	NS	6.72	0.004
Fatigue	11.00	8.00	NS	5.87	0.0001

*p-values from independent t-tests, adjusted based on Benjamini and Hochberg (1995) criteria to account for experimentwise error ($p = 0.0075$)

The HMP group reported greater levels of in-group leader direction and innovation and less need for outside direction as compared with other work groups on the Earth, suggesting that the HMP crewmembers were working under fairly high autonomous conditions. Additionally, they reported more cohesion as compared with mission control groups. They also reported being more fatigued than mission control personnel, but this might have been an artifact of their experience working in northern Canada during the constant daylight of summertime, where work conditions are more physically taxing than sitting in front of a mission control console. Taken together, these results suggest that people working under conditions of relatively high autonomy are not adversely affected. They may even be more creative and more likely to bond together cohesively as compared with other work groups.

12.4 Mars500 105-Day Study

Our most recent study involved the Mars500 Program, which is designed to study performance and interactions of a group of six individuals engaged in a simulated mission to Mars. The crew was confined for 520 days in a simulator at the Institute for Biomedical Problems (IBMP) in Moscow, from June 3, 2010 to November 4, 2011. The lower floor contained living and laboratory modules for the crew, and the upper floor contained a mock-up of the Mars surface on which the crew simulated geological and other planetary activities.

As a trial pilot study for this mission, a 105-day simulation was conducted at the IBMP from March 31 to July 14, 2009, which is discussed in this chapter. This pilot study involved a crew of six men: two Russian cosmonauts (one of whom was the commander), a Russian medical doctor, a Russian sports physiologist, a German mechanical engineer, and a French airline pilot. The crew simulated a number of scenarios related to launch, outbound journey to Mars, and working on the Martian surface. During the first 10 weeks of the mission, the crewmembers interacted in real time with outside monitoring staff ("mission control") during low autonomy conditions. In contrast, during the last 5 weeks they had more responsibility for monitoring and planning their own activities and also experienced a 40-min two-way Mars-like communication delay with the outside, the high autonomy condition.

Table 12.2 Crewmember mean scores during low and high autonomy (Cohen's $d \geq 0.6$)

Subscale	Low autonomy mean score	High autonomy mean score	Effect size (Cohen's *d*)
Depression-dejection	0.56	0.38	0.77
Vigor-activity	16.55	15.57	0.78
Anger and aggression	1.30	1.10	0.63
Self discovery	5.09	5.46	0.60
Leader control	4.96	4.38	0.81
Supervisor support	7.73	6.55	0.94
Work pressure	2.88	4.08	1.06

As before, both crew and mission control subjects completed weekly POMS, GES, and WES measures. An additional WES subscale, Autonomy, was added, along with 6 questions investigating the impact of the autonomy condition on individual and group performance. Our subject sample included all 6 crewmembers and 18 of the19 members of the monitoring staff in mission control (10 men and 9 women of medical and engineering backgrounds, the majority of whom were Russian employees of the IBMP). Debriefings with three members of the monitoring staff (conducted by Dr. Kanas) and of the crewmembers (by agency personnel ["Mission Accomplished: 105-day Mars Mission Simulation Ends in Moscow" 2009]) suggested that crew morale and cohesion were high throughout the mission and that mission goals were accomplished.

Due to the small subject sample sizes, we decided to focus our data analyses of crewmember low versus high autonomy differences using effect sizes, expressed in terms of Cohen's d values (where values of 0.5 are seen as defining a moderate difference, and values of 0.8 or above a large difference) (Cohen 1988). Out of a total of 21 subscales, 7 had effect sizes of 0.60 or higher (none were in the 0.46–0.59 range), and these are shown in Table 12.2.

The results suggested that the crewmembers found high autonomy to be a positive experience that allowed them to depend less on direction from their commander or outside supervisors than in the low autonomy condition.

The elevated Work Pressure score during high autonomy was largely influenced by data from the four Russian subjects (whose effect size was 4.00), suggesting that they perceived more of an increase in work stress than the Europeans in going from the low autonomy condition to the less-structured high autonomy condition (Fig. 12.2). This finding may have reflected the influence of Russian national characteristics or space agency organizational factors, where more direction is the norm.

One of the six new autonomy questions dealing with the experience of freedom among crewmembers in planning their own work schedule produced a higher score in the high (versus low) autonomy condition, with an effect size of 0.66. Given that this question was used as a fidelity check on the two autonomy conditions, this finding is congruent with our predictions. Even though the WES Autonomy subscale and the other five autonomy questions resulted in equal or higher overall mean scores in the high versus low autonomy condition, none of these comparisons

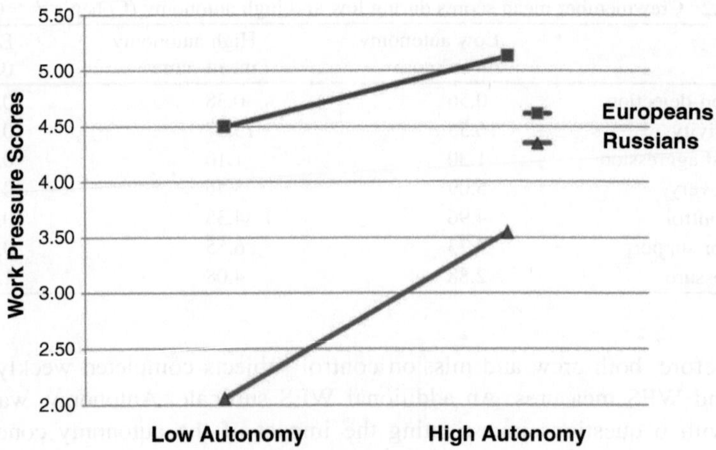

Fig. 12.2 Work Pressure under low and high autonomy

Table 12.3 Russian and European mean scores during the entire mission (Cohen's $d \geq 0.6$)

Subscale	Russian mean score	European mean score	Effect size (Cohen's d)
Task orientation	8.80	8.29	2.73
Order and organization	8.70	7.75	1.75
Managerial control	6.70	6.11	1.25
Leader support	8.37	7.32	1.36
Independence	6.17	6.64	0.85
Confusion-bewilderment	1.65	2.11	0.83
Anger and aggression	0.77	2.18	1.99
Leader control	4.44	5.46	1.15
Work pressure	2.58	4.68	1.50

produced an effect size greater than 0.40, although Russians scored marginally higher and Europeans marginally lower in work performance efficiency and accuracy during high autonomy.

Because our previous on-orbit and pilot work suggested that there may be differences in response between Russian and non-Russian crewmembers, we examined our data in terms of overall mean scores between Russian and European subjects. The differences that resulted in effect sizes at or above 0.60 are shown in Table 12.3. It is important to be mindful that we studied only four Russian and two European crewmembers, so what we report is highly specific to this particular small sample of individuals.

The results suggested that during this mission, the Russian crewmembers, as compared with the Europeans, generally felt more oriented to the task, perceived higher levels of organization and direction from management, and acknowledged

Table 12.4 Russian and European mean scores during low and high autonomy (eta^2 values $\geq$0.290)

Subscale	Low autonomy mean scores		High autonomy mean scores		eta^2
	Russian	European	Russian	European	
Total mood disturbance	−10.02	−10.44	−10.10	−6.45	0.530
Fatigue- inertia	0.84	0.99	0.85	1.80	0.344
Anger- hostility	0.46	0.41	0.35	1.38	0.293
Leader control	4.51	5.87	4.30	4.55	0.590
Leader support	8.53	7.15	8.05	7.75	0.549
Expressiveness	3.97	4.74	3.45	5.50	0.724
Cohesion	8.51	8.61	8.65	8.40	0.396
Autonomy	6.78	6.01	6.75	7.48	0.423

more support from the Russian commander. They also scored lower in independence than their European colleagues. The Europeans acknowledged having more overall confusion and group aggression, while at the same time perceiving more direction from the commander than the Russians. Similar to the findings from our on-orbit studies that Russian crewmembers scored lower on work pressure than their American counterparts (Kanas and Manzey 2008; Boyd et al. 2009), the Russians here scored lower than the Europeans throughout the mission (i.e., during both the high and low autonomy periods taken together, Fig. 12.2). Many of these findings are to be expected in a study conducted in Moscow with Russian mission control personnel and a crewmember majority that is Russian (including the commander).

We also looked for possible nationality x condition interaction effects in our subscales using a mixed model analysis of variance procedure that allowed us to find partial eta squared (eta^2) values as an indication of how much of the variance was accounted for by the interaction. According to Cohen (1988), values above 0.14, accounting for 14% of the variance, are "large" effects. Since our sample sizes were so small (four Russians and two Europeans), we report only those subscales with an eta^2 value of 0.29 or higher, and we present these findings as suggestive outcomes needing confirmation from larger samples. These results are shown in Table 12.4.

The results showed that the European crewmembers reacted to the high autonomy condition with higher scores on three measures of dysphoria (Total Mood Disturbance [see Fig. 12.3], Fatigue-Inertia, and Anger- Hostility), and they experienced a greater reduction in direction from the Russian commander than the Russian crewmembers. The Europeans also showed an increase in support from the commander during high autonomy, whereas the Russians perceived a drop. Expressiveness declined for the Russians but rose for the Europeans during high autonomy. The Russians perceived a rise and the Europeans a drop in cohesion during high autonomy. Finally, the Europeans experienced a rise in their Autonomy scores from the low to high autonomy condition, which was to be expected, whereas the Russians did not report much difference in the two conditions.

The reactions of the Russian mission control subjects to different autonomy conditions are summarized in Table 12.5. During high as compared with low

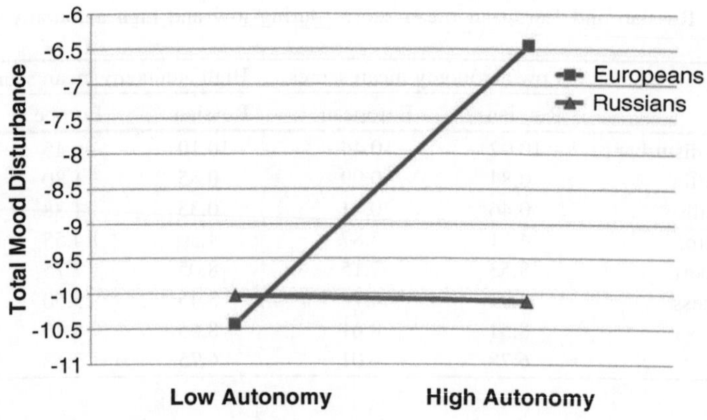

Fig. 12.3 Total mood disturbance under low and high autonomy

Table 12.5 Mission control mean scores during low and high autonomy (Cohen's $d \geq 0.6$)

Subscale	Low autonomy mean score	High autonomy mean score	Effect size (Cohen's d)
Tension-anxiety	9.17	9.61	0.79
Confusion-bewilderment	6.24	6.60	0.68
Task orientation	7.49	7.17	1.66
Depression-dejection	7.72	7.04	2.25
Cohesion	6.98	7.17	0.92

autonomy, they experienced more tension and confusion and less task orientation. These findings may have resulted from their having vague work assignments and fewer tasks to perform. The mission control subjects also reported a lower depression score but a higher group cohesion score during the high autonomy period.

Partly as a validity check and partly to compare our findings with those from our on-orbit studies, we examined the overall mean scores on our subscales between the crewmembers and mission control personnel. The significant results are shown in Table 12.6. As was the case in our Mir and ISS studies (Kanas and Manzey 2008; Boyd et al. 2009), crewmembers reported scores consistent with higher positive mood states than mission control personnel. Crewmembers also reported being more cohesive, receiving more support from their leader, and being better organized and oriented to their tasks than mission control personnel. Most of these significant differences were replicated in separate analyses conducted under low and under high autonomy conditions. These findings more likely reflected the characteristics of people volunteering to work in an isolated and confined environment versus a mission control setting than to effects related to low or high autonomy.

Table 12.6 Crewmember and mission control mean scores during the entire mission (significant subscales only)

Subscale	Crewmember mean score	Mission control mean score	p^*
Total mood disturbance	−9.75	21.39	0.000
Tension-anxiety	2.66	8.92	0.000
Depression-dejection	0.50	7.32	0.009
Confusion-bewilderment	1.80	6.10	0.000
Anger-hostility	0.51	7.13	0.004
Anger and aggression	1.24	2.71	0.046
Fatigue-inertia	0.98	5.69	0.000
Cohesion	8.55	6.86	0.009
Leader support	8.02	6.68	0.049
Order and organization	8.39	6.58	0.004
Task orientation	8.63	7.17	0.009

*p-values from independent t-tests, adjusted based on Benjamini and Hochberg (1995) criteria to account for experimentwise error ($p = 0.05$)

We examined our data to look for evidence of displacement of negative emotions to the outside and a relationship between task and support roles of the leader with group cohesion, all of which we found in our on-orbit studies. We have defined displacement operationally as the negative association of perceived outside Supervisor Support with six subscales measuring in-group dysphoria (Kanas and Manzey 2008; Kanas et al. 2007b). Using a mixed model linear regression analysis, we found the predicted negative association ($p < 0.05$) for all subjects for four of these subscales: Total Mood Disturbance, Tension-Anxiety, Anger-Hostility, and Anger and Aggression. The negative correlations just missed significance for the other two: Work Pressure ($p = 0.054$) and Depression-Dejection ($p = 0.071$). For all subjects, we did not find the predicted positive association between scores on the Leader Control subscale and Cohesion, but we did find it between Leader Support and Cohesion ($p = 0.000$). These effects were stronger for mission control subjects alone than crewmembers alone (where there were no significant correlations), which was likely due to the small number of subjects in the latter group.

12.5 Conclusions

All of the above findings were the results of pilot studies intended to evaluate the safety and feasibility of high autonomy and to test the use of our measures under low versus high autonomy conditions. The sample sizes were too small to evaluate many of the results using standard statistical methods, so they are presented descriptively or with effect sizes. One must be cautious in generalizing these results to other space analogue settings having subjects of different nationalities. In presenting these findings, we tended to be generous in reporting what we could, since we wanted to use this pilot work to suggest some possible conclusions that could inform a future hypothesis-testing study on-board the ISS.

Note that the concept of autonomy is a relative one, ranging from conditions where crewmembers have minimal input into their activities to situations where they must completely plan their work and rest schedule, with minimal input from the outside. In our pilot studies, autonomy in the NEEMO 13 and Mars500 105-day missions focused primarily on non-emergency work activities, and a Mars-like communication delay with the outside was instituted. In the HMP 2008 study, the crew had more independent control over their activities throughout the day, with minimal outside contact. An actual expedition to Mars will be similar to the Mars 500 situation, except that the crewmembers will need to deal with emergencies themselves, whereas mission control would have likely intervened if a true emergency occurred during the 105-day simulation.

Debriefings of the subjects after the missions were completed suggested that the high autonomy condition was safe and well-received by the crewmembers, and there were no adverse effects. Mission goals were generally accomplished. Our findings suggested that high autonomy offered some advantages for crewmembers over low autonomy, especially in terms of improved mood, personal discovery, and innovation.

But not everyone reacted the same. For example, during the Mars500 105-day study, the four Russian crewmembers reported a greater rise in work pressure from low to high autonomy than the two Europeans. In contrast, the European crewmembers reported a greater rise in dysphoric mood in going from low to high autonomy, whereas the Russians' emotional state remained the same or slightly improved. These findings suggest the influence of cultural factors related to national or agency organizational differences between Russians and non-Russians. However, it should be noted that the European findings were based on only two subjects participating in a Russian-operated study, so it is difficult to generalize these findings to European participants in other isolated and confined environments, such as the ISS. Furthermore, in the Mars500 pilot study and in our on-orbit work involving Russians and Americans, the overall level of perceived work pressure was low for the Russians as compared with other national groups. Perhaps decades of experience working with space stations have allowed Russians to develop unique coping strategies that protect them from stress on the job. These possible Russian/non- Russian cultural differences need to be explored further in future space studies.

As found in our previous on-orbit studies, crewmembers selected to work in isolated and confined environments on the Earth may react more positively than those who are monitoring them in mission control. In addition, during periods of high crewmember autonomy, mission control personnel may experience confusion concerning their work role. Management needs to consider the impact of high crew autonomy on the activities of monitoring staff in mission control and to prepare and train them to deal with changes in their activities during these periods of time.

These conclusions are very tentative, but they do suggest issues to be considered in a future hypothesis-testing study. We believe that it is now time to study the effects of low versus high crew autonomy with larger subject samples during actual space missions on-orbit using the ISS. The results of such a study will inform training and support strategies for crews and their mission control colleagues who will be involved with future exploratory missions to Mars and beyond, where high autonomy will be the norm.

Acknowledgements Figures 12.2 and 12.3 and parts of this chapter were presented at the Workshop on Human Behaviour and Performance in Analog Environments and Simulations, ESA/ESTEC, Noordwijk, The Netherlands, December 7–8, 2009. The work reported in this chapter was funded by NASA (Contract numbers NCC-9-0161 and NNJ05HG01A) and the National Space Biomedical Research Institute (Contract number NCC-9-58-NBPF00005). The authors would like to thank the Institute for Biomedical Problems in Moscow for its support of our participation in the Mars500 Program's 105-day study.

References

Benjamini, Y., Hochberg, Y.: Controlling the false discovery rate: a practical and powerful approach to multiple testing. J. R. Stat. Soc. **57**, 289–300 (1995)

Bioastronautics Roadmap: http://bioastroroadmap.nasa.gov (Jan 2010). Accessed 25 Mar 2012

Boyd, J.E., Kanas, N.A., Salnitskiy, V.P., Gushin, V.I., Saylor, S.A., Weiss, D.S., Marmar, C.R.: Cultural differences in crewmembers and mission control personnel during two space station programs. Aviat. Space Environ. Med. **80**, 532–540 (2009)

Cohen, J.: Statistical Power Analysis for the Behavioral Sciences, 2nd edn. Lawrence Earlbaum Associates, Hillsdale (1988)

Kanas, N., Manzey, D.: Space Psychology and Psychiatry, 2nd edn. Springer/Microcosm Press, Dordrecht/El Segundo (2008)

Kanas, N., Salnitskiy, V., Grund, E.M., Gushin, V., Weiss, D.S., Kozerenko, O., Sled, A., Marmar, C.R.: Social and cultural issues during Shuttle/Mir space missions. Acta Astronaut. **47**, 647–655 (2000a)

Kanas, N., Salnitskiy, V., Grund, E.M., Gushin, V., Weiss, D.S., Kozerenko, O., Sled, A., Marmar, C.R.: Interpersonal and cultural issues involving crews and ground personnel during Shuttle/Mir space missions. Aviat. Space Environ. Med. **71**(Suppl. 9), A11–A16 (2000b)

Kanas, N., Salnitskiy, V., Weiss, D.S., Grund, E.M., Gushin, V., Kozerenko, O., Sled, A., Bostrom, A., Marmar, C.R.: Crewmember and ground personnel interactions over time during Shuttle/Mir space missions. Aviat. Space Environ. Med. **72**, 453–461 (2001a)

Kanas, N., Salnitskiy, V., Grund, E.M., Weiss, D.S., Gushin, V., Kozerenko, O., Sled, A., Marmar, C.R.: Human interactions during Shuttle/Mir space missions. Acta Astronaut. **48**, 777–784 (2001b)

Kanas, N., Salnitskiy, V., Grund, E.M., Weiss, D.S., Gushin, V., Kozerenko, O., Sled, A., Marmar, C.R.: Human interactions in space: results from Shuttle/Mir. Acta Astronaut. **49**, 243–260 (2001c)

Kanas, N., Salnitskiy, V., Grund, E.M., Weiss, D.S., Gushin, V., Bostrom, A., Kozerenko, O., Sled, A., Marmar, C.R.: Psychosocial issues in space: results from Shuttle/Mir. Gravit. Space Biol. Bull. **14**(2), 35–45 (2001d)

Kanas, N., Salnitskiy, V., Grund, E.M., Gushin, V., Weiss, D.S., Kozerenko, O., Sled, A., Marmar, C.R.: Lessons learned from Shuttle/Mir: psychosocial countermeasures. Aviat. Space Environ. Med. **73**, 607–611 (2002)

Kanas, N.A., Salnitskiy, V.P., Ritsher, J.B., Gushin, V.I., Weiss, D.S., Saylor, S.A., Kozerenko, O.P., Marmar, C.R.: Human interactions in space: ISS vs. Shuttle/Mir. Acta Astronaut. **59**, 413–419 (2006)

Kanas, N.A., Salnitskiy, V.P., Ritsher, J.B., Gushin, V.I., Weiss, D.S., Saylor, S.A., Kozerenko, O.P., Marmar, C.R.: Psychosocial interactions during ISS missions. Acta Astronaut. **60**, 329–335 (2007a)

Kanas, N.A., Salnitskiy, V.P., Ritsher, J.B., Gushin, V.I., Weiss, D.S., Saylor, S.A., Kozerenko, O.P., Marmar, C.R.: Crewmember and mission control personnel interactions during International Space Station missions. Aviat. Space Environ. Med. **78**, 601–607 (2007b)

Kanas, N., Saylor, S., Neylan, T., Boyd, J., Weiss, D., Marmar, C.: Effects of high crewmember autonomy in two space analog conditions. In: Proceedings of the 17th International Academy of Astronautics 'Humans in Space' Symposium, Moscow, 7–11 June 2009

Kanas, N., et al.: High versus low crewmember autonomy in space simulation environments. Acta Astronaut. **67**, 731–738 (2010)

McNair, D.M., Lorr, M., Droppleman, L.F.: Profile of Mood States Manual (Revised). Educational and Industrial Testing Service, San Diego (1992)

Mission accomplished: 105-day Mars mission simulation ends in Moscow, 14 July 2009. http://www.esa.int/SPECIALS/Mars500/SEMIS47CTWF_0.html. Accessed 25 Mar 2012

Moos, R.H.: Group Environment Scale Manual, 3rd edn. Consulting Psychologists Press, Palo Alto (1994a)

Moos, R.H.: Work Environment Scale Manual, 3rd edn. Consulting Psychologists Press, Palo Alto (1994b)

Ritsher, J.B.: Cultural factors and the International Space Station. Aviat. Space Environ. Med. **76** (Suppl. 6), B135–B144 (2005)

Roma, P.G., Hursh, S.R., Hienz, R.D., Emurian, H.H., Gasior, E.D., Brinson, Z.S., Brady, J.V.: Autonomous versus scheduled mission management: behavioral and biological effects in simulated space- dwelling groups under communication constraints. In: Proceedings of the 17th International Academy of Astronautics 'Humans in Space' Symposium, Moscow, 7–11 June 2009

Saylor, S., Kanas, N., Baskin, P., Neylan, T.C., Boyd, J.E., Weiss, D.S., Marmar, C.: Effects of high vs. low autonomy on space crewmember performance: preliminary findings from an undersea submersible study. In: Proceedings of the 59th International Astronautical Congress, Glasgow, 29 Sept–3 Oct 2008, #IAC-08-A1.1.16, International Astronautical Federation, Paris

Chapter 13
Effects of Autonomous Mission Management on Crew Performance, Behavior, and Physiology: Insights from Ground-Based Experiments

Peter G. Roma, Steven R. Hursh, Robert D. Hienz, Zabecca S. Brinson, Eric D. Gasior, and Joseph V. Brady

Abstract Technical constraints during long-duration space expeditions will limit the ability for Earth-based management of astronaut crews and will thus increase the prevalence of autonomous operations. To provide experimentally-derived insights on the effects of crew autonomy, we utilized a laboratory-based simulation model to assess crew performance effectiveness and biopsychosocial adaptation under rigid schedule-based management of crew activities by Mission Control versus more flexible, autonomous management of activities by the crews themselves. In 2 separate experiments, 33 research volunteers formed 11 long-term 3-person mixed-gender crews that were extensively trained over several months in an interdependent computer-based planetary surface exploration task. Following training, in Experiment 1, three crews each completed two different types of 3–4 h test missions: Scheduled missions, in which they were directed by Mission Control according to a strict topographic and temporal region-searching sequence, and Autonomous missions, in which the crews received minimal baseline support from Mission Control and were free to establish their own protocols for exploring the planetary surface. In Experiment 2, eight 3-person crews were trained in identical fashion as Experiment 1, except these crews were each tested under four different scenarios, with the Scheduled and Autonomous sessions completed either under normal conditions of full communications capabilities or following the

P.G. Roma (✉) • S.R. Hursh
Institutes for Behavior Resources and Johns Hopkins University School of Medicine, Baltimore, Maryland, USA
e-mail: PeteRoma@gmail.com

R.D. Hienz • J.V. Brady
Johns Hopkins University School of Medicine and Institutes for Behavior Resources, Baltimore, Maryland, USA

Z.S. Brinson • E.D. Gasior
Institutes for Behavior Resources, Baltimore, Maryland, USA

D.A. Vakoch (ed.), *On Orbit and Beyond*, Space Technology Library 29,
DOI 10.1007/978-3-642-30583-2_13,

unexpected loss of audio and text-messaging functions. Overall, autonomous mission management led to improved task performance (more high-valued geologic samples were retrieved), increased subjective self-reports of positive mood, fewer references to negative emotions, greater use of socially-referent language in unstructured debriefing logs, and attenuated physiological stress reactivity. Consistent with observational field research, these laboratory-based investigations provide powerful experimental evidence supporting a causal relationship between crew autonomy and improved performance, enhanced psychosocial adaptation, and sustained biobehavioral health. Future work should systematically examine interactions with culture and personality in diverse multinational crews, assess the effects of other operationally-relevant stressors such as heavy workload and circadian disruption, and evaluate the effects of bounded autonomy in space analogue environments and on the International Space Station using rigorous experimental methods. The controlled laboratory data presented herein contribute to an emerging empirical knowledge base on crew autonomy which the international astronautics community may build upon for future research and ultimately draw upon when designing and managing long-duration exploratory missions.

13.1 Introduction

Even the most seemingly mundane act of space exploration is a truly astonishing accomplishment for our species, representing the culmination of millennia worth of accumulated knowledge and the extraordinary efforts of those dedicated to the acute tasks necessary for realizing the dream of human spaceflight. As has been apparent since the beginning of formalized efforts, space exploration is an exceptionally complex, expensive, and inherently dangerous endeavor, and so not surprisingly, very little is ever purposely left to chance by those responsible for mission management, including the activities of the space travelers themselves.

Historically, with manned expeditions going no farther than the Moon, the constant presence of "Mission Control" as an immediate resource and proactive contributor to Astronauts' daily operations has been a feature taken for granted in space exploration. However, as the international astronautics community moves toward manned missions to increasingly distant targets such as near-Earth asteroids (NEAs) and Mars, the technical constraints imposed by the vast space between Earth and these targets will inherently limit Mission Control's ability to directly manage their astronaut crews. For example, depending on where they are in their respective orbits, the time required for round-trip electronic transmissions between Earth and Mars ranges from 6 to 44 min, functionally nullifying real-time communication as a viable management technique (Jaap et al. 2007). Accordingly, the recent emphasis on long-duration missions has generated increased interest in the effects of "bounded autonomy"—that is, the opportunity for individuals or teams to internally decide how best to accomplish pre-assigned mission objectives (Kanas and Manzey 2008; Kanas et al. 2013; Langfred and Moye 2004; Stepanova et al. 2009).

Longer-term goals of recasting the International Space Station (ISS) as a test bed and launch pad for deep-space exploration, establishing permanent lunar or Martian settlements, and other independent space tourism endeavors only heighten the importance of building a reliable empirical knowledge base on how crew autonomy affects mission performance and biopsychosocial adaptation in small interdependent groups operating in long-term isolation and confinement.

Since the space community's interest in autonomy is relatively recent, there is limited small-group and/or space-specific research directly addressing the topic, most of which is presented or otherwise summarized or touched upon in the present volume. However, the empirical human industrial/organizational (I/O) psychology research literature indicates that autonomy generally improves task performance and subjective mood (e.g., (Gardell 1971; Hackman and Lawler 1971)), the latter of which is also supported by some emerging space analogue research by Kanas and colleagues (Kanas et al. 2010). Moreover, a lack of control at work is generally viewed as a psychosocial stressor precipitating a physiological stress response (Calnan et al. 2000; Maina et al. 2009), which under chronic conditions is an unwelcome addition to the myriad physical health and neuropsychiatric risks already inherent to long-duration space exploration (Kanas and Manzey 2008; Clément 2011). The following report describes experimental work in our laboratory explicitly exploring some behavioral and biological effects of autonomous versus highly scheduled mission management protocols in long-term three-person teams performing a laboratory-based computerized planetary exploration task. Adopting a multidisciplinary approach, multiple measures of performance, psychosocial adaptation, and stress physiology were obtained in response to highly scheduled versus autonomously managed mission plans in two separate experiments. The first experiment focused on the effects of autonomous mission management under otherwise baseline conditions, with the second experiment determining the generality of the effect when confronted with an operationally relevant stressor, in this case, the unexpected loss of audio and text-messaging communication. Scientists and subject matter experts can address the psychosocial issues relevant to long-duration space exploration in a variety of ways, including thoughtful reflections on personal experience (e.g., Hawley 2010; Pettit 2010), careful theoretical and logical analysis (Kanas and Manzey 2008; Brady 2005), and finally empirical observation, which itself may take the form of post-hoc analysis and synthesis of existing relevant data (Kanas and Manzey 2008; Suedfeld et al. 2011; Shea et al. 2009), targeted prospective observational study (Dinges et al. 2009), or rigorous experimental manipulation using carefully constructed models based on critical features that define the "real world" scenario(s) of interest. All of these approaches have their relative strengths and weaknesses, but should ultimately converge to provide clarity and guidance on the issue(s) at hand. The purpose of the ground-based work described below is to explore and elucidate some fundamental effects of team task autonomy by utilizing the logical power of controlled experimentation, with the ultimate goal of contributing to the evidence base upon which space exploration missions are designed, managed, and optimized.

13.2 Method

13.2.1 Participants

A total of 33 individuals from the local community voluntarily consented to participate as subjects. This included 14 males and 19 females ranging in age from 22 to 34 year (mean = 27 year). Although no special skills or background were required beyond a commitment of availability for recurrent training and experimental participation over several months, all volunteers were fluent English speakers and high school graduates, 66% of whom were college-educated through at least the bachelor's level, with an additional 30 % having completed some bachelor's-level coursework or 2-year degrees. The volunteers were assigned based on mutual availability into three-person mixed gender "crews" which remained intact throughout all training sessions and their respective experiments (9 individuals comprising 3 crews in Experiment 1 and 24 individuals comprising 8 crews in Experiment 2). All training, testing, and experimental procedures took place between 0900 and 1800 h under normal room lighting in a climate-controlled laboratory facility. Each participant received $20 per hour for participation plus performance bonuses for retrieving "bio-positive" geological samples (described below). All procedures described herein conformed to US and UN regulations governing the treatment of human research subjects, and all protocols were approved by the US Department of Health and Human Services-registered Institutional Review Board at the Institutes for Behavior Resources.

13.2.2 Planetary Exploration Simulation

Our development and early application of the computer-generated planetary exploration simulation task (PES) employed for the present studies has been described in previously published reports (Brady et al. 2004; Hienz et al. 2005, 2007, 2008; Roma et al. 2011). In the PES, three-person crews are trained and participate in experimental "Missions" during which the crew engages in a computerized geologic exploration expedition on a simulated planetary surface. Each crewmember is assigned to a different simulated exploration vehicle—either an "Orbiter," "Lander," or "Rover"—that is operated via an individual workstation that is physically, visually, and acoustically isolated from the other crewmembers' workstations. Each workstation contains one computer for navigating the planetary surface and collecting geologic samples and a second computer for electronically communicating via audio, video, "white board," and text messaging with the other crewmembers and with Mission Control (research staff monitoring the simulated mission). The Orbiter can remotely sense and display geologic sample locations on the surface for others to see and can analyze readings from collected samples to determine their geologic grade values; by design, the Orbiter continuously circles the planet and is alternately in and out of "communication range" with the surface vehicles in 9-min increments. The Rover can move across the surface, collect and store samples, and transmit geologic information to the Orbiter for further detailed analysis. The Lander provides logistic support for the Rover on the planetary surface such as additional fuel and sample storage space, can dock with the Rover to transport it to other locations on the planetary surface, and can also collect and store samples under certain scenario-specific conditions. Exploratory missions occur in up to six distinct regions on the planet's surface (e.g., crater, open plain, mountain), and each region contains randomly distributed geologic samples of varying grade value that are distinguished by a range of sizes, shapes, colors, densities, biological life signs, and combinations of these characteristics.

Each PES mission is 153 min of total "flight time" with a short break at mid-session. At the start of each mission, Mission Control assigns the task of collecting high-grade geologic samples as part of routine planetary exploration/survey work. A surface area map on the navigation computer of each crewmember displays those regions available for collecting samples. A number of geologic samples are randomly dispersed within each region, and each sample varies in size (small, medium, large), shape (round, oval, triangular, rod, square), color (red, green, blue, orange, yellow), density (low, medium, high), magnetite presence indicator (none, bio-negative, bio-positive), and grade value (0–120 arbitrary units). The Lander and Rover maneuver across the simulated planetary surface and collect samples for subsequent analysis. Once a vehicle collects a sample, its primary characteristics (size, shape, color) are displayed in the vehicle's collection tray. The remaining characteristics (e.g., density, grade value) are known only after the vehicle transmits the sample's data to the Orbiter, who then completes the analysis and transmits the analysis back to the surface vehicle. The immediate goal of the crew is to determine the five rules of each region (the specific combinations of size, shape, and color associated with positively valued samples) based upon information derived from the collection and analysis of samples, and collect those samples that have positive grade values to maximize crew point total. For all zero-valued samples, primary characteristics are assigned randomly and do not match any of the five specified rules within a region. Other features of the scenario impose realistic operational constraints on performance effectiveness, such as limits on fuel, vehicle sample storage capacity, vehicle speed, region boundaries, sensor ranges, forced "time-outs" for simulated radiation storms, and potential vehicle damage due to collisions and radiation exposure. Figure 13.1 below shows representative screen shots of the PES terrain map (left), operations "dashboard" (middle), and communications panel (right).

Upon crew formation, all members have the opportunity to perform the roles of Orbiter, Lander, and Rover during their initial orientation/training sessions, but role assignments remain fixed throughout subsequent training and testing once mutually agreed upon by the crew and Mission Control. Training to establish familiarity and comfort with the PES paradigm consists of weekly missions under the guidance of Mission Control, with decreasing levels of proactive intervention across missions eventually defaulting to a reactive support role. A crew is considered to have achieved baseline performance competency upon scoring at least 50% of the maximum possible grade value of geological samples collected for two consecutive missions. This final mission score includes samples collected minus "damage"

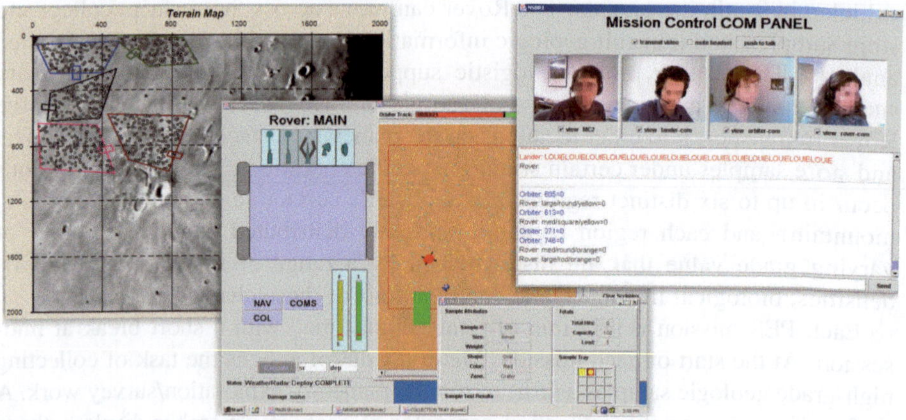

Fig. 13.1 Screen shots of the computer-based planetary exploration simulation task (PES) terrain map (*left*), operations "dashboard" (*middle*), and communications panel (*right*)

points for either improperly stowing gear during the simulated radiation storms or for crashing vehicles into obstacles. The crew competency criterion is typically accomplished in novice groups within 12 sessions over the course of 3–4 months.

13.2.3 Experimental Manipulations

Each "Scheduled" mission served to model the extremes of time pressure and performance demands seen in traditional mission management, and was based on pre-determined, randomly generated topographic and temporal region-searching sequences for the two surface-bound vehicles, which were directed and enforced by Mission Control via text and audio. The Scheduled missions consisted of 20 pre-determined surface exploration intervals ranging from 1 to 12 min in duration and summing to the standard 153-min mission length. As exemplified in Table 13.1, the Lander and Rover have their own respective mission plans. By contrast, crews in the "Autonomous" condition were permitted to explore the planet's surface regions however they chose within the confines of the standard mission length and other standard technical parameters. The crews were informed of how their missions would be managed during their pre-mission briefing, and Mission Control provided the same level of reactive support service throughout all missions regardless of management condition. For Experiment 1, exposure to the Scheduled and Autonomous mission types was counterbalanced across crews on non-consecutive days, with two of the three crews operating under the Scheduled condition first. For Experiment 2, exposure to the four mission types (Scheduled + Full Comms, Scheduled−Audio & Text, Autonomous + Full Comms, Autonomous−Audio & Text) followed a pseudo-Latin square design with the first four crews each completing the missions in a different sequence on non-consecutive days; this pattern was then repeated for the final four crews upon their recruitment and training.

Table 13.1 Example search sequence imposed during Scheduled missions in the planetary exploration simulation (PES) task

	Lander		Rover	
Interval	Location	Duration	Location	Duration
1	6	8	3	8
2	1	6	4	4
3	5	5	4	9
4	4	12	4	4
5	6	7	6	11
6	1	12	5	5
7	4	12	6	7
8	3	6	3	8
9	5	5	1	11
10	3	12	5	9
11	3	7	5	6
12	4	6	2	8
13	2	8	2	7
14	6	6	6	12
15	6	3	4	11
16	5	6	1	3
17	5	11	6	10
18	2	3	2	9
19	1	6	3	8
20	5	12	6	3

The two ground-based vehicles (Lander and Rover) were each assigned a mission plan consisting of 20 search intervals; each interval was comprised of a randomly generated combination of which region to survey (1–6) and duration of the search in minutes (1–12). Interval transitions were directed and enforced by Mission Control. During autonomously managed missions, the crews were free to conduct the geological survey however they saw fit under otherwise identical conditions to those of the Scheduled mission.

13.2.4 Performance, Behavioral, and Physiological Measures

The simulation program automatically recorded crew mission performances and all communication exchanges. Crew performance for any given mission was quantified by the total point value of all samples collected (minus damage as described above) expressed as a % of that mission's maximum possible points. In addition to this overall group-level performance measure, several acute individual measurements were taken immediately before the start of each mission (pre-mission), immediately after min 72 (mid-mission), and immediately after min 153 (post-mission).

Psychosocial adaptation was assessed via computerized visual analog scale (VAS) self-report questionnaires administered pre-, mid-, and post-mission, with the present studies focusing on the final post-mission assessments. The VAS battery consisted of 12 individual items; for each item, a subject responded by using the mouse cursor to select a point along a line displayed on the screen that best reflected his or her current state (anchored by "Not at All" on one end to "Extremely" at the other). VAS responses were recorded in mm along the response line, and for analyses were converted to % of the response line's length. The 12 items that defined the VAS battery were grouped and averaged to reflect Positive (e.g., "Happy"), Negative (e.g., "Stressed"), and Physical (e.g., "Fatigued") responses to the mission (see Hienz et al. (2008) for additional details).

All crewmembers also completed a debriefing logbook entry at the end of each mission where they provided their candid opinions on the mission in an otherwise unguided and unstructured text-based format. The individual debriefing logs were corrected for grammar and spelling and then processed through the 2007 version of the Linguistic Inquiry and Word Count program (LIWC; (Pennebaker and King 1999)). The software assigned the content of the debriefings into various non-mutually exclusive linguistic categories according to a ~4,500 word English dictionary and hierarchical taxonomy. LIWC data are expressed as the percentage of the entire debriefing log (all words) accounted for by words associated with any given linguistic category. The categories we focused on for the present studies were total Word Count, First Person Singular ("I"), First Person Plural ("we"), Positive Emotion (e.g., "happy," "love"), Negative Emotion (e.g., "anxiety," "hate"), Achievement (e.g., "earn," "win"), and Social Processes (e.g., "listen," "share").

Finally, to yield additional objective insights on relevant neurobiological mechanisms and physiological consequences of autonomous mission management, non-invasive saliva samples were collected from each subject at the same pre-, mid-, and post-mission points as the VAS assessments. Subjects first rinsed their mouths out with filtered water to remove any residual food or beverage that may interfere with the assay, and then were instructed to provide saliva samples via passive drool through a plastic straw directly into a 200-μl cryovial. Individuals who had difficulty producing samples were offered a piece of plain Trident sugar-free chewing gum to promote salivation. All vials were placed in a −20 °C freezer within 5 min of sampling, where they remained in storage until shipped on dry ice for processing. For analysis, given the well-documented relevance of the physiological stress system (hypothalamic-pituitary-adrenal axis) to long-term physical and behavioral health, mood regulation, cognition, and psychosocial interactions (Byers and Yaffe 2011; Kirschbaum and Hellhammer 1994; Kudielka et al. 2009; Lloyd and Nemeroff 2011; Plessow et al. 2012), all saliva samples in Experiment 1 were assayed for the stress hormone cortisol. All samples in Experiment 2 were also assayed for cortisol, with a subset assayed for the sympathetic nervous system biomarker alpha-amylase (Ali and Pruessner 2012). Finally, a subset of samples from Experiment 2 was assayed for the steroid hormone testosterone, which is classically associated with sexual behavior and aggression, but which recent experimental evidence suggests is critical for behaviors maintaining social order and status (Eisenegger et al. 2010, 2011). Total cortisol (μg/dL), alpha-amylase (U/mL), and testosterone (pg/mL) values were determined by a commercial firm (Salimetrics, College Park, PA, USA, www.salimetrics.com) via enzyme immunoassay operating under inter- and intra-assay coefficients of variation $<10\%$.

13.2.5 Statistical Analysis

All crews participated in all planned missions for both Experiments 1 and 2, thereby generating the full suite of performance and psychosocial data. Adequate saliva samples for cortisol assay were obtained from 6 participants in Experiment 1 and all 24 participants in Experiment 2. Alpha-amylase and testosterone were assayed from 18 and 9 participants in Experiment 2, respectively. Cortisol and testosterone were assayed in duplicate, with the average value used for analysis of any given sample; alpha-amylase was determined by standard singlet assay. For Experiment 1, variables with sample sizes appropriate for inferential statistics were subjected to paired-samples *t* tests comparing the missions conducted under Scheduled versus Autonomous mission management. One-tailed tests were applied to variables with hypothesized effects as outlined previously (Roma et al. 2011) (two-tailed otherwise and as noted). All variables from Experiment 2 were subjected to two-tailed Analyses of Variance (ANOVA) with between-groups factors of mission management (Scheduled vs. Autonomous) and communications status (Full Comms vs. Audio & Text Out); significant interaction effects were followed by Fisher's LSD post-hoc tests. Additional analyses are described in detail below. Statistical significance for all analyses was set at $\alpha = .05$. Unless otherwise noted, all data are presented as mean ± SEM.

13.3 Results

13.3.1 Experiment 1

13.3.1.1 Crew Performance

As hypothesized, PES task performances improved under the autonomous condition, increasing by at least 9 percentage points in all three crews (mean ± SD: 43 ± 21 vs. 60 ± 12; Cohen's $d = 0.98$). Although the crews differed in their baseline performances and sensitivity to the effect, the consistency of autonomy's benefit across three independent groups is noteworthy (see Fig. 13.2).

13.3.1.2 Subjective Self-Reports

As illustrated in Fig. 13.3, paired-samples *t* tests of the post-mission VAS data revealed that compared to the Scheduled condition, the Autonomous mission management condition significantly increased Positive self-reports (38 ± 8% vs. 47 ± 6%; $t(8) = 2.22$, $p < .05$) and significantly decreased Negative self-reports (39 ± 6% vs. 23 ± 4%; $t(8) = 2.22$, $p < .01$), but had no effect on Physical self-reports (41 ± 7% vs. 37 ± 8%; $t(8) = 0.60$, $p > .50$, two-tailed).

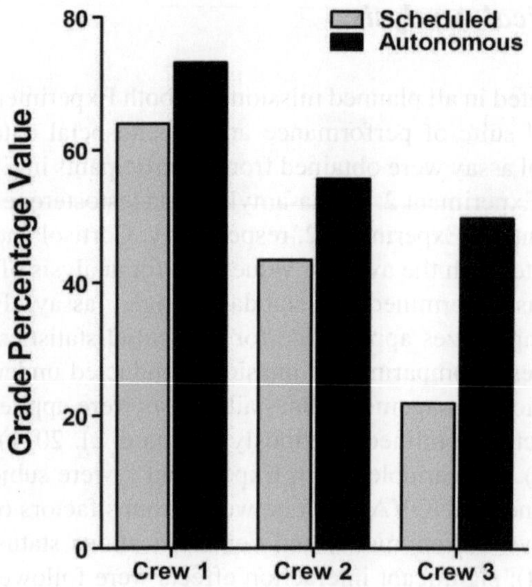

Fig. 13.2 Improved task performance under Scheduled and Autonomous mission management conditions in three PES crews (Cohen's d = 0.98)

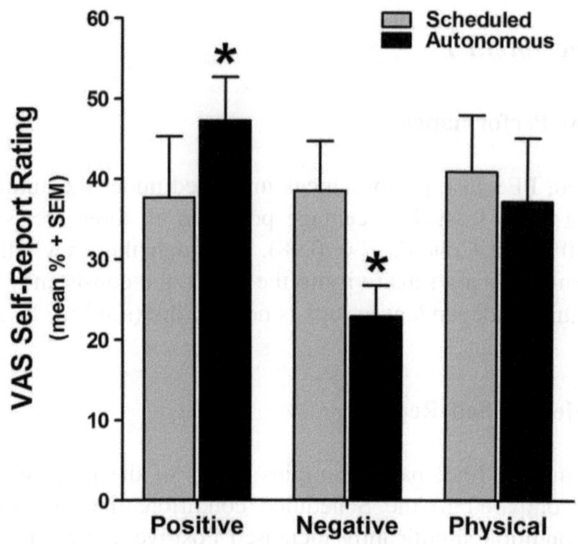

Fig. 13.3 Visual analog scale (VAS) self-report ratings following 3–4 h PES missions operating under Scheduled or Autonomous mission management. Data are expressed as mean + SEM % of maximum (n = 9, *p < .05 Scheduled vs. Autonomous)

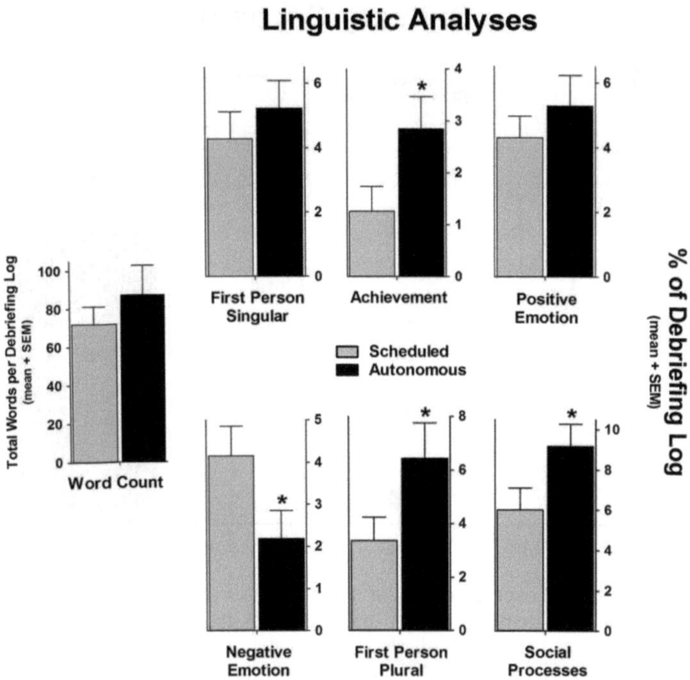

Fig. 13.4 Linguistic Inquiry and Word Count (LIWC) analyses of unstructured debriefing logs following 3–4 h PES missions operating under Scheduled or Autonomous mission management. Note that the y-axes vary in scale for each linguistic category. Data from each category are expressed as the mean + SEM % of all words in the debriefing log ($n = 9$, *$p < .05$ Scheduled vs. Autonomous)

13.3.1.3 Linguistic Analyses

The autonomous condition significantly increased the use of First Person Plural (3 ± 0.9% vs. 7 ± 1.3%; $t(8) = 1.96, p < .05$) and references to Social Processes (6 ± 1% vs. 9 ± 1%; $t(8) = 2.20$, $p < .05$) and Achievement (1 ± 0.5% vs. 3 ± 0.6%; $t(8) = 2.20$, $p < .05$) while significantly decreasing expressions of Negative Emotion (4 ± 0.7% vs. 2 ± 0.7%; $t(8) = 2.20$, $p < .05$). Autonomy did not affect total Word Count of the debriefing log entries (72 ± 9 vs. 87 ± 16 words; $t(8) = 1.04$, $p > .30$, two-tailed) nor did it alter the use of First Person Singular (4 ± 0.9% vs. 5 ± 0.9 %; $t(8) = 1.96, p = .09$, two-tailed) or expressions of Positive Emotion (4 ± 0.7% vs. 5 ± 0.9%; $t(8) = 1.96, p > .10$; see Fig. 13.4).

13.3.1.4 Physiology

Regardless of time point or session type, raw cortisol values ranged from 0.04 to 1.18 μg/dL. To control for expected individual differences in baseline hormonal activity, cortisol values at each time point from each mission were transformed to

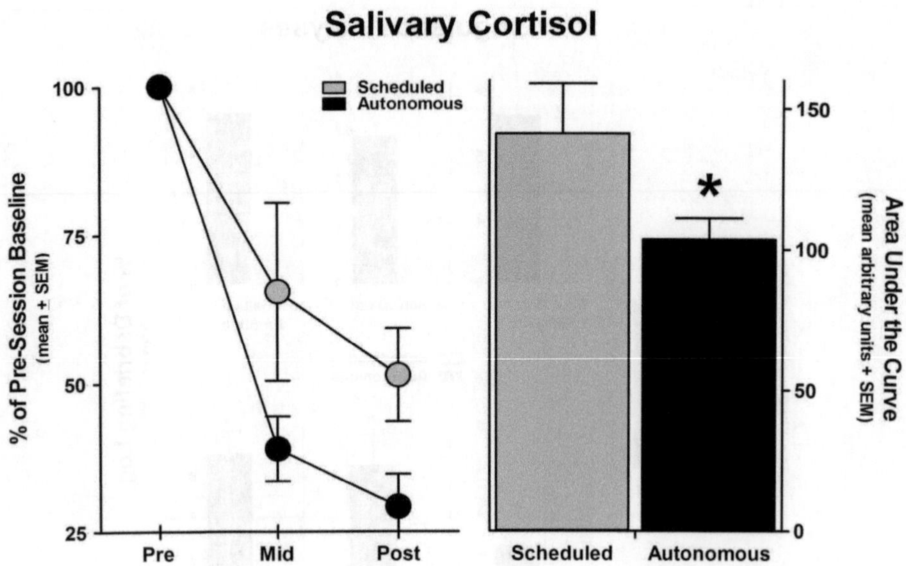

Fig. 13.5 Salivary cortisol over the course of 3–4 h PES missions operating under Scheduled or Autonomous mission management. The left panel presents the mean ± SEM % of pre-session baseline before, during, and after each mission. These values were calculated for each subject and subjected to area under the curve (AUC) analyses, the mean + SEM of which were directly compared as seen in the *right* panel (n = 6, *$p < .05$ Scheduled vs. Autonomous)

the % of that mission's pre-session baseline, and these values were plotted and subjected to individual area under the curve (AUC) analyses. Mean AUC values over the course of the Scheduled and Autonomous missions were then compared via paired-samples *t* test. This analysis confirmed that overall cortisol production was significantly lower during the Autonomous missions relative to the Scheduled missions ($t(6) = 2.59$, $p < .05$; see Fig. 13.5).

13.3.2 Experiment 2

Encouraged by the positive results of Experiment 1, we then utilized the same basic parameters in a more complex design with more crews in Experiment 2, the purpose of which was to (1) determine the reliability of the performance and biopsychosocial benefits of autonomy via systematic replication, (2) begin testing the limits of autonomy's effects when confronted with operationally-relevant stressors, in this case, unexpected and uncontrollable communications constraints, and (3) extend our understanding of concomitant stress physiology and neurobiology with the inclusion of analyses for salivary alpha-amylase and testosterone output.

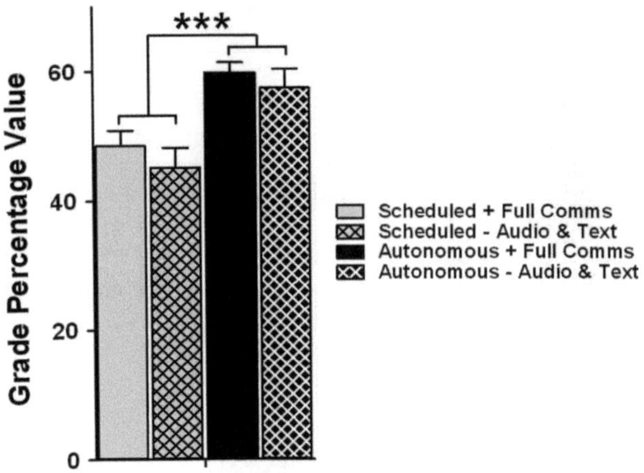

Fig. 13.6 Improved PES task performance under Scheduled and Autonomous mission management conditions despite the unexpected and uncontrollable loss of audio and text-messaging capabilities. A significant main effect of autonomy is indicated by brackets (***$p < .001$)

13.3.2.1 Crew Performance

Consistent with Experiment 1, crew PES performances improved under autonomous conditions, with grade values averaging 47 ± 2% in scheduled missions versus 59 ± 2% in autonomous missions ($F(1,32) = 20.5$, $p < .001$). Interestingly, there were no main or interaction effects of communications constraints on task performance (Fs (1,32) < 1.3, $ps > .20$; see Fig. 13.6).

13.3.2.2 Subjective Self-Reports

Replicating the findings of Experiment 1, mean Positive self-reports significantly increased during Autonomous versus Scheduled missions (56 ± 3% vs. 45 ± 2%; $F(1,96) = 9.0, p < .01$). However, as seen in Fig. 13.7, there were no significant effects of autonomy or communications constraints on Negative or Physical self-reports (Fs(1,96) < 3.2, $ps > .07$).

13.3.2.3 Linguistic Analyses

Analysis of post-mission debriefings revealed content of 90 ± 4 words per individual log. Consistent with Experiment 1, autonomous mission management significantly reduced the average expression of Negative Emotion (3.5 ± 0.3% vs. 2.7 ± 0.3%; $F(1,96) = 4.0, p < .05$); moreover, autonomy also increased the average expression of Social Processes (7.7 ± 0.5% vs. 9.3 ± 0.5%; $F(1,96) = 4.8$, $p < .05$). As depicted in Fig. 13.8, the only effects of communications constraints were main effects revealing unexpected but significant increases in First Person Plural

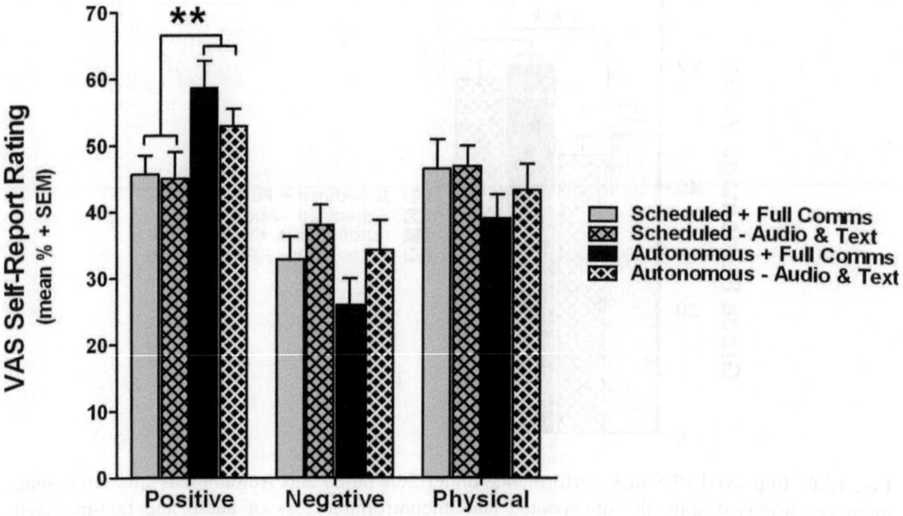

Fig. 13.7 Visual analog scale (VAS) self-report ratings following 3–4 h PES missions operating under Scheduled or Autonomous mission management with full communications or following the unexpected and uncontrollable loss of audio and text-messaging capabilities. Data are expressed as mean + SEM % of maximum (n = 24). A significant main effect of autonomy is indicated by brackets (**$p < .01$)

($3.6 \pm 0.3\%$ vs. $4.8 \pm 0.5\%$; $F(1,96) = 4.3$, $p < .05$) and Social Processes ($7.5 \pm 0.4\%$ vs. $9.4 \pm 0.6\%$; $F(1,96) = 6.9$, $p < .05$). There were no significant effects of mission management or communications constraints on total Word Count or the First Person Singular, Positive Emotion, and Achievement linguistic categories (Fs $(1,96) < 3.7$, $ps > .05$).

13.3.2.4 Physiology

Regardless of time point or session type, raw cortisol values ranged from 0.02 to 1.45 μg/dL, raw alpha-amylase values ranged from 1.7 to 219.4 U/mL, and raw testosterone values ranged from 11.58 to 331.95 pg/mL. Mean AUC values calculated based on raw shifts from pre-session baseline and analyzed via $2 \times 2 \times 2$ ANOVA (including Gender) revealed a significant interaction between autonomy and communications constraints in alpha-amylase ($F(1,36) = 4.4$, $p < .05$). Consistent with Experiment 1, post-hoc analyses confirmed that physiological stress reactivity was attenuated by autonomous mission management under baseline technical conditions, but the loss of audio and text messaging capabilities elicited a stress response that was equivalent to the highly scheduled mission yet not reversible by autonomy (Autonomous + Full Comms vs. all other conditions $p < .05$). There were no significant main or interaction effects of gender, crew autonomy, or communications outages on cortisol or testosterone (Fs$(1,72) < 3.5$, $ps > .06$; see Fig. 13.9).

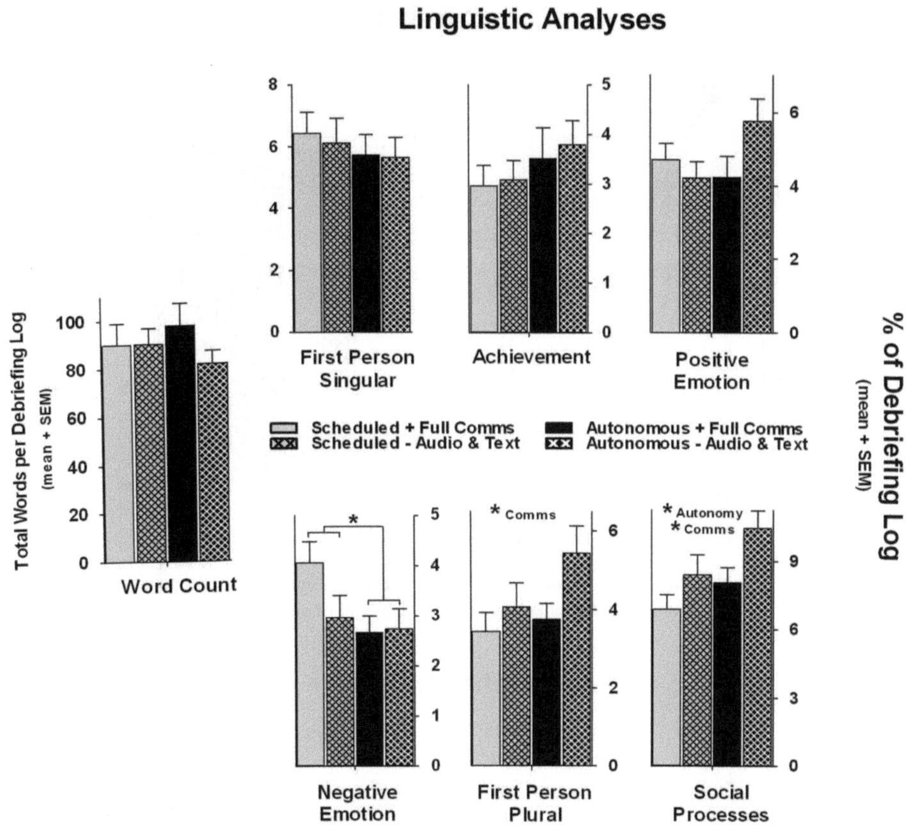

Fig. 13.8 Linguistic Inquiry and Word Count (LIWC) analyses of unstructured debriefing logs following 3–4 h PES missions operating under Scheduled or Autonomous mission management with full communications or following the unexpected and uncontrollable loss of audio and text-messaging capabilities. Note that the y-axes vary in scale for each linguistic category. Data from each category are expressed as the mean + SEM % of all words in the debriefing log ($n = 24$). Significant main effects of autonomy or communications outages are indicated by brackets or text (*$p < .05$)

13.4 Discussion

The results of the present study clearly revealed several apparent benefits of autonomous mission management on crew performance and individual behavioral and biological indices of psychosocial adaptation within a long-term laboratory-based planetary exploration simulation. In Experiment 1, crew autonomy enhanced task performance in all three groups, decreased subjective self-reports of negative emotional states, reduced references to negative emotions and greater use of social language in the unstructured debriefing logs, and attenuated physiological stress reactivity across the 3–4 h missions. Although these results are clear and make a

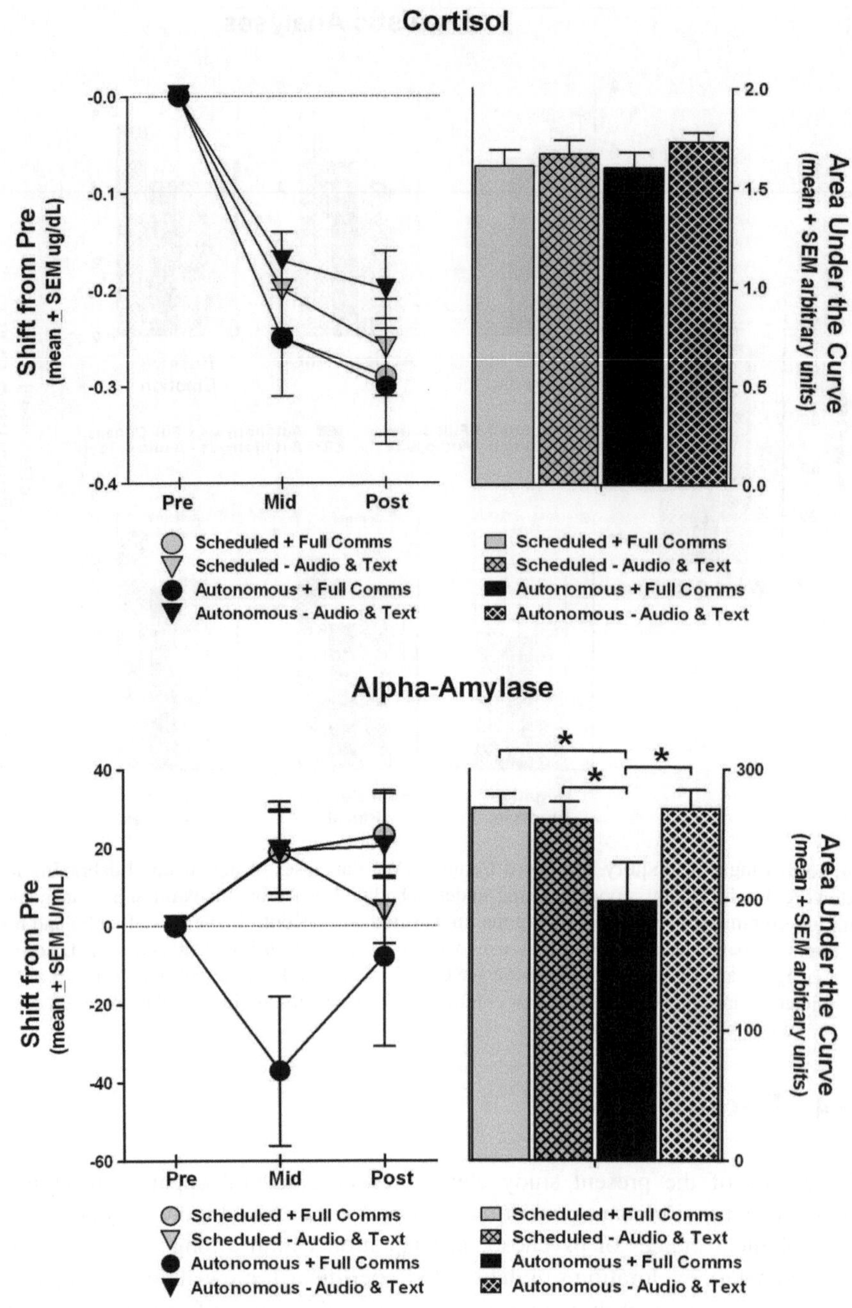
Cortisol
-0.0
-0.1
-0.2
-0.3
-0.4
Shift from Pre
(mean ± SEM ug/dL)
Pre
Mid
Post
2.0
1.5
1.0
0.5
0.0
Area Under the Curve
(mean + SEM arbitrary units)
Scheduled + Full Comms
Scheduled - Audio & Text
Autonomous + Full Comms
Autonomous - Audio & Text
Scheduled + Full Comms
Scheduled - Audio & Text
Autonomous + Full Comms
Autonomous - Audio & Text
Alpha-Amylase
40
20
0
-20
-40
-60
Shift from Pre
(mean ± SEM U/mL)
Pre
Mid
Post
*
*
*
300
200
100
0
Area Under the Curve
(mean + SEM arbitrary units)
Scheduled + Full Comms
Scheduled - Audio & Text
Autonomous + Full Comms
Autonomous - Audio & Text
Scheduled + Full Comms
Scheduled - Audio & Text
Autonomous + Full Comms
Autonomous - Audio & Text

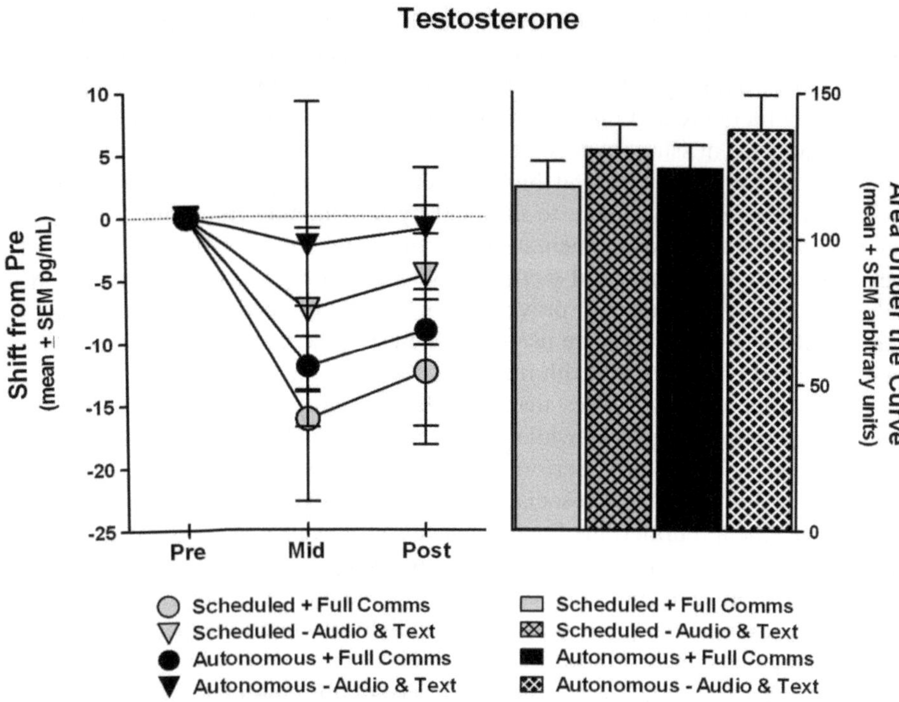

Fig. 13.9 Salivary cortisol (n = 24), alpha-amylase (n = 18), and testosterone (n = 9) over the course of 3–4 h PES missions operating under Scheduled or Autonomous mission management with full communications or following the unexpected and uncontrollable loss of audio and text-messaging capabilities. The *left* panels present the mean ± SEM shift from pre-session baseline before, during, and after each mission. These values were calculated for each subject and subjected to area under the curve (AUC) analyses, the mean + SEM of which were directly compared as seen in the *right* panels. Significant differences between specific sessions are indicated by brackets (*$p < .05$)

promising first step in the experimental analysis of autonomous mission management for the international astronautics community, there are still several issues worth considering for future research and application. Among them is the direction and consistency of the effects of autonomy across different conditions of relevance to space exploration. For example, one of the reasons for investigating autonomy is that limited communications and unpredictable technical difficulties will require independent action outside of standard operating procedures and crew routines. To test this, we designed and implemented Experiment 2 as a follow-up study to explore the effects of autonomy in teams confronted by the unexpected and uncontrollable operational stressor of communications outages, and a remarkably similar pattern of benefits emerged under autonomous mission management. In addition to communications constraints, heavy workload and endogenous biological rhythms often converge to impair performance, psychological adaptation, and interpersonal interactions in many operational settings (Roma et al. 2011; Mallis and

DeRoshia 2005; Mollicone et al. 2008), and data from recently completed studies in our laboratory are now being analyzed to test the interaction between autonomy and circadian phase during extended 12-h duty periods. To those familiar with the I/O psychology literature and other workplace research, it should come as no surprise that autonomous mission management generally proves beneficial at least in terms of psychosocial adaptation and biobehavioral stress response. The experiments presented here now contribute to the field by highlighting individual benefits of autonomous mission management in well-trained small groups engaged in interdependent long-term team-based work. Whether autonomy's benefits are limited to relatively benign conditions or prove resilient to various operational and biobehavioral challenges, all insights are useful as interest in the management of psychosocial factors and behavioral health increases.

Understanding the variables that alter the trajectory of autonomy's effects is important, but it is also worthwhile to explore the breadth of those effects. In the present studies, crew task performances as well as subjective and objective measures of individual psychosocial adaptation improved under autonomous mission management. Further, the linguistic analyses revealed an unsolicited increase in the use of socially-referent language, suggesting a potential enhancement of group affiliation and cohesion. Given the importance and natural interactions between neurobiology, behavioral health, individual cognitions/perceptions, and social interactions (including work), these findings are of special interest to many in the space research community since the matter of building, monitoring, and maintaining long-term group cohesion is a high priority issue (Salas et al. 2008; Wilson et al. 2007), particularly in the age of increasingly diverse international and mixed-gender crews. Whether in the laboratory, field, or space itself, systematic assessment of autonomy's effects on the formation, maintenance, and repair of group task and social cohesion should remain a high priority (Schmidt et al. 2009) along with the development and inclusion of inherently group-level metrics.

Regardless of the factors surrounding this study of autonomous mission management, perhaps the broadest consideration is that of generalizability of results from a population of community volunteers that meet in a laboratory for several missions a month to real astronaut crews that constantly live, train, and work together. Unlike many team-based experimental paradigms (Langfred and Moye 2004; DeShon et al. 2004), the simulation task employed in our laboratory does require a standing commitment lasting months, with some groups working together for years, and it seems highly unlikely that the kind of autonomous mission management that benefitted these subjects would negatively affect the truly exceptional individuals that constitute any nation's astronaut corps. Nonetheless, basic laboratory experiments such as the ones reported here provide a sound logical basis and "proof of concept" foundation to warrant scalable tests of autonomy in even higher-fidelity settings such as the fully immersive environment of a residential laboratory like that employed by Brady and colleagues (Brady et al. 1975; Emurian and Brady 2007; Emurian et al. 2009) or the Russian Mars500 isolation capsule (e.g., Kanas et al. 2010), for much-needed controlled field experiments in isolated and operationally demanding analogue environments (Shea et al. 2009; Binsted et al. 2009), and even for evaluation in active astronaut crews in space. Although the knowledge gaps are considerable, the increasingly multinational and collaborative nature of space exploration—including major contributions expected from China (China's Space Activities in 2011)—should also provide exciting opportunities for laboratory, analogue, and space research as well as development of cross-cultural if not truly universal tools and techniques for assessing the effects of autonomy and maintaining performance, behavioral health, and biopsychosocial adaptation in small groups on long-duration missions. The collective results of such efforts can provide empirical insight on autonomy as both an inherent feature of future missions and another potential prophylactic measure against the psychosocial rigors of life aloft (cf. Carter et al. 2005; Craske et al. 2009).

In conclusion, when considering the dynamic interplay between mission management, environmental influences, neurobiology, and psychosocial factors, it is important to consider the nature of autonomy itself. The present study focused on intra-task autonomy in that the crews engaged in a single (albeit complex) task with a single goal bound by a fixed temporal framework, with the mission management approach governing how the crew pursued that goal within that framework. This approach differs from autonomous conditions that take the form of crew involvement in the sequencing and timing of multiple distinct tasks, social activities, and rest periods over an extended temporal framework. Even when task autonomy per se does not apply in light of optimal performance training, this higher-order type of autonomy is of known interest to astronauts the world over (Jaap et al. 2007; Stepanova et al. 2009). However, given the tremendous investment of resources committed to even the most routine space-bound activities, concerns still linger over crew autonomy as increasing the risk of compromised performance, frequent insubordination, or even mutiny (V.I. Gushin, 2009, personal communication). In this regard, as the space research community moves forward with its study of psychosocial factors, it is equally important to define not only what autonomy is, but what it is *not*. Autonomy is not anarchy. Particularly in highly disciplined and extraordinarily well-trained astronaut crews, there is a difference between autonomy within a pre-assigned task, autonomy in selecting and sequencing required tasks over the course of a duty period, and complete freedom to chose whether or not to even work towards mission goals or any other end. Fortunately, the present study suggests that regardless of immediate improvements in performance, incorporation of autonomous mission management strategies does not lead to anarchy, and may in fact contribute to critically important long-term benefits on individual and group behavioral health and well-being. Given the inevitability of autonomous crew operations in future missions, acknowledging this operational reality and building an evidence-based understanding of how autonomy works may be the most responsible course of action to protect space agencies' investments in human and other resources. With a secure empirical knowledge base established, leaders of the international astronautics community and all interested stakeholders may make better informed decisions regarding crew autonomy to universal benefit when managing ongoing projects and designing long-duration missions beyond Earth's orbit.

Acknowledgements This work was supported by the US National Space Biomedical Research Institute through NASA NCC 9-58-NBPF01602 and Directed Research Project NBPF00008. Portions of this work were previously published by the authors as part of a peer-reviewed scientific journal article (Roma et al. 2011), and were used with permission from Elsevier. The authors have no interests that may be perceived as conflicting with the research presented herein, and were entirely responsible for the design of the studies, the collection, analysis, and interpretation of data, the preparation of the manuscript, and the decision to submit for publication.

References

Ali, N., Pruessner, J.C.: The salivary alpha amylase over cortisol ratio as a marker to assess dysregulations of the stress systems. Physiol. Behav. **106**(1), 65–72 (2012)

Binsted, K.A., Kobrick, R.L., Bishop, S., O'Griofa, M., Lapierre, J.: An overview of the human factors research conducted on a four-month mars simulation in the Canadian high arctic, with recommendations for future analogue research. Paper presented at the biannual meeting of the International Academy of Astronautics' Humans in Space Symposium, Moscow (2009)

Brady, J.V.: Behavioral health: the propaedeutic requirement. Aviat. Space Environ. Med. **76**, B13–B24 (2005)

Brady, J.V., Bigelow, G.E., Emurian, H.H., Williams, D.M.: Design of a programmed environment for the experimental analysis of social behavior. In: Carson, D.H. (ed.) Man-Environment Interactions: Evaluations and Applications. 7: Social Ecology, pp. 187–208. Environment Design Research Association, Milwaukee (1975)

Brady, J.V., Hienz, R.D., Hursh, S.R., Ragusa, L.C., Rouse, C.O., Gasior, E.D.: Distributed interactive communication in simulated space-dwelling groups. Comput. Hum. Behav. **20**, 311–340 (2004)

Byers, A.L., Yaffe, K.: Depression and risk of developing dementia. Nat. Rev. Neurol. **7**, 323–331 (2011)

Calnan, M., Wainrwright, D., Almond, S.: Job strain, effort-reward imbalance and mental distress: a study of occupations in general medical practice. Work Stress **14**, 297–311 (2000)

Carter, J.A., Buckey, J.C., Greenhalgh, L., Holland, A.W., Hegel, M.T.: An interactive media program for managing psychosocial problems on long-duration spaceflights. Aviat. Space Environ. Med. **76**, B213–B223 (2005)

China's Space Activities in 2011.: Information Office of the State Council, Beijing. http://www.scio.gov.cn/zxbd/wz/201112/t1073727.htm. Accessed 25 Mar 2012

Clément, G.: Fundamentals of Space Medicine, 2nd edn. Microcosm/Springer, El Segundo/Dordrecht (2011)

Craske, M.G., Rose, R.D., Lang, A., Welch, S.S., Campbell-Sills, L., Sullivan, G., Sherbourne, C., Bystritsky, A., Stein, M.B., Roy-Byrne, P.P.: Computer-assisted delivery of cognitive behavioral therapy for anxiety disorders in primary-care settings. Depress. Anxiety **26**, 235–242 (2009)

DeShon, R.P., Kozlowski, S.W., Schmidt, A.M., Milner, K.R., Wiechmann, D.: A multiple-goal, multilevel model of feedback effects on the regulation of individual and team performance. J. Appl. Psychol. **89**, 1035–1056 (2004)

Dinges, D.F., Mollicone, D., Ecker, A.: Psychomotor vigilance self test on ISS (reaction self test on expeditions 21 & 22). NASA Technical Report # JSC-CN-18714, NASA Johnson Space Center, Houston (2009)

Eisenegger, C., Naef, M., Snozzi, R., Heinrichs, M., Fehr, E.: Prejudice and truth about the effect of testosterone on human bargaining behaviour. Nature **463**, 356–359 (2010)

Eisenegger, C., Haushofer, J., Fehr, E.: The role of testosterone in social interaction. Trends Cogn. Sci. **15**, 263–271 (2011)

Emurian, H.H., Brady, J.V.: Behavioral health management of space dwelling groups: safe passage beyond earth orbit. Behav. Anal. Today **8**, 25–41 (2007)

Emurian, H.H., Canfield, K., Roma, P.G., Gasior, E.D., Brinson, Z.S., Hienz, R.D., Hursh, S.R., Brady, J.V.: Behavioral systems management of confined microsocieties: an agenda for research and applications. In: Proceedings of the 39th International Conference on Environmental Systems, SAE International, Warrendale (2009), paper number 2009-01-2423

Gardell, B.: Technology, alienation and mental health in the modern industrial environment. In: Levi, L. (ed.) Society, Stress and Disease, vol. 1, pp. 148–180. Oxford University Press, London (1971)

Hackman, J.R., Lawler III, E.E.: Employee reactions to job characteristics. J. Appl. Psychol. **55**, 259–286 (1971)

Hawley, S.A.: Mission to Mars: Risks, challenges, sacrifices and privileges. One Astronaut's perspective. J. Cosmol. **12**, 3517–3528 (2010)

Hienz, R.D., Brady, J.V., Hursh, S.R., Ragusa, L.C., Rouse, C.O., Gasior, E.D.: Distributed communication and psychosocial performance in simulated space dwelling groups. Acta Astronaut. **56**, 937–948 (2005)

Hienz, R.D., Brady, J.V., Hursh, S.R., Banner, M.J., Gasior, E.D., Spence, K.R.: Communication constraints, indexical countermeasures, and crew configuration effects in simulated space-dwelling groups. Acta Astronaut. **60**, 362–378 (2007)

Hienz, R.D., Brady, J.V., Hursh, S.R., Gasior, E.D., Spence, K.R., Emurian, H.H.: Effects of incentives on psychosocial performances in simulated space-dwelling groups. Acta Astronaut. **63**, 800–810 (2008)

Jaap, J., Meyer, P., Davis, E., Richardson L.: In-space crew-collaborative task scheduling. NASA Document ID 20070013716 (2007)

Kanas, N., Manzey, D.: Space Psychology and Psychiatry, 2nd edn. Microcosm/Springer, El Segundo/Dordrecht (2008)

Kanas, N., Saylor, S., Harris, M., Neylan, T., Boyd, J., Weiss, D.S., Baskin, P., Cook, C., Marmar, C.: High versus low crewmember autonomy in space simulation environments. Acta Astronaut. **67**, 731–738 (2010)

Kanas, N., Sandal, G.M., Boyd, J.E., Gushin, V.I., Manzey, D., North, R., Leon, G.R., Suedfeld, P., Bishop, S.L., Fiedler, E.R., Inoue, N., Johannes, B., Kealey, D.J., Kraft, N.O., Matsuzaki, I., Musson, D., Palinkas, L.A., Salnitskiy, V.P., Sipes, W., Stuster, J., Wang, J.: Psychology and culture during long-duration space missions. In: Vakoch, D.A. (ed.) On Orbit and Beyond: Psychological Perspectives on Human Spaceflight. Springer, Heidelberg (2013)

Kirschbaum, C., Hellhammer, D.H.: Salivary cortisol in psychoneuroendocrine research: recent developments and applications. Psychoneuroendocrinology **19**, 313–333 (1994)

Kudielka, B.M., Hellhammer, D.H., Wüst, S.: Why do we respond so differently? Reviewing determinants of human salivary cortisol responses to challenge. Psychoneuroendocrinology **34**, 2–18 (2009)

Langfred, C.W., Moye, N.A.: Effects of task autonomy on performance: an extended model considering motivational, informational, and structural mechanisms. J. Appl. Psychol. **89**, 934–945 (2004)

Lloyd, R.B., Nemeroff, C.B.: The role of corticotropin-releasing hormone in the pathophysiology of depression: therapeutic implications. Curr. Top. Med. Chem. **11**, 609–617 (2011)

Maina, G., Palmas, A., Bovenzi, M., Filon, F.L.: Salivary cortisol and psychosocial hazards at work. Am. J. Ind. Med. **52**, 251–260 (2009)

Mallis, M.M., DeRoshia, C.W.: Circadian rhythms, sleep, and performance in space. Aviat. Space Environ. Med. **76**, B94–B107 (2005)

Mollicone, D.J., Van Dongen, H.P., Rogers, N.L., Dinges, D.F.: Response surface mapping of neurobehavioral performance: testing the feasibility of split sleep schedules for space operations. Acta Astronaut. **63**, 833–840 (2008)

Pennebaker, J.W., King, L.A.: Linguistic styles: language use as an individual difference. J. Pers. Soc. Psychol. **77**, 1296–1312 (1999)

Pettit, D.: Mars landing on earth: an astronaut's perspective. J. Cosmol. **12**, 3529–3536 (2010)
Plessow, F., Kiesel, A., Kirschbaum, C.: The stressed prefrontal cortex and goal-directed behaviour: acute psychosocial stress impairs the flexible implementation of task goals. Exp. Brain Res. **216**(3), 397–408 (2012)
Roma, P.G., Hursh, S.R., Hienz, R.D., Emurian, H.H., Gasior, E.D., Brinson, Z.S., Brady, J.V.: Behavioral and biological effects of autonomous versus scheduled mission management in simulated space-dwelling groups. Acta Astronaut. **68**, 1581–1588 (2011)
Salas, E., Wilson, K.A., Murphy, C.E., King, H., Salisbury, M.: Communicating, coordinating, and cooperating when lives depend on it: tips for teamwork. Jt. Comm. J. Qual. Patient Saf. **34**, 333–341 (2008)
Schmidt, L.L., Keeton, K., Slack, K.J., Leveton, L.B., Shea, C.: Risk of performance errors due to poor team cohesion and performance, inadequate selection/team composition, inadequate training, and poor psychosocial adaptation. In: McPhee, J.C., Charles, J.B. (eds.) Human Health and Performance Risks of Space Exploration Missions: Evidence Reviewed by the NASA Human Research Program, pp. 45–84. NASA Johnson Space Center, Houston (2009)
Shea, C., Slack, K.J., Keeton, K.E., Palinkas, L.A., Leveton, L.B.: Antarctica meta-analysis: psychosocial factors related to long-duration isolation and confinement. Final Report submitted to the NASA Behavioral Health and Performance Element (2009)
Stepanova, S.I., Nesterov, V.F., Saraev, I.F., Galichiy, V.A., Savchenko, E.G., Lavrentieva, I.N., Rudometkin, N.M.: The ISS crew work/rest schedule (WRS): open issues and satellite challenges. Paper presented at the biannual meeting of the International Academy of Astronautics' Humans in Space Symposium, Moscow (2009)
Suedfeld, P., Wilk, K.E., Cassel, L.: Flying with strangers: postmission reflections of multinational space crews. In: Vakoch, D.A. (ed.) Psychology of Space Exploration: Contemporary Research in Historical Perspective, pp. 143–175. National Aeronautics and Space Administration, Washington, DC (2011)
Wilson, K.A., Salas, E., Priest, H.A., Andrews, D.: Errors in the heat of battle: taking a closer look at shared cognition breakdowns through teamwork. Hum. Factors **49**, 243–256 (2007)

Chapter 14
Near-Term Extended Solar System Exploration

Harvey Wichman

Abstract Any discussion of long-duration space flight is of necessity a multidisciplinary discussion. It is of little value, for example, to discuss psychological issues as if they existed independently of engineering, sociological, physiological, and even political and economic issues. Simply put, there is no sense in discussing flights to the outer planets to be conducted at this time because it is not psychologically or physiologically possible. As an example, recently the Jet Propulsion Laboratory (JPL), in conjunction with the European Space Agency (ESA), sent the Cassini/Huygens probe to the planet Saturn and its moon, Titan. Given the resources available at the present time, that journey took 7 years. That would mean a 14-year round trip for a human space crew with no time at the destination. This is not physiologically possible, and it would be psychologically absurd. Yet humankind dreams of exploring the entire solar system, planets, moons, asteroids, and comets. Given that all of the disciplines mentioned above overlap in their involvement with human space flight, this chapter examines the promise of near-future long duration spaceflight from a multidisciplinary point of view.

This chapter explores psychology's contributions to understanding human performance in space, with emphasis on the role of the autonomic nervous system. We consider what data we have about how long explorers in situations of high risk, crowding, lack of privacy, the inability to leave, and sometimes great discomfort, can be away from home, family, and life as we normally know it, and recover to live normal lives. After finding evidence to support a psychological benchmark duration, the chapter examines technology now in serious development that would make such a flight possible, using trips to Mars and Saturn as examples.

But even meeting the psychological benchmark does not address the physiological debilitation of such an extended period of weightlessness. The chapter thus examines a potential physiological "deal breaker," bone demineralization, as well as current active research that suggests that this critical problem is in the process of being

H. Wichman (✉)
Claremont McKenna College and Claremont Graduate University, Claremont, USA
e-mail: hwichman@cmc.edu

D.A. Vakoch (ed.), *On Orbit and Beyond*, Space Technology Library 29,
DOI 10.1007/978-3-642-30583-2_14, © Springer-Verlag Berlin Heidelberg 2013

understood and solved. After proposing two promising solutions for bone demineralization, the chapter suggests that near-term spaceflight to the outer planets is reasonable to expect. The chapter closes by considering the advantages of returning to the Moon as a step toward exploratory missions to Mars, as well as the value of including humans in future missions to asteroids.

14.1 Introduction

This book is titled *On Orbit and Beyond.* The preponderance of the book is about living on orbit. Most of what we know about spacefaring has been derived from orbital flight. It is a fascinating story that began when the Soviets successfully orbited a spacecraft carrying the dog, Laika, demonstrating that mammals can live in space. Soon after, the flights of Yuri Gagarin and John Glenn showed that humans could live and perform successfully in space too. From then on, we became frequent orbital spacefarers.

However, this book is also about living in space *beyond* Earth orbit. That began in the Apollo program when astronauts on Apollo flights 8, 10 and 13 went out and orbited the Moon. On flights 11, 12, 14, 15, 16 and 17 astronauts landed and spent time exploring the Moon. U.S. astronauts departed the Moon in December 1972 and no one has since flown beyond Earth orbit, a period of 40 years.

This chapter emphasizes living in space *beyond* Earth orbit. The expectation is that the long hiatus will soon end. We will return to the Moon to colonize it and continue on to such exciting destinations as Mars, asteroids, comets and Lagrange Points. To do so, we have to consider two major issues. First, if we have not gone beyond Earth orbit with human crews in 40 years, what makes us think we can do it now, or at least soon? Second, how is life beyond orbit different from life on orbit?

14.2 Why Human Spaceflight Beyond Orbit May Soon Be Possible

The greatest threat in spaceflight beyond orbit and the protection of the Earth's Van Allen Radiation Belts is twofold: the increased *level* of cosmic radiation encountered and the extended *duration* of that exposure. Using current technology, a one-way trip to the Moon takes about 3 days. A trip out to Mars, the nearby planet we most want to visit, requires around 7 months. A trip to Saturn takes 7 years.

The Moon trip has been done. The Mars trip stretches credulity, and the trip to Saturn is simply ridiculous. The radiation problem can be solved or significantly ameliorated passively, by developing better insulating materials, or actively, by cleverly employing magnetic fields around spacecraft. Progress is being made but is limited, in the case of magnetic fields by the huge demands for electrical power required. More on this later. Clearly, for the moment, duration is the greatest challenge to travel beyond Earth orbit.

Duration is also critical physiologically. It is not currently feasible for a human to tolerate several months in weightlessness without crippling deficits in bone density. Duration can also be a limiting factor psychologically. Just how long can people can survive periods away from their normal social environments and loved ones while enduring lack of privacy, crowding, imminent danger and the inability to leave the situation and still recover and live normal lives?

This is an answerable question. Studies described earlier in this volume involving sailors on tiny vessels in the time of Captain Cook and Columbus often spent 2½ years away from home. Shackleton's Antarctic expedition, frozen in the ice, and many other examples of exploration show that 2½ years is tolerable. So, for now, all other things being equal, let us accept the duration away from Earth of 2½ years to be the maximum to which we would want to expose spacefarers psychologically. The question then becomes whether we can find a way in the next few years to fit some long distance explorations into this 2½ year time frame. If not, the discussion is over and there is no need to go further.

So, for the sake of discussion let us consider trips to Mars and the moons of Saturn.

A round trip to Mars would take about 14 months. With 6 months on station that would be 30 months or 2½ years. This is possible psychologically but not physiologically. A round trip to Saturn would take 14 years with no time on station. This is out of the question under any circumstances. What then gives us any hope of drastically reducing these travel times in the near future?

14.3 The VASIMR Rocket Motor

This would be the end of this chapter were it not for some very heartening news. Former NASA astronaut Franklin Chang Díaz (Fig. 14.1) and his team of scientists and engineers at Ad Astra Rocket Company are well along in developing a new type of electric propulsion rocket called Variable Specific Impulse Magneto Rocket, or VASIMR (Fig. 14.2) (Chang Díaz 2000; Longmier et al. 2011; Bering et al. 2011). The propulsive force for this ingenious motor is supplied by plasma. Plasma is a state of matter in which atoms have been so energized by heating that they can no longer hold themselves together, so they come apart into their respective subatomic particles. Plasma is what makes up the solar wind emanating from the sun.

To explain how the motor works, let us use hydrogen as an illustrative fuel because it is the simplest atom. Hydrogen is heated in the rocket with microwave energy, similar to the heating that occurs in a microwave oven. The hydrogen is heated to a million degrees Fahrenheit. It then comes apart into its constituent charged particles, electrons and protons. Normally these particles would melt the rocket and that would be the end of it. Here is where Chang Díaz's genius comes in. He has found a way to create powerful, precisely shaped magnetic fields that entrap the particles and keep them from touching the chamber walls. They are then accelerated out the rocket nozzle at velocities approaching the speed of light, in the process providing thrust.

Fig. 14.1 Franklin Chang Díaz (Photo credit: NASA)

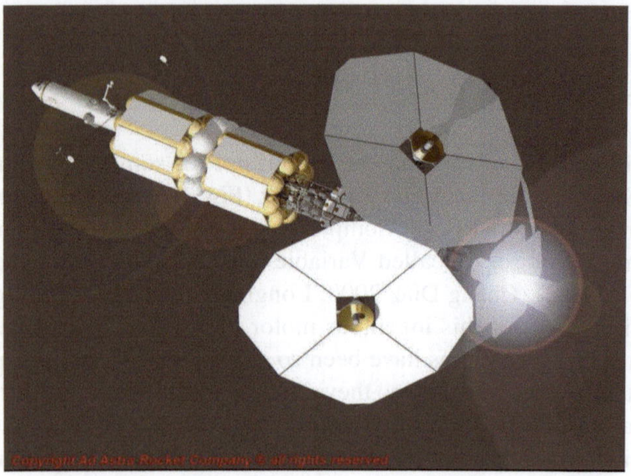

Fig. 14.2 Concept for piloted mission to Mars powered by four VASIMR engines (Illustration credit: Ad Astra Rocket Company)

A common measure of motor efficiency is specific impulse, which is defined as the pounds of force generated per pound of propellant expended per second. For comparison purposes, the specific impulse of arguably our best chemically fueled rocket engines, the main engines on the Space Shuttles, is 453 in a vacuum.

The specific impulse of the VASIMR is an extremely efficient 4,900. What differentiates these engines is that the thrust of the Shuttle engines is very high, 400,000 lb at liftoff, whereas the VASIMR at high electric power may produce about 40 lb of thrust. Clearly a chemical rocket motor would be required to get from Earth to space and into orbit. But once there, a plasma motor needs little thrust to accelerate a space vehicle to very high speeds in an environment where there is no friction.

The Ad Astra team has calculated the time it would take to fly to Mars as about 39 days, when Mars and Earth are in favorable orbital positions (Ilin et al. 2011). This compares with a flight time of about 7 months for a current chemical rocket. The team has calculated the time for a round trip to Saturn, with 6 months on station, as being 2½ years. A flight to Saturn with 6 months to explore its moons fell within our 2½ year psychological limit.

Is this all too good to believe? Admittedly, there are serious critics of using electric rockets such as VASIMR, such as Robert Zubrin, founder of The Mars Society and author of *The Case for Mars* (Zubrin and Wagner 2011) and *How to Live on Mars* (Zubrin 2008). One of Zubrin's criticisms of the VASIMR is the huge amount of electrical energy it requires to achieve practical thrust levels. He believes it would require a nuclear power unit to be practical. (We will return to nuclear power in the following section.)

On the other hand, NASA plans to mount a VASIMR motor on the International Space Station (ISS) in 2015, just 2 years from now. There, it can be tested and further developed in a space environment (Squire et al. 2011). It will have a practical use for the space station. The ISS needs to be accelerated periodically in order to maintain its altitude. That task used to be performed by a Space Shuttle using its orbital engines while docked to the station. The Shuttles are now retired, so a new source of thrust has to be found. On the ISS, VASIMR will use solar panels for electrical power. An electric thruster is ideal because its exhaust produces no chemical residue to coat the optics and windows on the ISS. Clearly, knowledgeable people at NASA consider VASIMR a viable electric rocket.

14.4 The Use of Nuclear Power in Space

Sources of electrical power are critical for all phases of space flight and extraterrestrial activity. Batteries and fuel cells are fine for low power demands and over short durations. Currently, the most popular sources of electricity in space are solar panels and devices called Radioisotope Thermoelectric Generators, or RTGs. In these generators an isotope such as plutonium-238 decays and in doing so produces heat that is transformed into electricity by devices called thermocouples. The heat is also often used to keep electronics from freezing in cold places like Mars.

Solar panels are an excellent source of electricity when one is as close to the Sun as the Earth and Moon are. The ISS, with its huge truss holding acres of solar panels, can supply a lot of electrical power. However, in every 90-min orbit of the

Earth, the ISS is in darkness half of the time, so power generation is not constant. Once one gets as far away from the sun as the orbit of Mars, the utility of solar panels is much diminished. Beyond Mars they are not practical.

Even though RTGs utilize a radioactive fuel source, they are safe enough to be launched into space at the present time. Thus they are regularly used in modest power applications such as the Cassinni/Huygens probe that was recently sent to Saturn and its moon Titan.

In February 2011 a conference on space nuclear power was held in Albuquerque, New Mexico. It was jointly sponsored by the American Nuclear Society and the American Institute for Aeronautics and Astronautics to bring all relevant parties together to discuss the cutting edge of this technology. The final conclusion of the eight Jet Propulsion Laboratory (JPL) delegates was, "O'Keefe [former NASA Administrator] Had It Right: The Next Major Step In Space Exploration Is Going To Require Space Nuclear Power." (From their PowerPoint slide number 23).

Because of the threat of a deadly disaster resulting in environmental pollution, we do not send nuclear power generators into space. However, the U.S. and Russia have many years of experience with mobile nuclear power generators in surface ships and submarines. These units have accumulated excellent records of reliability and safety. There is a general renewal of optimism in the nuclear power field that the time is near for using nuclear reactors in space. These reactors could serve as sources of propulsion in nuclear rockets and as sources of electrical power in nuclear generators. We are approaching the end of the era of exploratory orbital spacefaring and quick, daring trips to spend a few hours on the Moon in manned spaceflight. In robotic spaceflight the huge successes with tiny probes have brought us to the point where it is necessary to send more elaborate robotic units to explore and analyze on a much grander scale. The electrical requirements for such units will demand nuclear power generators.

The reason it takes 7 years to get to Saturn is not because Saturn is so far away. It is because our current chemical rockets cannot get space vehicles up to the high speeds necessary to make the long trip in times compatible with human needs. Consider, for example, the flight of the Cassinni/Huygens probe to Saturn and its moon Titan. With great ingenuity the flight managers at JPL launched the spacecraft and then sent it to Venus where it was accelerated in a gravity assist maneuver—a sort of "crack the whip" event that sent it back toward Earth. They did this multiple times, back and forth between the two planets, before sending it off on a long journey to Jupiter for an even bigger boost, then finally sending the spacecraft on its way to Saturn. That huge journey took 7 years.

Much work is being done in a variety of countries including Japan, China, India, Russia, the U.S. and the consortium in the European Space Agency to safely use nuclear power in space. Nuclear-powered rockets were designed more than 30 years ago in the early stages of the U.S. space program. Experience with nuclear power generators is extensive. Now that the time is ripe to utilize space nuclear power, clever ideas are being put forth for safe ways to put radioactive components into space separately and then bring them together when they are safely away from Earth. Many of the pieces are already in place to begin exploiting this resource as soon as there is the financial and political will to motivate it.

Thus the problem of very long times spent going to and from distant locations in the solar system, which would have been physically and psychologically disabling, may now be solvable in the near future. This means it is reasonable to begin planning extended human space explorations now.

14.5 Solving the Problem of Physical Deterioration from Weightlessness

Up until now the practical limit for time spent in weightlessness has been 6 months. That is the usual tour of duty for crewmembers on the ISS. It may be some time before this can be stretched to the 2½ year psychological span that we have empirically determined is possible. But is there any hope of improving the present situation? The answer may be yes in two ways. One is to find a practical way to produce a one-g artificial gravity. The other is somehow to treat crew persons, medicinally or physically, to prevent bone demineralization.

14.5.1 Artificial Gravity

Most people will remember the great rotating spaceship in the movie *2001: A Space Odyssey*. We are not yet ready to construct that scale of spaceship as a means of creating artificial gravity. However, Robert Zubrin asserts that the solution is rather simple (Zubrin 2008). Once a crew in its capsule is on its way to Mars, a tether 1,500 m long could be trailed out with a small rocket at the end. When the tether became taught, the capsule-tether-rocket would become a mechanical system. Firing the rocket would cause the system to rotate. A slow rotation rate of 2.6 RPM would produce an artificial gravity about equal to that on Earth. If this works, it could be a relatively simple solution for the major physiological problem of long duration space flight.

A demonstration experiment probably should have been conducted long ago. However, a one-g environment was not needed for the quick round trips to the Moon, and a rotating structure is not practical for space stations festooned with solar panel arrays, docking stations, etc. But it could be the answer for flights involving long rides out and back to distant destinations. Since the Apollo program ended, NASA has been preoccupied with constructing the ISS for manned flight and sending robotic probes to far distant locations. But NASA is now conducting serious research addressing the issues related to long duration human spaceflight (NIH selects another investigator for the International Space Station 2011).

Consider the following research project that could be conducted now rather inexpensively. It would test Zubrin's idea for a rotating capsule-tether-rocket system and more.

First, a small-scale crew capsule would be flown to prove out all the subsystems and demonstrate the viability of the principle proposed by Zubrin. It would rotate to produce a one-g environment. Then two more of the capsules would be flown with small animal payloads—probably rats. Half of the animals would be male and half female to check for sex differences. Half of the animals would be in cages with running wheels, and half would be without, to check for exercise effects. But most importantly, one group would be rotated to simulate the Moon's gravitational force (one-sixth that on Earth) and one group would live under conditions equaling the Martian gravity (0.38 Earth's gravity). We would then know how mammals (and probably humans) would respond to long periods on the Moon and on Mars. These are critical data required before any human colonizing is begun. These are experiments that cry out to be conducted now because if the results supported Zubrin's proposal, they would provide a large motivational impetus to continue space flights beyond Earth orbit.

14.5.2 Preventive Treatment for Bone Deterioration

Research on how to prevent bone deterioration in weightlessness is now being vigorously conducted and is beginning to show encouraging results. There are both chemical and physical approaches to solving this problem. We will briefly review one approach that at first appears to be physical, but probably is electrochemical due to the physical activity. The important thing is that it illustrates the ingenuity being brought to bear on problems considered unsolvable a couple of decades ago.

In the early days of space venturing it was assumed that one must be in a one-g environment to avoid deconditioning, hence the idea of spinning space stations. However there was a problem with this type of thinking. The same type of deterioration occurs in long-term bed-rest patients and they are in a one–g environment. Clearly something other than just being in a one-g environment is responsible for maintaining healthy bones.

Bones are made up of calcium crystals of a specific type called piezoelectric crystals. Such crystals have a pair of interesting characteristics. First, if a physical load is applied to the crystal and slightly deforms it, the crystal generates a miniscule electric current. Second, if, conversely, a small current is applied to the crystal, the crystal will slightly deform. In bone crystals the miniscule currents serve to distribute calcium ions throughout and create strong bones. But to create the currents, the bones have to be physically loaded so the force deforms the crystalline bones. That is what walking does normally on Earth and what does not happen in bed rest and weightlessness. It is only during the act of deforming and during elastic reforming that the currents are generated. Adults typically walk between 4 and 10,000 steps a day (2–5 miles). This produces two pulses per step (pressure on—pressure off). Perhaps by providing something like 5,000 pressure pulses (10,000 electrical pulses) per day to space crews, the bone demineralization problem could be solved or significantly ameliorated. As a wonderful spinoff for rehabilitation medicine, this could work for bed-rest patients as well.

These pulses would not likely have to be uniformly distributed throughout the day. Probably several periods of rapid vibration would be sufficient. Some experiments sponsored by NASA that are now in progress are using pleasant vibration rates of 90 cycles per second. Looking at the vibrating platform one cannot even see its motion. Animal studies with mice, rats and turkeys are providing early data that are very encouraging. Another type of treatment that corroborates the piezoelectric effect is inserting a limb with a damaged bone in a cylindrical electromagnet and generating pulsing magnetic fields to move the calcium ions about in the damaged bone. This technique is proving quite effective in facilitating bone healing.

14.6 Now Is the Time to Begin Planning Human Exploration of the Solar System

Space flight beyond Earth orbit is not much different from within Earth orbit except for the lack of protection of the radiation belts and the duration of flights, which can be very much longer. Space activities are different when crews have landed on another object in the solar system, because astronauts then spend much more time outside the spacecraft in a protective suit. When on a moon or planet, crews will no longer be weightless, and that could be a great advantage.

The beauty of the situation is that we are close to conquering the difficulties discussed above. We have many reasons to be optimistic that these and other solutions, some not yet thought of, are soon to be found, just as these unlikely solutions were discovered. Since most of what we know about spaceflight within Earth orbit also applies to flight beyond, we start with a huge fund of accumulated knowledge and experience.

The scientists, engineers, astronauts and cosmonauts who introduced the Space Age half a century ago did not have this luxury. They started from scratch. Yet once they were intensely focused on the task at hand, they were hugely successful. We in the next generation now stand on the shoulders of giants. From our vantage point we should now look to the farther horizons and prove ourselves worthy to follow in the footsteps of those who set us on our way.

14.7 Mind-Body Interaction in Space Psychology

Much of the very useful information in this volume falls strictly within the discipline of psychology. It can be parsed into the various sub-disciplines of the field such as social, clinical, organizational and human factors psychology. The information has been derived from laboratory experiments, natural experiments and years of practical experience with human spaceflight.

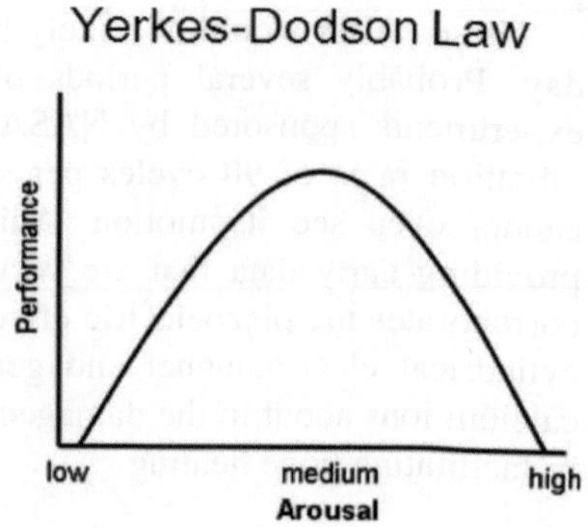

Fig. 14.3 According to the Yerkes-Dodson Law, performance is highest at an optimal level of arousal, while performance is lower at levels of arousal that are either lower or higher than this optimum

Most people think of psychology as being about the mind. But mind is the activity of the brain. And the brain is part of the body. Mind and body do not function independently of one another. Bodily functioning can both facilitate and interfere with effective mental activity. By the same token, mental activity can greatly facilitate or inhibit bodily functioning, including destroying it, as in the case of suicide. Thus, we would be remiss in establishing a complete context for comprehending this information if we did not reflect upon what is happening in parallel—physically and mentally—when we apply psychological principles.

14.8 Performance

A good way to examine the link between the physical and the mental is to consider performance. Performance is important to spaceflight because that is what space crews do—they perform, and their effective performance is critical to the success of any mission. Performance occurs when a mind wants to accomplish something; the body is what does the accomplishing. Even if the mind wants to accomplish a *mental* task, the brain has to do that task.

In psychology, the Yerkes-Dodson Law helps characterize the interplay of mind and body to performance (Yerkes and Dodson 1908). This law states that performance is an optimum function of arousal. An optimum function is portrayed graphically as an in inverted-U shaped curve (Fig. 14.3). This means that when one's arousal level is low, one's performance will be poor. As arousal level increases, so will performance, up to a point: the *optimum*. From that point on, as arousal increases, performance continually decreases. Nearly all living systems operate according to the optimum principle.

To take an example that is intuitively obvious, plotting an organism's state of health against the amount of food it consumes produces an optimum curve. As food intake continues to increase, state-of-health diminishes, ultimately to death at the opposite end.

Although the inverted-U is the general form of the Yerkes-Dodson Law, the optimal level of arousal is not the same for all activities. For example, the optimal level of arousal for complex or intellectual activities is lower than for physical activities.

14.9 The Autonomic Nervous System

To define what is meant by arousal, it is necessary to describe the system of nerves that operate our bodies—usually without conscious control—called the *autonomic nervous system*. It consists of two branches, the *sympathetic* and the *parasympathetic* systems. Most bodily organs receive nerves from both systems, and generally they have opposite effects. For example, sympathetic nerves to the heart speed it up, and parasympathetic nerves slow it down. However, parasympathetic nerves to the stomach increase its activity, and sympathetic nerve impulses decrease its activity. Thus, in an emergency a huge outflow of sympathetic neural impulses will speed up the heart to cope with the threat and slow down stomach activity to allow blood to be diverted to muscles. (Digesting a sandwich doesn't matter if one is caught by the tiger.)

In a general sense, the basic distinction between the sympathetic and parasympathetic systems is that the sympathetic system prepares the body to cope with demands from the environment, and the parasympathetic system prepares the body to rehabilitate itself from the bodily expenditures and damage from coping activities. Thus, the term arousal in the Yerkes-Dodson law refers to arousal of the *sympathetic* branch of the autonomic nervous system. Too little arousal to cope leads to poor performance, but too much arousal also leads to poor performance—thus the notion of optimal arousal levels to achieve the best performances. It is important to note that while both physical activity and mental activity are optimal functions of arousal, generally the optimum level of arousal for physical activity is greater than it is for mental activity.

Arousal of the sympathetic nervous system is typically a two-factor phenomenon, with one factor being physical and the other mental. Some cognitive states are occurring in parallel with the heightened physical activity usually associated with sympathetic arousal. These cognitive states can be either positive or negative. For example, a bride-to-be preparing for her wedding can be highly aroused and delighted, while a person being led to the gallows can be highly aroused and in a cognitive state of dread. We call people in such high states of arousal *stressed*. The stress can be cognitively positive or negative. Persons who feel stressed often say they are *tense*, indicating that the sympathetic nervous system and the hormones that facilitate arousal, such as adrenalin, are causing muscle groups to be partly in contraction, prompting the changes in posture, gestures and facial expressions that we associate with persons under stress.

Both branches of the autonomic nervous system are always functioning. If one puts sensors on someone to detect the state of arousal of the sympathetic system, one would see that during meditation there is a dramatic reduction in sympathetic activity (hence the utility of meditation in stress management). In contrast, simply asking the person to state his or her name produces a significant increase in sympathetic activation. Both sympathetic and parasympathetic activity vary constantly throughout the day. At times one or the other system may be dominant as the mind/body system strives to maintain equilibrium (homeostasis) as well as meet the challenges of the environment. To manage human behavior, it is important to understand that this interplay of mind and body is constantly taking place. For example, to keep pilots of multiengine aircraft from being paralyzed with fear if an engine should fail, we have them practice engine failure procedures, so that when an engine failure does occur, the pilot will not perceive it as catastrophic. Thus, his or her cognitive state will not overexcite the sympathetic system. Rather, the sympathetic system would be at the optimal level of arousal for coping with the challenge. Instead, if the pilot were *paralyzed with fear*, his or her sympathetic system would be so hyperaroused that it would cause all skeletal muscles to be in a state of contraction at the same time.

What all of this tells us is that a person is not a single entity, but rather a system of interacting elements, both physical and mental. Only if the professional teams that design flight hardware and crew performance programs operate from this point of view will they be able to create effective long-duration spaceflight environments.

While it is true that there is much similarity between spaceflight on orbit and beyond, there are two important differences. The farther one gets from Earth orbit, the less likely it is that a rescue mission can be accomplished, and the longer it takes to communicate back and forth. In all of the spaceflights undertaken to date, the flight crews enjoyed the support of large ground crews constantly available to them for advice and problem solving. The interaction between the ground and flight crew following the explosion on Apollo 13 stands as a shining example. The 2-s time delay for radio signals to traverse from Earth to Moon was insignificant. However, the time delay for communication with Mars varies between 10 and 20 min each way, depending on the relative positions of the two planets. At Saturn, the one-way time delays would vary between 68 and 84 min. These delays are significant. Clearly, spacefarers on deep space missions are going to have to be much more autonomous than space crews to date. This will post a huge challenge for deep space program designers. However, as suggested in Chap. 6 of this volume, we can begin to meet that challenge right here on Earth in spaceflight simulators.

14.10 Where to Next for Human Spaceflight Beyond Earth Orbit?

During the heyday of the Cold War between the Soviet Union and the allied countries of the West, the space programs of the Soviets and the United States conducted a competition unprecedented in the history of science and technology. This intense competition produced a huge surge in knowledge and large numbers of spinoffs of great value to non-space related areas of human endeavor. Great moments of national pride accompanied each new achievement as the two countries see-sawed back and forth attempting to be the leader in space.

But in parallel with the intense desire to be first and best, something very interesting was happening among those who were active players in the competition. It turns out that science often trumped politics, and in spite of national allegiances, the simple quest for knowledge turned out to be a superordinate goal that motivated the Cold War enemies to become comrades in a higher cause—the pure search for knowledge. International meetings of scientists of all sorts continued taking place, albeit with sensitivity to military issues. But scientists did meet and inspire one another, and many apolitical friendships quietly formed.

Then as the Cold War wound down, we suddenly found cosmonauts learning English and riding in Shuttles, while astronauts were learning Russian and riding in Soyuz capsules to Russian space stations. Eventually, what started out to be the U.S. space station *Freedom* became what President Reagan had originally envisioned, the International Space Station. If the ISS achieves nothing else, it will stand forever as a monument to the possibilities of international cooperation.

While it is true that science and technology are still rife with competitiveness, and the new players in spaceflight such as Japan, India and China are eager to prove themselves as major spacefaring nations, the groundwork for much international cooperation has been solidly established. The melding of the forces of friendly competition and cooperation is certain to accelerate the rate of progress in space science and technology. This, in addition to the achievements cited earlier that have been accomplished or are on the threshold of coming to fruition, should convince us that the continuation of flight beyond Earth orbit is imminent.

If so, it makes sense to think about where we will go first in opening this new era of *beyond*.

It will almost certainly be to return to the Moon. The equipment we used to do that before now only exists in museums. Returning to the Moon will be a good way to test the new equipment we will create for flight beyond Earth orbit and to train a new generation of spacefarers to use that equipment.

An analogy with the settlement of the United States is informative. The American West was opened first by daring frontierspeople. These were followed by hardy trappers and hunters who found ways to exploit the riches of the West. Next came the development of settlements, which ultimately turned into permanent communities providing security, comfort and economic development. So too it will take many flights to the Moon to develop, by successive approximations, a permanent human presence there. The experiences involved in colonizing the Moon will be good preparation for taking the much bigger step of sending an exploratory team to Mars. In fact, the kind of research described earlier could be undertaken during the period of Moon colonization.

As an aside, the term colonialism has fallen into disrepute in modern times. There are serious international legal and political problems involved in such undertakings, and already we see concrete and sustained deliberations about these issues, as the following two examples illustrate. First, in Antarctica the international scientific community has an effective working model for permanent research communities sponsored by different countries. As such, Antarctica serves as an analogue for space. The agreements worked out there greatly inform us about issues that we may need to address on moons, planets and asteroids. Second, organizations such as the Denver-based United Societies in Space have been addressing such issues for nearly two decades. United Societies in Space was founded in 1994 by Oleg Alifanov of the Moscow Aviation Institute and Declan O'Donnell, an American attorney. The organization convenes conferences, develops white papers, and publishes a journal, *Space Governance*. It has recently incorporated a subsidiary, International Space Development Authority Corporation, to broaden the scope of its deliberations. The point is that legal, political and economic issues regarding space exploration and exploitation are not issues to be first faced in the future. A foundation for serious international discussions has been developing for some time.

We may find that building ever larger and more powerful chemical rockets to lift heavy loads out of Earth's deep gravity well is less effective than carrying smaller pieces up to the ISS and assembling them there. Such an approach will be made economically feasible by the development of safe, totally reusable space launch systems with low operational costs. This approach was pioneered in the early 1990s with the McDonnell Douglas, *Delta Clipper*, single-stage-to-orbit program. Such research is currently the focus of several entrepreneurial groups. Bigelow Aerospace in Las Vegas is currently flying two inflatable habitats (*Genesis I* and *II*) on Earth orbit. Larger, more sophisticated units are forthcoming. The relatively cramped habitats on the ISS had to be sized to fit in the cargo bay of the Shuttles. Bigelow Aerospace's work with inflatables gives hope of solving the spacecraft crowding problem. Again, this is not futuristic dreaming but extant research. These tightly packaged flexible structures can be carried to space on current chemical rockets and then inflated in space to greatly increase their volume.

There are other places of interest to visit in the solar system besides planets and moons. One example is the five Lagrange points in the Moon-Earth gravitational system. These points in space, also called libration points, were discovered by the French mathematician, Louis Lagrange. They are locations at which all the gravitational forces balance out to zero. Theoretically an object of insignificant gravity itself, compared to the Earth and Moon, would remain "parked in space" if placed in such a location. This could be very useful as a storage depot in space.

This chapter will end by considering one other possible destination beyond Earth orbit for humans to visit: an asteroid. In January 2009 the International Academy of Astronautics published a volume titled *Dealing with the Threat to Earth from Asteroids and Comets* (Bekey 2009). The document reflects the work of experts from many disciplines who came together during three international conferences and two planetary defense workshops between 2001 and 2007, where they discussed the cutting edge of issues relating to comets and asteroids. There are many more reasons to be interested in comets and asteroids other than their threat of crashing into the Earth, and of necessity the volume deals with many of them. However their threat and what we might do about it is the focus of the book.

Near Earth Objects, or NEOs, have always been a threat, but it is only now that we really might be able to detect them early enough and then be able to do something to mitigate the threat. While this may sound more like science fiction than something we can realistically begin planning for in the near future, consider the following.

Fig. 14.4 Near Earth Asteroid Rendezvous (NEAR) flew by asteroid 433 Eros in December 1998 and began orbiting the asteroid in February 2000. The following year, the mission concluded with a safe landing of the orbiter on 433 Eros' surface (Illustration credit: NASA)

In 1996 the United States sent a spacecraft named NEAR Shoemaker 200 million miles to intercept a large asteroid named 433 Eros (Fig. 14.4). NEAR is an acronym for Near Earth Asteroid Rendezvous, and the craft was also named to honor planetary scientist Eugene M. Shoemaker. Eros is approximately 8 × 8 × 20 miles in size. It rotates as it whizzes through space. The task of the flight managers was to catch Eros, and then, because it is massive enough to have significant gravity, go into a slow orbit around the asteroid. Because of the long time required for radio signals to make a round trip between Earth and the asteroid, the spacecraft had to be pre-programmed to operate quite autonomously.

The project was successful, and much was learned about this asteroid. The plan did not call for the spacecraft to attempt to land on Eros, and thus the craft had no landing gear. But when the project was completed, rather than simply shutting down the spacecraft and letting it drift, the flight managers decided to attempt a landing on Eros. They had the spacecraft approach the asteroid at about 5 miles per hour—bumper-bumping speed—and a successful landing was accomplished. This event took place over a decade ago: in February 2001.

Recently a Japanese spacecraft *Hayabusa* (peregrine falcon) advanced the science of asteroid exploration even further. It landed on an asteroid named 45143 Itokawa (to honor the Japanese rocket pioneer, Hideo Itokawa) and sampled the dust on the asteroid's surface (Fig. 14.5). After landing and taking off from Itokawa five times, it returned to Earth orbit and discharged its reentry package of space particles, which safely entered our atmosphere and parachuted to a landing in the Australian outback. As this chapter was being written, the returned material was being studied in Japanese laboratories.

Thus we see that the groundwork has already been laid for intercepting and landing on asteroids. We would always expect to precede piloted missions with robotic missions that pave the way without risk to humans. There are thousands of asteroids circling the Sun, many in the outer reaches of the solar system in an area called the Kuiper belt. Occasionally these objects bump together, knocking one another out of their orbits, which may put them at risk for collision with the Earth. It is rare that the Earth is struck by an asteroid large enough to produce a worldwide catastrophe that could destroy civilization. But it has happened in the past, and it will probably happen again. If it lies within our power to intervene and prevent such a calamity, it makes

Fig. 14.5 Hayabusa deploying a marker used to guide the spacecraft during its descent to the surface of the asteroid 45143 Itokawa (Illustration credit: MEF/JAXA · ISAS)

sense to prepare to do so. Along the way we would greatly advance our spaceflight technology and learn much about our solar system and its origin.

Grandiose science fiction stories have been written and movies made about how one might protect the Earth from collision with a comet or asteroid with such things as atomic bombs. But consider this. Everything in the solar system travels in a curved path around the Sun. An asteroid on a collision course with Earth means that its curved path crosses that of the Earth. All that is required to keep the Earth and the asteroid from reaching the same point in space at the same time is to speed up or slow down the asteroid, even just a little bit, and the two objects will not collide. A wide variety of ways to do this have been proposed and several are discussed in the International Academy of Astronautics book. But think of the rapid development that is occurring in electric propulsion discussed earlier in this chapter. (The Japanese Hayabusa spacecraft used xenon ion engines for propulsion).

Here is a proposal for an experiment that might call for sending a human team to an asteroid. A series of international missions, so that costs and technologies are shared, are planned to a selected asteroid. First, robotic probes are sent to test all the navigational procedures, management programs, hardware and redundancies required for safe human flight. Then, all of the necessary equipment is flown to the asteroid and stored there. This would include guidance electronics to enable the spacecraft with a human crew to land in close proximity to their supplies. Finally, the human crew would be sent to attach the electric generator and electric rocket motor to the asteroid. From that point on, many experiments could be conducted after the crew left, to teach us how to maneuver large space objects. These activities could lead to mining asteroids for valuable minerals and retrieving water ice from comets and transferring it to colonies on the Moon or at Lagrange points.

14.11 Conclusion

The orbital flight space frontier has been opened. Very soon now, space tourists will be taking suborbital rides to the edge of space. It is not a long trip, only about 100 km or 60 miles. But it is straight up, and that makes all the difference. It will not be long before companies will have employees at work in orbiting laboratories and production facilities. Orbiting space hotels will not be far behind. The new space frontier is human flight beyond Earth orbit. This chapter has elaborated on a variety of the issues, related to opening this new frontier to human experience. Humankind has been dreaming of such things for a long time. But the dream remained just that: a dream. The material presented in this chapter suggests that the dream is over. The dawn of the new era of human exploration of the solar system is upon us. What we are experiencing now, historians one day will call the beginning of human space flight beyond Earth orbit. What an adventure that is going to be. It is time to get on with it.

Acknowledgments The author thanks Dr. Franklin Chang Díaz, Dr. Robert Vondra and Ad Astra Rocket Company for calculating trip times with the VASIMR rocket motor. The author also thanks Dr. William Gaubatz for his invaluable review of this manuscript.

References

Bekey, I. (ed.): Dealing with the Threat to Earth from Asteroids and Comets. Indian Space Research Organization [ISRO], Bangalore (2009)

Bering, III E.A., Longmier, B.W., Ballenger, M., Olsen, C.S., Squire, J.P., Chang Díaz, F.R.: Performance studies of the VASIMR® VX-200. Paper presented at the 49th AIAA Aerospace Sciences Meeting and Exhibit, Orlando (2011)

Bruce Hammer.: NIH selects another investigator for the International Space Station, 19 Oct 2011. Available at www.nasa.gov/mission_pages/station/research/news/NIH_Hammer.html. Accessed 8 Jan 2012

Chang Díaz, F.R.: The VASIMR Rocket. Sci. Am. **283**(5), 90–97 (2000)

Ilin, A.V., Cassidy, L.D., Glover, T.W., Chang Díaz, F.R.: VASIMR® human mission to Mars. Paper presented at the Space, Propulsion & Energy Sciences International Forum, College Park (2011)

Longmier, B.W., Cassidy, L.D., Ballenger, M.G., Carter, M.D., Chang Díaz, F.R., Glover, T.W., Ilin, A.V., McCaskill, G.E., Olsen, C.S., Squire, J.P., Bering III, E.A.: VX-200 magnetoplasma thruster performance results exceeding fifty-percent thruster efficiency. J. Propul. Power **27**(4), 915–920 (2011)

Squire, J.P., Olsen, C.S., Chang Díaz, F.R., Cassidy, L.D., Longmier, B.W., Ballenger, M.G., Carter, M.D., Glover, T.W., McCaskill, G.E.: VASIMR® VX-200 operation at 200 kW and plume measurements: future plans and an ISS EP test platform. Paper presented at the 32nd International Electric Propulsion Conference, Weisbaden (2011)

Yerkes, R.M., Dodson, J.D.: The relation of strength of stimulus to rapidity of habit-formation. J. Comp. Neurol. Psychol. **18**, 459–482 (1908)

Zubrin, R.: How to live on Mars: a trusty guidebook to surviving and thriving on the red planet. Three Rivers Press, New York (2008)

Zubrin, R. with Wagner, R.: The case for Mars: the plan to settle the red planet and why we must, rev. and updated. Free Press, New York (2011)

Chapter 15
From Earth's Orbit to the Outer Planets and Beyond: Psychological Issues in Space

Nick Kanas

Abstract Current planning for the first interplanetary expedition to Mars envisions a crew of six or seven people and mission duration of around 2.5 years. However, this time frame is much less than that expected on expeditions to the outer solar system, where total mission durations of 10 years or more are likely. Although future technological breakthroughs in propulsion systems and space vehicle construction may speed up transit times, for now we must realistically consider the psychological impact of missions lasting for one or more decades.

Available information largely deals with on-orbit missions. In research that involved Mir and ISS missions lasting up to 7 months, our group and others have studied the effects of psychological and interpersonal issues on crewmembers and on the crew-ground relationship. We also studied the positive effects of being in space. However, human expeditions to the outer planets and beyond will introduce a number of new psychological and interpersonal stressors that have not been experienced before. There will be unprecedented levels of isolation and monotony, real-time communication with the Earth will not be possible, the crew will have to work autonomously, there will be great dependence on computers and other technical resources located on board, and the Earth will become an insignificant dot in space or will even disappear from view entirely.

Adaptation of Kanas (2011), reprinted with permission from Elsevier. Parts of this chapter were presented at the 6th IAA Symposium on Realistic Near-term Advanced Scientific Space Missions, Aosta, Italy, July 6–9, 2009.

N. Kanas (✉)
University of California, San Francisco and Veterans Affairs Medical Center, San Francisco, CA, USA
e-mail: nick.kanas@ucsf.edu

D.A. Vakoch (ed.), *On Orbit and Beyond*, Space Technology Library 29,
DOI 10.1007/978-3-642-30583-2_15, © Springer-Verlag Berlin Heidelberg 2013

Strategies for dealing with psychological issues involving missions to the outer solar system and beyond will be considered and discussed, including those related to new technologies being considered for interstellar missions, such as traveling at a significant fraction of the speed of light, putting crewmembers in suspended animation, or creating giant self-contained generation ships of colonists who will not return to Earth.

15.1 Introduction

The first expedition to Mars will likely involve a crew of six or seven people and a mission duration of around 2.5 years (Kanas and Manzey 2008). Being confined with the same individuals for so long at great distances from home might create psychological and interpersonal stress for the crew and impact on their ability to carry out mission goals. However, this time frame is less than what might be expected on expeditions to the outer solar system. Here mission durations of 10 or more years are likely. Technological breakthroughs in propulsion systems and space vehicle construction may speed up transit times in the future, but for now we must realistically consider the psychological impact of missions lasting over 10 years.

People have been on-orbit for as long as 14 months with no apparent negative sequelae. Although emotional adjustments and psychosomatic reactions have occurred in space, severe psychotic episodes have not been reported, probably due to the careful psychiatric selection and support of crewmembers (Kanas and Manzey 2008). However, the amount of psychosocial research information from space is small and has been limited to the near-Earth environment.

15.2 Psychological Issues During Long-Duration Near-Earth Space Missions: What Do We Know?

Kanas et al. (2000, 2001, 2007; Boyd et al. 2009) conducted two major NASA-funded international on-orbit studies of psychological and interpersonal issues affecting crewmembers and mission control personnel during missions to the Mir and the International Space Stations. Four areas were studied: possible changes in mood and group social climate over time, displacement of group tension to outside monitoring personnel, cultural differences, and leadership roles. The Mir study sample involved 5 American astronauts, 8 Russian cosmonauts, and 42 American and 16 Russian mission control personnel. The ISS study sample involved 8 American astronauts, 9 Russian cosmonauts, and 108 American and 20 Russian mission control personnel. Subjects completed a weekly study questionnaire that included items from the Profile of Mood States, the Group Environment Scale, the Work Environment Scale, and a Critical Incident Log.

The findings from the two studies were similar, and there was evidence for national (American vs. Russian), occupational (crewmember vs. mission control), and organizational (NASA vs. Russian Space Agency) cultural differences. Overall, there was little support for the presence of changes in levels of mood and group interpersonal climate over time, and no evidence for the so-called "3rd quarter phenomenon", where mood and cohesion levels are thought to worsen after the halfway point of the mission (Kanas and Manzey 2008; Kanas et al. 2007). However, both studies provided evidence for the displacement of negative emotions in the group to outside individuals, with the effect being stronger for the isolated crewmembers than the mission control subjects. Crewmembers scored lower (i.e., were happier) on a number of dysphoric mood subscales than ground personnel. However, both groups scored lower on the dysphoric subscales compared to normative samples of people who worked in non-space related jobs on Earth. In both studies, the support role of the leader was related to group cohesion among crewmembers, and both the task and support roles of the leader were related to cohesion among mission control personnel. Crewmembers scored higher on a measure of cultural sophistication than mission control personnel, and Russians reported greater language flexibility than Americans (Kanas and Manzey 2008; Boyd et al. 2009). Americans scored higher on a measure of work pressure than Russians. For the crewmembers, the astronauts scored one standard deviation above published norms on this measure, whereas the cosmonauts scored below the norm. Both national and organizational factors were implicated in accounting for this difference.

Other investigators have studied similar phenomena. For example, Gushin et al. (1997; Gushin 2003) found that space crews and other isolated groups showed decreases in the scope and content of their communications and a tendency to filter what they said to outside personnel, a phenomenon Gushin called psychological closing. Crewmembers also showed preferences to interact with some individuals in mission control and not others. In part, this was felt to be due to a process of autonomization, where isolated crews become more egocentric and perceive outsiders as opponents, which could lead to mutual tension and misunderstanding. How these behaviors relate to the displacement of negative emotions found by the Kanas team remains to be evaluated in future studies.

Surveys of personnel involved with space missions also have shed light on important psychosocial issues in space. Tomi et al. (2007) conducted a survey of 75 astronauts and cosmonauts and 106 mission control personnel aimed as assessing intercultural issues that that could lead to misunderstandings and conflict during space missions. Both crew and ground subjects rated coordination problems between member organizations that were involved with the missions as the biggest issue, followed by communication difficulties due to simple misunderstandings. Other mentioned difficulties related to differences in language and work management styles and communication problems between mission control personnel and their support teams. The most important countermeasure for dealing with these problems involved cross-cultural training of astronauts and mission control personnel, with over 83% of the survey respondents stating that this training should involve key members of both groups training together in order to encourage team-building. Nechaev et al. (2007) conducted a survey of 11 cosmonauts that was aimed at gaining information regarding human factors issues concerning an expedition to Mars. Factors seen as potentially causing psychological problems included isolation and monotony, communication delays with the Earth due to the long distances involved, leadership issues, differences in management style among the involved space agencies, and cultural problems related to the international makeup of the crew. Careful crewmember selection and in-flight psychological support were seen as important psychological counter-measures. Finally, Sandal and Manzey (2009) conducted a survey of 576 employees of the European Space Agency aimed at exploring important cultural issues impacting on performance, both within ESA and between ESA and other space agencies. They found a link between cultural diversity and challenges related to human interactions that could interfere with efficiency. Especially important were factors related to leadership issues and decision-making processes. The authors concluded that cross-cultural training was an important countermeasure to deal with such diversity, both within ESA and between ESA and other space program teams.

Isolated and confined environments can be growth-enhancing and positive for participants as well. In an attempt to characterize the positive aspects of space flight, Ihle et al. conducted a study involving 39 astronauts and cosmonauts who had participated in at least one space mission and completed their Positive Effects of Being in Space questionnaire (Ihle et al. 2006). Every respondent reported at least some positive change as a result of flying in space. The items clustered into eight subscale categories: Perceptions of Earth, Perceptions of space, New possibilities, Appreciation of life, Personal strength, Changes in daily life, Relating to others, and Spiritual change. Of these, only one was significantly different from the others: Perceptions of Earth. One of the items in this subscale, "I gained a stronger appreciation of the Earth's beauty," had the highest mean score, with the average rating indicating a "great degree" of positive change. The results of this study suggest that space travel is a meaningful experience for the participants and that they especially value observing the Earth.

15.3 Missions to Mars

Although these findings have relevance for future on-orbit missions and even expeditions to the Moon, caution must be used in extrapolating them to missions beyond the Earth's neighborhood, such as an expedition to Mars. There are several reasons for this. A Mars crew will be millions of miles away from Earth on a mission lasting some 2.5 years in duration. This will create a situation where crewmembers will experience a severe sense of isolation and separation from the Earth, with maximum two-way communication delays of up to 44 min (average = 25 min), infrequent resupply, and impossibility of evacuating home during a medical or psychiatric emergency. Consequently, the crewmembers will be much more autonomous from mission control than during near-Earth missions, and they will need to plan their own schedule and deal with their own problems. Support based on real-time communication with family, friends, and counselors on Earth will be much less possible than during near-Earth missions. The crew will be more dependent on on-board and planetary resources for food, water, and fuel. The increased isolation and mission duration may impact on their psyches and produce feelings of monotony and homesickness. Finally, no human being has ever experienced the Earth as an insignificant dot in the heavens, the so-called "Earth-out-of-view phenomenon" (Kanas and Manzey 2008). As mentioned above, gazing at the Earth has been rated as the major positive factor of being in space, and the absence of this factor during an expeditionary mission on crewmember psychology may result in increased isolation, homesickness, dysphoria, or possibly suicidal or psychotic thinking.

Countermeasures for dealing with these stressors can be categorized in terms of crew selection, pre-launch training, mission monitoring and support, and post-mission readaptation to Earth. Crews should be selected to include people who are mindful of both psychological and interpersonal issues and who can problem-solve around psychosocial difficulties. They must feel comfortable working alone on a project when necessary as well as interacting with their teammates and valuing teamwork in general. Commanders should be selected who have shown skill in using both task and support leadership characteristics in accordance with the needs of their team. They also should be sensitive to the impact of psychological and cultural factors (both national and organizational) on individual and crew behavior.

Crewmembers and mission control personnel should receive pre-launch psychosocial education training aimed at recognizing and dealing with important psychological and interpersonal issues. Specific topic areas for this training include: psychological and interpersonal issues involving isolated and confined groups, the relationship between crewmembers and mission control personnel (including displacement), the impact of national and organizational cultural differences on the mission, and the appropriate use of leadership. For expeditionary missions, ways to deal with autonomy and time delays in communication should be discussed and incorporated into training, using both didactic and experiential techniques. Some of this training needs to involve both crew and ground personnel working together.

During the mission itself, crewmembers and mission control personnel should receive computer-based psychosocial education training refresher courses to remind them of key issues discussed prior to launch. For crewmembers, such in-flight training may be done after entering space and mid-way through the mission, and perhaps even more frequently during multi-year expeditions. Crew and ground personnel should be sensitive to the stressors involved in each other's jobs and periodically take time to discuss the mission and their mutual ability to communicate with each other clearly and openly. Such "bull sessions" may especially be useful for crews on expeditionary missions to help them deal with autonomy and festering interpersonal problems. Crews also may value an on-board

Table 15.1 Characteristics of planetary missions involving human crews

Planet	Average distance from Earth (10^6 km)	Average 2-way communication time (min)	Round-trip mission duration (years)
Mars	229	25	1.3
Jupiter	778	87	4.5
Saturn	1,427	159	8.3
Uranus	2,870	319	16.8
Neptune	4,497	500	26.3
Pluto	5,906	657	34.5

telescope with which to see the Earth in real time. Strategies need to be developed to allow crewmembers to communicate efficiently with people on Earth during time-delayed conditions (e.g., e-mail messages that include suggested responses at the end to which the recipient may reply). Families at home need to be supported during the mission, both informally (e.g., peer-led groups) and formally (e.g., counseling).

Finally, post-return readaptation debriefings and supportive activities (e.g., individual counseling, time alone with families) need to be employed to assist the crewmembers in readjusting to life on Earth. Especially for high-profile on-orbit missions or for early missions to the Moon and Mars, ways to cope with issues of fame and glory should be discussed, and the privacy needs of both the astronauts and their family members must be respected.

15.4 Missions to the Outer Planets

In addition to the above, new psychosocial factors and countermeasures should be considered when one begins to talk about missions to the outer planets. Table 15.1 lists some important characteristics of such missions.

The average distance of each planet from the Earth is given in the second column. Based on this distance the approximate time it would take for a two-way audio communication from the Earth is given in the third column (calculated by dividing the average distance by a transmission time equal to the speed of light, then doubling the final figure). In the last column, the approximate travel time for a round-trip mission (excluding time spent exploring the planet and its vicinity) is shown, calculated by extrapolating from the flight time to and from Mars to account for the increased distance to each planet. These figures assume that the method of travel to the outer planets will be similar to that currently being planned for a Mars mission, without any special technological innovations or breakthroughs.

It is notable that the figures in the last column of the table are of the same order of magnitude as the actual travel times observed during the Voyager 2 "Grand Tour" mission through the solar system. Launched on August 20, 1977, and taking advantage of a favorable planetary alignment that allowed for gravity assist, it took

Voyager 2 just under 2 years to make a one-way trip to Jupiter, 4 years to reach Saturn, nearly 8.5 years to reach Uranus, and 12 years to reach Neptune. Doubling these figures closely approximates the round-trip durations listed in column four.

Examining this table, it is apparent that there are dramatic increases in distance from the Earth as one proceeds to the outer planets using conventional propulsion systems. In a similar manner, the average two-way communication times and round-trip mission durations increase almost geometrically. Certainly, many of the same psychological and interpersonal issues discussed above that are relevant to a Mars mission (e.g., isolation from Earth, periods of monotony, increased crewmember autonomy, reliance on on-board resources, "Earth-out-of-view phenomenon") would apply here as well, but in addition there are some new issues.

Consider a mission to Saturn. The two-way communication time of over 2.5 h would make real-time discussions with mission control or family and friends realistically impossible, thus accentuating the isolation and loneliness felt by the crewmembers. The Earth would appear as an even more insignificant dot in space, adding to the sense of isolation. In addition, what would one do to occupy time during the 4 years it would take to get to Saturn (and the 4 years it would take to get back), and how would the crewmembers get along with each other during this time? A 40-year old astronaut undertaking this mission would be nearly 50 when he or she returned, with children having grown up and family and friends aging and perhaps dying in the interval. These factors would create a selection problem for the crew—what kind of people would want to go on such a mission? Time and aging effects would also increase the chances of crewmembers developing a medical or psychiatric illness.

It is hard to imagine that such missions to the outer planets would occur under current methods of travel, and new concepts would likely be employed, such as velocities representing a significant fraction of the speed of light and suspended animation. Both of these issues will now be discussed.

15.5 Interstellar Missions

In this discussion, interstellar missions will include travel to the Oort Cloud as well as to the nearest neighboring stars in our Galaxy. The Oort cloud is a broad spherical shell of comet nuclei that is located 0.63–0.94 light years away from our Sun. The reason to go there includes gaining knowledge of comets and their composition, mining the nuclei for valuable minerals, and looking for extra-terrestrial life, as has been suggested by Freeman Dyson (2003). The nearest star to our Sun is Proxima Centauri, which is 4.2 light years away, followed by Alpha Centauri A and B, located 4.4 light years away. A light year means that if one could travel at the speed of light, it would take this amount of time to reach these stars. (Faster than light speeds or travel through worm-holes are ideas that belong more to science fiction than to science fact, and they will not be considered here.)

Using current technology, interstellar travel is highly unlikely. For example, Sheffield (2003) estimates that it would take Voyager 2 some 497,000 years to reach the star Sirius, which at 8.7 light years away is just over twice as far away as Proxima Centauri. In order to travel interstellar distances, new propulsion systems will be necessary, such as nuclear pulse rockets, ramjets, and beamed power starships (Forward 2003). These systems are capable of achieving very high speeds by present standards, but how much would this help? For example, at 5% the speed of light, it still would take 88 years to reach Alpha Centauri (Sheffield 2003), which would be longer than the expected lifetime of the crewmembers. Bracewell (1976) has devised a scenario where a round trip to Proxima Centauri could be made in 11 years, assuming a 1-year acceleration to near-light speed, followed by a 3.5-year coast in deep space, and then a 1-year deceleration to the star (with a similar scenario for the return). However, traveling at near-light speed presents difficult technological problems, and the effects of the resultant space-time changes that occur at relativistic speeds on the human body are unclear. For example, Semyonov (2009) has pointed out that the oncoming flow of interstellar gas and cosmic rays impacting on a space ship traveling at relativistic speeds could present a unique radiation hazard that would require special windward shielding in order to protect the crew.

In addition to increasing the speed of travel, two other options have been suggested for interstellar travel. The first, which is also feasible for travel in the outer solar system, is to put the crew in suspended animation during much of the mission. In this scenario, after the critical activities involving the launch and the setting of the course have been accomplished, the crew would be put to sleep, and their physiological functions would be slowed down until such time when they would be near their destination and would be awakened to perform landing and exploration activities. The space vehicle would be on autopilot during the bulk of the mission, and computers would handle life support and navigation. The technology to do this safely has yet to be developed. But in addition, it would raise a number of psychosocial questions. How would the crewmembers feel about their lives being dependent upon computer-controlled machines for such a long period of time? What would it be like to suddenly awaken after several years so far away from home, knowing that your family and friends have aged or even died, and everything that you have ever loved is impossibly far away? What would the return home be like trying to reintegrate with people after such a long time. Finally, how much interpersonal stimulation and support could be achieved during such a long mission by a crew that would likely consist of less than a dozen people?

A second scenario that has been advocated for interstellar expeditions has been the use of giant self-contained generation ships that would be full of people who would plan to colonize space rather than return to Earth. For example, Terra (1996) has discussed the use of moving habitats containing some 25 individuals each that would cluster together in communities living in the Oort Cloud. Strong (1965) has envisioned giant 100-megaton star ships containing 100–150 people that would be equipped for a century-long journey to the stars. Woodcock (1996) imagines even larger one million metric ton star ships the length of 11 football fields that would carry 10,000 people to the stars. Accelerating to a maximum velocity of 0.15 the speed of light, then decelerating to reach a star some 10 light years away, such a behemoth would complete its journey in about 130 years.

The psychosocial impact of a permanent divorce from the home planet is unclear. Certainly, the effects on the first generation of colonists would be profound, since they would vividly remember the Earth and would have left friends and associates behind. Subsequent generations would likely be less affected, since their total existence and reference point would be the generation ship itself. Nevertheless, images and stories of the Earth would no doubt be preserved in such star ships and would likely be the subject of future lore and longings as the colony evolved over time. Colonist selection also would be an issue. Who would be selected? How many extended family members would be allowed? How much cultural and religious diversity could be tolerated? Would there be one common language, and if so, which one? How would a diverse enough gene pool be assured to minimize the appearance of dangerous recessives and maximize the stability of future generations? What kind of social structure would be optimal?

Moore (2003) has discussed some of the ethical and demographic issues involving a multi-generational crew of 150–180 people on a 200-year mission to Alpha Centauri. In order to establish a stable population and maximize the productivity of the group, and based on his computer modeling techniques, he has suggested several social engineering principles, such as starting with a crew of young, childless married couples and asking them to postpone parenthood until late in the women's reproductive life. This would maintain genetic variation by lengthening the generations, and it would result in smaller sibships, with more time available for mission tasks and less needed for childrearing. The result would be the production of well-defined demographic echelons over time, with people clustering into discrete age groups. In this way, the number of non-productive young and old people would not increase, and the social network would remain stable. But, how will the participants in such a mission accept such social engineering? Will later generations simply see these echelons as a normal part of life, or will they rebel? Will there be unforeseen consequences of such social engineering that would threaten the life of the colony? No one knows for sure.

15.6 Conclusions

A summary of important psychological issues related to travel to the outer planets and beyond is given in Table 15.2. Many of these issues can be ameliorated by prelaunch psychological, interpersonal, multicultural, and team-building training and preparation (Kanas and Manzey 2008). Other issues need to be dealt with during the mission itself, and all crewmembers should be sensitized to recognize and deal with these issues as they come up, since the link to support staff on Earth will be tenuous or non-existent, and the crew will experience unprecedented levels of autonomy and need for self-regulation. It is not too early to study the effects of increased autonomy on space crews, perhaps using the International Space Station as a research platform. In this way, countermeasures can be developed and tested that will allow future crews to function more independently from the Earth.

Table 15.2 Psychological issues in deep space

1.	Selection: who would want to be away from family and friends for years to decades?
2.	Unknown physical and psychological effects of traveling at a significant fraction of the speed of light
3.	Extreme feelings of isolation and loneliness
4.	Lack of support from Earth due to extreme distance and communication delays
5.	Earth-out-of-view phenomenon
6.	Filling leisure time over the years
7.	Increased risk for medical and psychiatric illness due to time away from home
8.	Lack of social contacts and novelty
9.	Returning home to a new world and aging or dead family and friends
	Additional issues related to suspended animation
10.	Unknown physical and psychological effects due to suspended animation technology
11.	Fear of trusting machines to maintain your life and wake you up
	Additional issues related to colonizing space
12.	Selection: who and how many could go given limited resources
13.	How much diversity (e.g., language, cultural factors, religion) would be tolerated or be seen as beneficial by the group
14.	Feelings of homesickness, especially for the first generation and their children who remember the Earth
15.	Psychological/ethical effects of social engineering (e.g., regulating coupling and birth rate)
16.	Keeping the social contract—rebellion by later generations who do not agree with colonizing space

Given the psychological impact of interstellar travel, one might ask if traveling beyond the Earth's neighborhood is worth the effort and risk. Kondo (2003) has suggested two reasons why we may want to travel to distant planetary systems: (1) to insure the long-term survival of the human race, should the Earth become inhospitable or should something happen to the Sun; and (2) to keep ourselves vibrant by exploring the unknown, which he feels is essential for any civilization to remain alive. Two additional reasons could be mentioned: the need to search for extra-terrestrial life, and the economic advantages of trade and mining activities involving distant mineral-rich planetary bodies. Kondo further speculates that over time numerous Earth-like planets will be discovered for us to go to, and he reasons that given the technological hurdles involved with interstellar travel, we should begin to address them now. We have discussed some psychological hurdles as well, and we have only begun to fully understand ways to cope with them. But as humans reach farther out in our own solar system, people will begin to experience the effects of living and working together in space far from the home planet, and this knowledge will help us to develop strategies of dealing with missions to the stars close to home and beyond.

Acknowledgments This chapter was supported by National Aeronautics and Space Administration Contract nos. NAS9-98093 and NCC-0161. The author would like to thank Ms. Stephanie Saylor; Astronaut Charles Camarda; and Professors Jennifer Boyd, Thomas Neylan, and Richard Wener for their helpful comments on this chapter.

References

Boyd, J.E., Kanas, N.A., Salnitskiy, V.P., Gushin, V.I., Saylor, S.A., Weiss, D.S., Marmar, C.R.: Cultural differences in crewmembers and mission control personnel during two space station programs. Aviat. Space Environ. Med. **80**, 1–9 (2009)

Bracewell, R.N.: The Galactic Club: Intelligent Life in Outer Space. San Francisco Book, San Francisco (1976)

Dyson, F.J.: Looking for life in unlikely places: reasons why planets may not be the best places to look for life. In: Kondo, Y., Bruhweiler, F.C., Moore, J., Sheffield, C. (eds.) Interstellar Travel and Multi-Generation Space Ships, pp. 105–120. Apogee Books, Burlington (2003)

Forward, R.L.: Ad Astra! In: Kondo, Y., Bruhweiler, F.C., Moore, J., Sheffield, C. (eds.) Interstellar Travel and Multi-Generation Space Ships, pp. 29–51. Apogee Books, Burlington (2003)

Gushin, V.I.: Problems of distant communication of isolated small groups. Human Physiol. **29**, 548–555 (2003)

Gushin, V.I., Zaprisa, N.S., Kolinitchenko, T.B., Efimov, V.A., Smirnova, T.M., Vinokhodova, A.G., Kanas, N.: Content analysis of the crew communication with external communicants under prolonged isolation. Aviat. Space Environ. Med. **68**, 1093–1098 (1997)

Ihle, E.C., Ritsher, J.B., Kanas, N.: Positive psychological outcomes of space flight: an empirical study. Aviat. Space Environ. Med. **77**, 93–101 (2006)

Kanas, N: From Earth's orbit to the outer planets and beyond: psychological issues in space. Acta. Astronautica. **68**, 576–581 (2011)

Kanas, N., Manzey, D.: Space Psychology and Psychiatry, 2nd edn. Microcosm Press/Springer, El Segundo/Dordrecht (2008)

Kanas, N., Salnitskiy, V., Grund, E.M., Gushin, V., Weiss, D.S., Kozerenko, O., Sled, A., Marmar, C.R.: Interpersonal and cultural issues involving crews and ground personnel during shuttle/Mir space missions. Aviat. Space Environ. Med. **71**(9 Suppl.), A11–A16 (2000)

Kanas, N., Salnitskiy, V., Weiss, D.S., Grund, E.M., Gushin, V., Kozerenko, O., Sled, A., Bostrom, A., Marmar, C.R.: Crewmember and ground personnel interactions over time during Shuttle/Mir space missions. Aviat. Space Environ. Med. **72**, 453–461 (2001)

Kanas, N.A., Salnitskiy, V.P., Boyd, J.E., Gushin, V.I., Weiss, D.S., Saylor, S.A., Kozerenko, O.P., Marmar, C.R.: Crewmember and mission control personnel interactions during International Space Station missions. Aviat. Space Environ. Med. **78**, 601–607 (2007)

Kondo, Y.: Interstellar travel and multi-generation space ships: an overview. In: Kondo, Y., Bruhweiler, F.C., Moore, J., Sheffield, C. (eds.) Interstellar Travel and Multi-Generation Space Ships, pp. 7–18. Apogee Books, Burlington (2003)

Moore, J.H.: Kin-based crews for interstellar multi-generation space travel. In: Kondo, Y., Bruhweiler, F.C., Moore, J., Sheffield, C. (eds.) Interstellar Travel and Multi-Generation Space Ships, pp. 80–88. Apogee Books, Burlington (2003)

Nechaev, A.P., Polyakov, V.V., Morukov, B.V.: Martian manned mission: what cosmonauts think about this. Acta Astronaut. **60**, 351–353 (2007)

Sandal, G.M., Manzey, D.: Cross-cultural issues in space operations: a survey study among ground personnel of the European Space Agency. Acta Astronaut. **65**, 1520–1529 (2009)

Semyonov, O.G.: Radiation hazard of relativistic interstellar flight. Acta Astronaut. **64**, 644–653 (2009)

Sheffield, C.: Fly me to the stars: interstellar travel in fact and fiction. In: Kondo, Y., Bruhweiler, F.C., Moore, J., Sheffield, C. (eds.) Interstellar Travel and Multi-Generation Space Ships, pp. 20–28. Apogee Books, Burlington (2003)

Strong, J.: Flight to the Stars. Hart, New York (1965)

Terra, R.P.: Islands in the sky: human exploration and settlement of the Oort Cloud. In: Schmidt, S., Zubrin, R. (eds.) Islands in the Sky: Bold New Ideas for Colonizing Space, pp. 95–116. Wiley, New York (1996)

Tomi, L., Kealey, D., Lange, M., Stefanowska, P., Doyle, V.: Cross-cultural training requirements for long-duration space missions: results of a survey of International Space Station astronauts and ground support personnel. Paper delivered at the Human Interactions in Space Symposium, Beijing, 21 May 2007
Woodcock, G.R.: To the stars! In: Schmidt, S., Zubrin, R. (eds.) Islands in the Sky: Bold New Ideas for Colonizing Space, pp. 183–197. Wiley, New York (1996)

Afterword

From the Past to the Future

Gro Mjeldheim Sandal
University of Bergen, Norway
Gloria R. Leon
University of Minnesota, USA

Abstract

Although NASA has been criticized for many years for neglecting psychological issues in research and operations, the past several years have witnessed an increased recognition of the importance of psychosocial and cultural factors in the success and safety of human space missions. The challenges associated with future long-duration missions involving extreme environments, isolation, and greater crew autonomy as the distance from Mission Control increases require effective countermeasures to mitigate the risk for behavioral health problems, clear psychiatric disorders, and impairments in effective crew interactions and task performance. International space missions also underscore the need to understand the potential safety implications of individual and cultural differences at a national, organizational, and professional level that involve both space crews and ground-based personnel. While the research literature on space psychology has increased over the last few years, many unanswered questions remain that require additional investigation.

Since the first solo flight of a human being into Earth orbit, human spaceflight has undergone significant changes in terms of crew composition, mission duration and complexity. Even the major achievement of the establishment of the International Space Station (ISS) put into service in 2000 must be regarded as just one further step towards a much bigger endeavor: human exploratory missions into

D.A. Vakoch (ed.), *On Orbit and Beyond*, Space Technology Library 29,
DOI 10.1007/978-3-642-30583-2, © Springer-Verlag Berlin Heidelberg 2013

outer space and the establishment of human outposts on other celestial bodies of the Solar System. This might include a return to the Moon, as well as flights of humans to asteroids and to our neighbor planet Mars. Technology is just one important aspect of such long-duration space missions. Beyond that, there is no doubt that different biomedical and psychological factors might represent important limitations to the success of the missions.

The chapters presented in this volume and publications elsewhere demonstrate advances in our understanding of challenges related to human spaceflights. Yet, a number of issues remain that require additional investigation. This is particularly the case in regard to long-duration exploratory space missions to the Moon and Mars, which to some extent can be expected to involve the same range of psychological issues and risks that have been reported from long-duration orbital flights, simulation studies and expeditions into analogue environments. Nonetheless, missions to Mars will add a new dimension to the history of human expeditions into *terrae incognitae* with respect to the distance and duration of travel. Such missions might not be comparable to any other undertaking humans have ever attempted because of the long distance of travel, the duration of permanent living under dependence of automated life-support systems, the degree of isolation and confinement, and the lack of short-term rescue possibilities in case of emergencies. Current knowledge about human adaptation under these conditions is very limited, but predictions about the emergence of certain psychological issues can be extrapolated from Earth-based analogues and studies and previous spaceflights. For example, historical expeditions to unknown parts of the world parallel some of the human requirements associated with future interplanetary missions. In addition, studies of personnel wintering over on Antarctic research stations, such as the Concordia station in which the European Space Agency (ESA) is involved, may provide highly useful data in such fields as telemedicine as well as psychophysiological reactions and group dynamics associated with isolation and lack of evacuation possibilities. Recently, a successful 500 day space simulation study aimed to model the living conditions of crews on future Mars missions was completed. While these Earth-bound analogues provide interesting platforms for research, a much more detailed understanding of the concrete scientific and operational demands of planetary exploration mission space crews and the design of their habitats is needed before the psychological issues associated with these missions can finally be assessed.

It is beyond doubt that political and commercial interests to a large extent will dictate crew composition for future missions, and these forces might introduce new factors that must be addressed as part of mission planning. An era of space tourism seems to be at hand, as highlighted by the study presented by Wichman in Chap. 6. Now that the Russian, European and American space agencies are equal partners on joint projects such as the ISS, they are faced with challenges related to collaboration among people with different cultural backgrounds. New space faring Asian nations are in a phase where they are in the process of building the capacity for manned spaceflight. Most notably, the progress made by the Chinese space program suggests that the country will be a key member of the international space community in the near future. For a long time, the impact of cultural variability seemed to be a neglected area in aerospace research. As demonstrated by several chapters in this volume, the last few years have witnessed an increased recognition of the potential safety risks associated with cultural heterogeneity in terms of nationality, organizational background and profession. Accidents like Challenger and Columbia have underscored the need for future studies to focus not only on people operating in space, but also on ground-based personnel.

Management of the safety culture may become even more complex in international space programs in which mission preparation and control often demand the coordinated effort of different space agencies. An interesting study by Tomi, Stefanowska, and Doyle involved data collection among ISS personnel from different agencies. The results demonstrated the need for training and intervention beyond the space crews, and the need to address differences in organizational cultures, in addition to those between national cultures. A similar conclusion was reached from another study involving 600 employees working for ESA that addressed challenges associated with collaboration with representatives from other agencies. Preliminary results indicate that the most prominent difficulties differed according to agency, and seem to reflect value differences anchored in organizational and national culture.

In contrast to the interest traditionally shown by the Russians in space psychology, now shared by European as well as Japanese space agencies, NASA has been criticized for neglecting psychological issues in research and operations. Many researchers and others observed that the NASA culture discouraged questions about the behavioral health of astronauts since they were assumed to have "the Right Stuff." Recently, these issues have gained more attention. In Chap. 1, Harrison and Fiedler describe some recent positive developments within the NASA establishment in moving away from "the marginalization of psychology." A more realistic recognition of stress and its consequences has led to a concern with prevention and countermeasures as a result of experiences during the Russian/American Mir/Shuttle program and the ISS.

Greater attention to the psychological health of astronauts is reflected in a NASA-sponsored behavioral health workshop held several years ago, followed by some increase in more formal operational attention to post-mission functioning of the astronaut and his/her family. However, there is still much planning, program implementation and change in organizational culture that needs to be carried out for these efforts to be more than a perfunctory gesture. A major problem in terms of astronaut and perhaps to some extent family behavioral health is the organizational factors that work against disclosure of psychological problems. The concern by astronauts that disclosure will result in non-assignment or removal from a flight assignment likely has considerable justification in reality. The "Right Stuff" culture and concerns about confidentiality feed into a system where personal problems might be denied or dealt with in dysfunctional ways such as substance abuse, impulsive or high risk behaviors, and family and other interpersonal conflicts. The findings of the 2007 NASA Astronaut Health Care System Review Committee indicated that none of those interviewed knew of an astronaut who had sought behavioral health care.

A focus on behavioral assessment and health should extend from the process of selection to training, in-flight and post-flight periods. Passing a psychological/psychiatric screening upon entrance to the astronaut corps does not predict the absence of psychological/behavioral problems that might occur at a later time. In considering crew selection for future long-duration lunar and planetary missions, a particularly thorny issue is the prediction of the later development of psychopathology in initially healthy individuals. This situation would affect not only the disabled person, but would also have a detrimental effect on overall crew performance, safety, and potentially, jeopardize the mission. Some disorders, for example, certain types of depression and anxiety disorders, as well as liability for schizophrenia have a considerable genetic component that might be detected by inquiries about family background. As the field of genetic testing and other statistical predictors of dysfunction develop over time, this information could be used in screening, selection, and later evaluation. However, in fairness to individual astronauts, it is very important to remain cognizant of the fact that "liability" is not the same as the development of a disorder. Moreover, in reality, it seems impossible to deal with possible deleterious behavioral health consequences in space simply by focusing on selection criteria, which thus far have been the most prominent countermeasure applied by space agencies. Therefore, the careful development and implementation of psychological, pharmaceutical, and other countermeasures to deal with these potential problems is extremely important.

In recent years, a number of reports, papers and books have identified and emphasized psychological problems related to long-duration spaceflights. Concerns have been raised about the stressors that crew members are exposed to, and the potential deleterious effects on health, group functioning, performance, and safety. More recently, Peter Suedfeld has brought a new perspective into the field by emphasizing the salutogenic, growth enhancing aspects of experiences in space as well as in other extreme environments and conditions. Rather than focusing on pathogenesis, this approach directs attention to the fostering of human strengths that act as buffers against psychological dysfunction and the adverse effects of stress. Chapter 10 by Suedfeld, Wilk, and Cassel presents a careful study of the change in values and coping mechanisms by majority and minority crew members from preflight, in-flight, to post-flight status, with somewhat different trajectories and value changes between the two groups. The in-flight increase by minority members in spirituality and family interest reflects positive growth; the decline in references to power and self-direction may be a reflection of a "host/guest" phenomenon. While these value changes were not sustained post-flight, it is possible that with the experience of an extended flight to the Moon or Mars, greater changes in values would occur that would be more stable and extended to the post-flight period. In general, a transformation of a value hierarchy might represent a coping strategy for maintaining motivation during long-duration missions. This transformation might also be reflected in terms of certain personality characteristics becoming more prominent or diminished, depending on the individual's space experiences.

The salutogenic aspects of space are also reflected in the study of self-initiated photography during leisure time in space and the larger number of photos taken as the mission progressed, described in Chap. 3 by Robinson and colleagues. The awesome, existential phenomenon of viewing the Earth from space has been commented on by many astronauts, and is the primary leisure time activity. The perhaps "hard-wired" pleasure of communing with nature and the feeling of spirituality that is often a part of this activity seems an important positive component of life in space, and is also a means of coping with negative aspects of space missions. For example, Russian cosmonauts have commented on the pleasure they experienced on space missions by growing plants, watering and tending to them and watching the plants thrive. A challenge for future long-duration missions to distant planets where the Earth and home will no longer be in view is to plan for other methods of viewing the cosmos and opportunities for leisure photography during planetary exploration.

While the transformation of values may be regarded as a constructive coping mechanism, these changes might also carry potential hazardous aspects. During a future expedition to Mars or beyond, a partial or complete loss of commitment to the usual (Earth-bound) system of values and behavioral norms could result which, in extreme cases, involve unforeseeable risks in terms of individual behavior, performance of mission tasks, and interpersonal interactions within the crew. This might make any external control and guidance of the crew extremely difficult. Reports from some military combat units indicate that micro-cultures that develop in isolated groups may diverge profoundly in values and behaviors from societal norms. Sharing a unique experience in an extremely isolated environment is likely to result in strong bonding between crew members. Although crew cohesion is generally regarded as positive factor for the psychosocial adaptation in space, the risk for groupthink in decision making must be considered. Groupthink is associated with lower quality performance since members are reluctant to critically evaluate of alternative ideas or viewpoints. Open-minded investigations may be necessary to understand group processes that are likely to occur in confined, and isolated crews, and in particularly under the condition of great autonomy. Space simulation studies on Earth have indicated that periods of high crew autonomy did not produce adverse effects or interfere with mission goals, and resulted in improved crewmember mood, creativity, and performance (see Chap. 9 by Kanas, and Chap. 8 by Sandal, Bye, and van der Vijver). However, the impact of high autonomy on crew performance and interaction has not yet been evaluated in space, but any breakdown of communication, co-operation and cohesiveness of a space crew must be considered as an important limiting factor to mission success and safety. One concern raised by several of the authors in this volume is that impairments of crew interactions and operations might be induced by cultural and gender differences. With long-duration missions, one would anticipate changes and fluctuations in interpersonal as well as task cohesion over time, and that both gender and culture could influence these relationships. Chapter 7 by Kring and Kaminski addresses the issue of gender composition and crew cohesion, also examining one-dimensional and multi-dimensional conceptions and types of cohesion, for example, interpersonal versus task cohesion. While there is a growing literature on mixed gender and multicultural groups in polar settings and space simulations, the information from space missions is primarily anecdotal. With respect to gender, this is partly because of the low number of women involved and the associated need to maintain confidentiality. Also, the small sample sizes make it difficult to isolate the effects of gender or culture from factors such as personality and professional training experiences. Most spaceflight experiences with multi-cultural space crews, thus far, stem from international Shuttle or Mir missions where cross-cultural aspects were inevitably confounded with "host/guest" differences, i.e., where the missions usually involved one "dominant culture" being the host for crew members from other countries, organisations and professions. Consequently, a different picture may emerge within a truly international facility with equal partners, as pointed out by Suedfeld, Wilk, and Cassel in Chap. 10. With more psychological data emerging from the International Space Station, a better understanding of the impacts of national and organisational cultures per se is within reach.

There is no doubt that national interests will continue to play a major role in decisions about the cultural mix of crews for specific missions. Within that context, decisions about crew composition ideally should be made on the basis of the compatibility of a group of persons possessing the range of skills required for the particular mission. This information can be obtained by behavioral observations of a larger group from which a particular crew will be selected. The use of a uniform assessment battery across countries to provide data to inform about selection for specific space missions would also be helpful. However, in reality this might be difficult to obtain. Space agencies have shown differences in preferences regarding measurements. Additionally, the same measure might not be equally applicable or valid for different national groups. In the context of long duration spaceflight such as a future Mars mission, there is still very little empirical data to inform on the ideal crew mix despite the many discussions that have ensued. Addressing this issue by conducting systematic research in analogue environments will require a significant commitment by space agencies in terms of priorities and funding. However, it is likely that existing basic research and studies from other fields on broader populations can provide important insights into group processes in spaceflights. For example, information about cooperation in multinational groups is available from fields such as anthropology, sociology, and cross-cultural psychology. Although one needs to be mindful about the fact that astronauts represent a highly selected group, findings from broader populations can to a larger extent be used to generate empirical testable hypothesis for research in analogue environments.

A comprehensive program in spaceflight behavioral health must be broad based, interdisciplinary, and address issues at the individual, small group, and organizational levels. In dealing with issues of relevance to multinational spaceflight, cross-cultural collaboration in research should be considered mandatory to prevent ethnocentrism in design and interpretations. Such collaboration, although often challenging, might provide research results that are more easily accepted and applicable across national agencies. Rigorous theories and thorough methodological designs are prerequisites for progress in the field. In Chap. 2, Bishop points out that one key methodological and validity issue is the added value of utilizing consistent measures across various expeditions, allowing a more valid comparison of individuals and teams across environments, including space. With awareness of this problem, researchers in the last several years have tended more often to use a common set of measures that facilitate comparison of results across studies. While this is a positive development, potential cultural bias in the assessment methodology must not be overlooked. For example, Ritscher has emphasized that cultural differences may affect the accuracy of methods for detecting distress in spaceflights so that specific methods will not work equivalently.

Robust and sensitive assessment methods for monitoring behavior and health in space are crucial with regards to the quality of research, but also for the early detection of behavioral health problems. Improved prediction, prevention, and treatment of distress will improve the safety of international long-duration space missions. With regard to both prevention and treatment, the development of countermeasures designed for autonomous crews become more important as we prepare for much longer exploration-class missions to Mars and beyond. In our view, giving a high priority to research programs in the area of behavioral health will prove to be highly beneficial for the safety of long-duration space missions.

References

Ritscher, J.B.: Cultural factors and the International Space Station. Aviat. Space Environ. Med. **76**(6), 135–144 (2005)

Sandal, G.M., Manzey, D.: Cultural determinants of co-working of ground personnel in the European Space Agency. Paper presented at the 58th international astronautical congress, Hyderabad, Sept 2007

Tomi, L., Stafanowska, P., Doyle, V.F.: Cross-cultural training requirements definition survey for the International Space Station. Paper presented at the 115th annual convention of the American Psychological Association, San Francisco, Aug 2007

About the Editor

Douglas A. Vakoch is a Professor in the Department of Clinical Psychology at the California Institute of Integral Studies, as well as the Director of Interstellar Message Composition at the SETI Institute. Dr. Vakoch is a licensed psychologist in the state of California, and his psychological research, clinical, and teaching interests include topics in psychotherapy and ecopsychology. As a corresponding member of the International Academy of Astronautics, Dr. Vakoch chairs that organization's Study Groups on Interstellar Message Construction and on Active SETI. Through his membership in the International Institute of Space Law, he examines policy issues related to interstellar communication. His books include *Communication with Extraterrestrial Intelligence (CETI)*; *Civilizations Beyond Earth: Extraterrestrial Life and Society*; *Ecofeminism and Rhetoric: Critical Perspectives on Sex, Technology, and Discourse*; and *Feminist Ecocriticism: Environment, Women, and Literature*. Dr. Vakoch serves as general editor of the book series Ecofeminist Theory and Practice, published by Berghahn Books, and he is a member of the editorial board of the journal *Ecopsychology*.

Dr. Vakoch has chaired numerous workshops and conference sessions, including several symposia at recent annual conventions of the American Psychological Association on the psychology of space exploration and on environmental psychology. While completing his M.A. in history and philosophy of science at the University of Notre Dame, he focused on both the history of astronomy and the history of psychology. He received his Ph.D. in clinical psychology with a quantitative concentration from the State University of New York at Stony Brook. After finishing his predoctoral clinical internship in health psychology at the University of Wisconsin-Madison Hospital and Clinics, Dr. Vakoch completed a National Research Service Award postdoctoral fellowship in the Department of Psychology at Vanderbilt University.

D.A. Vakoch (ed.), *On Orbit and Beyond*, Space Technology Library 29,
DOI 10.1007/978-3-642-30583-2, © Springer-Verlag Berlin Heidelberg 2013

Index

D.A. Vakoch (ed.), *On Orbit and Beyond*, Space Technology Library 29,
DOI 10.1007/978-3-642-30583-2, © Springer-Verlag Berlin Heidelberg 2013

MIX
Papier aus verantwortungsvollen Quellen
Paper from responsible sources
FSC® C105338

If you have any concerns about our products,
you can contact us on
ProductSafety@springernature.com

In case Publisher is established outside the EU,
the EU authorized representative is:
Springer Nature Customer Service Center GmbH
Europaplatz 3, 69115 Heidelberg, Germany

Printed by Libri Plureos GmbH
in Hamburg, Germany